Sensing Mars: Remote Sensing to Understand the Red Planet

Steven Hobbs

Sensing Mars: Remote Sensing to Understand the Red Planet

Springer

Steven Hobbs
Mars Society Australia
Clifton Hill, VIC, Australia

ISBN 978-3-032-24415-4 ISBN 978-3-032-24416-1 (eBook)
https://doi.org/10.1007/978-3-032-24416-1

This Springer imprint is published by the registered company Springer Nature Switzerland AG
The registered company address is: Gewerbestrasse 11, 6330 Cham, Switzerland

This book is dedicated to my awesome daughters Sabina and Nadia Hobbs who may one day reach for Mars. It is also dedicated to those who are passionate in learning more about Mars.

Foreword

Mars, the most Earth-like planet in our solar system, continues to captivate our collective imagination and scientific curiosity. Its mysteries—from the potential for life to its history of liquid water and its dynamic geological features—serve as a backdrop for unparalleled exploration beyond Earth. Remote sensing and robotics have emerged as indispensable tools for unraveling these mysteries, enabling us to transcend the challenges of distance, harsh conditions, and human absence on this red frontier.

Presented in 11 chapters, *Sensing Mars* offers a comprehensive journey through the active field of Martian exploration. Readers are introduced to a multidisciplinary approach, blending remote sensing techniques, engineering ingenuity, and planetary science to decode Mars' secrets. Its author, Steven Hobbs, has thoroughly researched the topic and brings a unique perspective gleaned from his professional background as a remote sensing expert as well as more than a decade of experience building and testing robots in Mars analog environments as a long-standing member of Mars Society Australia.

Accompanied by appropriate and well-chosen illustrations and examples, the book is divided in the following way:

Chapter 1 (Introduction) explores pre-space-age studies of Mars, historical spacecraft missions, and foundational knowledge about Mars' properties, moons, and formation. Key topics include early telescopic observations, 19th-century interpretations, and the advent of space-age missions like Mariner and Viking.

Chapter 2 (Space-Age Exploration and Technologies) highlights the engineering challenges of interplanetary travel and the rise of new more affordable and accessible technologies including CubeSats that may be essential for future missions. This chapter also offers an overview of spacecraft propulsion, navigation, and the operational design of missions.

Chapter 3 (Remote Sensing Considerations for Mars) introduces the principles of electromagnetic energy, sensor resolution, calibration, and data processing, all of which are essential for accurate planetary observations.

Chapter 4 (Optical Remote Sensing) covers imaging technologies such as multi-band and hyperspectral sensors and includes case studies on their applications and advancements for Martian exploration.

Complementary, non-imaging sensors are presented in Chap. 5 (Other Martian Remote Sensors) where tools including radar, thermal infrared, ultraviolet, and high-energy radiation sensors are described, emphasizing their role in studying Mars' subsurface, atmosphere, and geological history.

Chapter 6 (Mapping, Cartography and Data Processing for Mars Missions) examines methods such as Geographical Information Systems (GIS), map projections, digital elevation modeling (DEMs), and the application of artificial intelligence for Mars data analysis and visualization.

Physical properties of Mars are emphasized in Chaps. 7–9. Chapter 7 (Mars Unveiled: Atmosphere and Surface) presents an analysis of Martian atmospheric dynamics, elemental surface composition and mineralogy, and the use of these features in interpreting Mars' current and paleoenvironments. Chapter 8 (Mars Unveiled: Cratering, Volcanism and Aeolian Activity) discusses Martian impact craters, volcanic activity, and aeolian processes that shape the surface over geological timescales. And Chapter 9 (Mars Unveiled: Ice and Water) explores Mars' polar ice caps, glaciers, and evidence of historical water flow, with a focus on paleoclimate and hydrology.

Chapter 10 (Mars Unveiled: Astrobiology) considers the potential for past or present life on Mars, the utility of leveraging Earth analogues to help test and prepare for future missions, the study of methane as an exciting though ambiguous indicator of life, and cutting-edge astrobiological experiments.

And finally, Chap. 11 (Mars of the Twenty-First Century: A Complex and Dynamic Planet) reflects on Mars as a dynamic planet with unique processes. Here, the author discusses briefly the future of Mars exploration, including the implications of remote sensing for human missions and the broader benefits of Martian studies for Earth.

Sensing Mars is both an introductory textbook and a reference work for planetary scientists, covering the scientific, technological, and engineering aspects of remote Mars observation. For students, it offers an accessible introduction to the principles of remote sensing, GIS, and robotics applied to Mars exploration. For planetary scientists, it acts as a comprehensive reference, combining foundational concepts with current research and case studies, making it invaluable for professionals in the field. Mars exploration holds transformative potential for science, technology, and humanity. Insights from the Red Planet deepen our understanding of planetary evolution, inspire technological innovation, and encourage global collaboration in space science. The search for life and the quest to understand Mars' past also help us reflect on Earth's fragility and resilience. Through *Sensing Mars*, the author invites you to explore this interdisciplinary and ever-evolving field of planetary remote sensing. Whether you are a scientist, student, or simply a curious mind, this book offers a window into Mars' captivating story and the innovative methods enabling us to reveal it.

December 2024

Jennifer G. Blank, Ph.D.
Senior Research Scientist
Blue Marble Space Institute of Science
NASA Ames Research Center
in Silicon Valley
Seattle, USA

Acknowledgements

The author would like to thank Dr. Jennifer Blank for writing the foreword, Ms. Gabriela Hobbs, Dr. Colin Paine and Mr. John Haythorpe for their work in editing and proofreading this manuscript. Ms. Bharti Sharma is also thanked for providing summary edits and a content check, particularly for the technical aspects of this book. University of New South Wales, Canberra, is acknowledged for providing laboratory space, workshop assistance and instrumentation needed for developing spectral-based applications that appear in this book. Dr. David Paull and Dr. Andrew Lambert are acknowledged for their assistance in offering scientific and engineering expertise. Dr. Jonathan Clarke, former Mars Society Australia president, is thanked for his support in the research and development of many of the case studies appearing in this book. The mission designers and operators of the missions and returned remote sensing data are gratefully acknowledged for making this information available to the community, making this book possible. Except where images were produced by the author, NASA/JPL and ESA are gratefully acknowledged for producing publicly released data, imagery and concept artwork appearing in this book. Finally, the author would like to thank Springer and its associates for making publication of this book possible.

Contents

1 Introduction ... 1
1.1 Pre-space-Age Mars Studies ... 4
1.1.1 Early Observations ... 4
1.1.2 Nineteenth Century Interpretations ... 6
1.1.3 Pre-space-Age Twentieth Century ... 7
1.2 History of Mars Spacecraft Exploration ... 9
1.2.1 Additional Martian Orbiters ... 13
1.2.2 Surface Missions ... 15
1.3 Properties of Mars ... 19
1.3.1 Size, Mass, Distance from the Sun ... 19
1.3.2 Moons ... 20
1.3.3 Formation and Age of Mars ... 22
References ... 25

2 Space-Age Exploration and Technologies ... 29
2.1 Space Engineering Considerations for Mars ... 29
2.1.1 NASA Engineering Phases ... 30
2.1.2 Deep Space Operational Impacts on Spacecraft Engineering ... 33
2.2 Travel Between Earth and Mars ... 37
2.2.1 Navigation, Propulsion and Attitude Control ... 37
2.2.2 Propulsion ... 39
2.2.3 Presence of Significant Perturbation Forces ... 41
2.2.4 Arriving at Mars ... 41
2.2.5 Reaching the Surface ... 45
2.3 The Influence of Space 2.0 ... 47
2.3.1 Trade Study Analysis ... 49
2.3.2 Commercial-Grade Components ... 50
2.3.3 Mars Analogue Environment Application for Space 2.0 ... 50

2.3.4 Space 2.0 Case Study: Identifying the Utility of Low-Cost Remote Sensing for High Altitude Ballooning Applications 51
References 56

3 Remote Sensing Considerations for Mars 61
3.1 Electromagnetic Energy and Influences 61
3.1.1 The Electromagnetic Spectrum 61
3.1.2 The Influence of the Martian Atmosphere 65
3.2 Sensor Resolution, Processing and Calibration Considerations 66
3.2.1 Spatial Resolution 66
3.2.2 Spectral Resolution 68
3.2.3 Temporal Resolution 68
3.2.4 Optical Designs 68
3.2.5 Sensor Data Processing 71
3.2.6 Radiometric Calibration 74
References 77

4 Optical Remote Sensing 81
4.1 Missions and Sensors 81
4.1.1 Orbiter Imager Types 81
4.1.2 Popular Sensor Types 83
4.1.3 Surface Imagers 84
4.2 Multi-band Imagery 86
4.2.1 Multispectral Imagery 86
4.2.2 Multispectral Band Ratios 90
4.2.3 Hyperspectral Sensors 92
4.2.4 Case Study, Design and Development of a Low-Cost Hyperspectral Sensor for Deep Space Applications 97
References 103

5 Other Martian Remote Sensors 109
5.1 Sensor Types 109
5.1.1 Radar Remote Sensing 109
5.1.2 Factors Affecting Martian Radar Remote Sensors 110
5.1.3 Thermal Infra-Red 114
5.2 High Energy Radiation—UV, X-Ray and Gamma Rays 120
5.2.1 Flown UV Sensors 120
5.2.2 High Energy Detectors 121
References 125

6 Mapping, Cartography and Data Processing for Mars Missions 129
6.1 Image Processing and Mapping 129
6.1.1 Map Projections 129
6.1.2 Mars Ellipsoid 132
6.1.3 Mars Data Processing 133
6.2 Elevation Data 138
6.2.1 Stereoscopic Imagery 140
6.3 GIS Creation and Management 143
6.3.1 GIS Support in Landing Site Selection 147
6.3.2 Case Study: Using GIS for Solar and Nuclear Rover Landing Site Selection 149
6.4 Artificial Intelligence (AI) for Martian Applications 152
6.4.1 Image Segmentation 152
6.4.2 Edge-Based AI 158
References 160

7 Mars Unveiled: Atmosphere and Surface 165
7.1 Atmosphere 165
7.1.1 Atmospheric Column 166
7.1.2 Clouds and Storms 170
7.1.3 Clouds 172
7.1.4 Stability of Water on Present Mars 175
7.1.5 Present-Day Water Activity 176
7.2 Elemental Composition and Mineralogy 177
7.2.1 Elemental Abundances 177
7.2.2 Global Mineralogy 180
References 184

8 Mars Unveiled: Cratering, Volcanism and Aeolian Activity 189
8.1 Crater Processes 189
8.1.1 Degradation 194
8.1.2 Crater Count Dating 196
8.2 Volcanism 199
8.2.1 Mud Volcanism 207
8.2.2 Valles Marineris, Planet-Scale Faulting and Relationship to Volcanism 207
8.2.3 Case Study Alba Mons—A Unique Volcano on Mars 209
8.3 Aeolian Features 211
8.3.1 Wind Streaks, Ripples, Yardangs and Ventefacts 212
8.3.2 Dunes 213
References 223

9 Mars Unveiled: Ice and Water ... 229
9.1 Ice ... 229
9.1.1 Poles ... 229
9.1.2 Polar Layered Deposits ... 232
9.1.3 Martian Ice Landforms ... 232
9.1.4 Ice Flows ... 234
9.2 Water ... 241
9.2.1 Valley Networks ... 241
9.2.2 Outflow Channels (Hesperian) ... 246
9.2.3 A Case Study: The Kasai Valles System ... 247
9.2.4 Crater Lakes ... 251
9.2.5 Fluvial Deltas ... 251
9.2.6 Lakes ... 253
9.2.7 A Martian Ocean ... 253
9.2.8 Case Study: Remote Sensing Reveals Complex Multi-process Interactions in Noachis Terra ... 257
9.3 Small-Scale Processes: Gullies and Recurring Slope Lineae ... 263
9.3.1 Gullies ... 263
9.3.2 Recurring Slope Lineae ... 266
References ... 267

10 Mars Unveiled: Astrobiology ... 275
10.1 Direct Search for Life ... 275
10.2 Earth Analogues for Astrobiology ... 277
10.2.1 Case Study: Field Robotic and Laboratory Astrobiology Investigations ... 280
10.2.2 ALH84001—Mars Meteorite Analysed on Earth ... 283
10.3 Atmospheric Methane ... 284
References ... 287

11 Mars of the Twenty-First Century: A Complex and Dynamic Planet ... 289
11.1 The Importance of Mars Missions ... 289
11.2 Remote Sensing as an Enabler for Understanding Mars ... 290
11.2.1 Multiple Processes Shaping Mars ... 291
11.3 Benefit to Earth ... 291
11.4 Future Exploration ... 292
11.4.1 Future Plans for Human Habitation ... 293
References ... 294

Appendix A: List of Major Successful Mars Instruments ... 297

Appendix B: Acronym List ... 303

Index ... 307

Chapter 1
Introduction

Was Mars once or perhaps still an abode for life? Did water play an important role in shaping the Martian surface? These questions have driven the bulk of Mars funding and exploration. More space missions have flown past, orbited, crashed into or landed on Mars than any other planet, yet direct evidence of past or present life is yet to be found.

The action of surface water, long assumed to be abundant on the planet prior to the space age is more complex than previously thought. Small scale hillside gullies and recurring slope lineae, once considered to be markers of recent water activity but now thought to be formed by more exotic multi-agent processes are examples of changes in humanity's understanding of the Red Planet.

The terabytes of returned spacecraft data have also resulted in the development of remote sensing and mapping techniques designed for Martian conditions (Fig. 1.1).

Multi-wavelength sensors, often derived from terrestrial experimentation but adapted for the Martian environment have enabled development of custom cartography and projection standards. These systems evolved from painstakingly cutting and stitching photographs to global mosaics from the early flyby and orbiter missions. Digital systems, plus global topographic mapping conducted by an early form of light detection and ranging (LIDAR), now make it possible to correctly locate remotely sensed data on the Martian surface. In fact, global maps of Mars exceed the accuracy of terrestrial equivalents because Earth's oceans obscure most of our planet's surface.

In the absence of human field observations, studies of Mars rely exclusively on remotely sensed orbital data coupled with limited lander-based analysis. Decades of Earth observation satellite design has shaped increasingly sophisticated sensors on Martian equivalents. Most of the electromagnetic spectrum, from radar through to high-energy radiation is used to feed a multi-disciplinary approach to investigating Mars.

Subsurface measurements have clarified the presence of underground liquid water on Mars with corresponding implications for past habitability. Visible-light-based albedo measurements, used by early telescopic observers to infer the character of

S. Hobbs, *Sensing Mars: Remote Sensing to Understand the Red Planet*,
https://doi.org/10.1007/978-3-032-24416-1_1

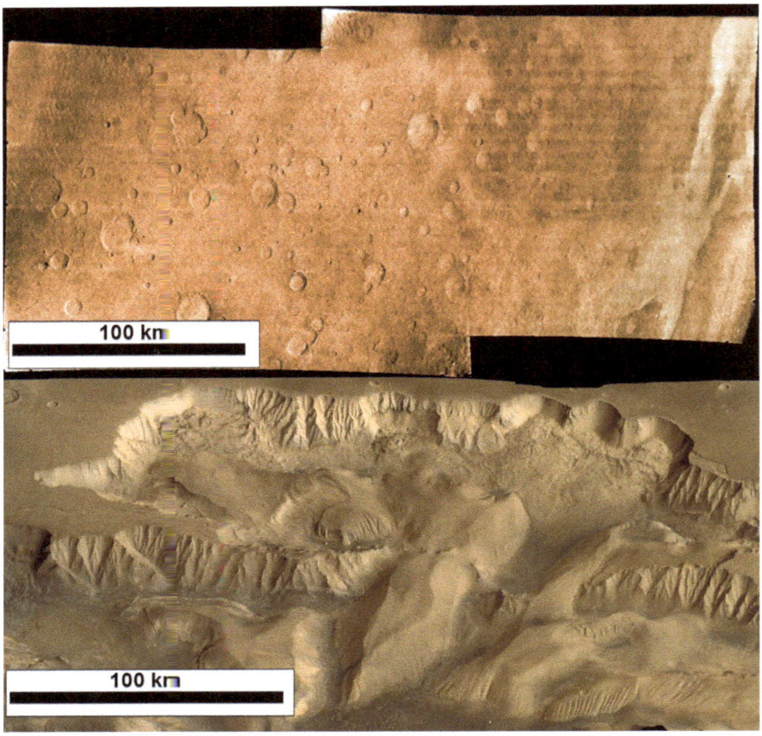

Fig. 1.1 Extraordinary progress has been achieved in humanity's understanding of Mars. The top mosaic shows the best imagery available at the time of the Apollo Moon landings. Scientists mistakenly believed Mars was flat, cratered and geologically dead. Obscure features to the extreme right of the image were later identified as part of the largest canyon system in the Solar System, Valles Marineris (bottom image, PIA000045). *Image credits* NASA/JPL

the Martian surface is now used to track global movement of dust and atmospheric modelling to assist landing on the planet. Multispectral and recently hyperspectral sensors are now providing insight into detailed mineralogic composition of the surface and interaction with geologic processes. As an example, the Opportunity Mars Exploration Rover (MER) was targeted at a site where hydrated minerals were identified by these methods. Spectral analysis combined with X-Ray diffraction (XRD) and other science data has now demonstrated beyond a reasonable doubt the past presence of long-standing liquid water in areas of the planets.

Thermal and high energy wavelengths are also used to help understand the Martian environment. Measuring the rates at which surface features heat up and cool down provides useful information of underlying geology. Gamma ray spectroscopy identified the presence of subsurface ice, later confirmed by the 2008 Phoenix polar mission. Additional high energy photon measurements in the ultraviolet (UV) range are showing researchers how Mars has lost most of its atmosphere to the solar wind from historic depletion of its magnetic field and implications for past volatile stability on the surface.

All Martian remote sensing methods rely on calibration methods designed for the Martian environment and atmosphere that differ significantly from Earth. Tailored calibration targets, and adaptation of techniques derived from Earth observation satellites are needed to convert collected data into scientifically useful information. These have placed constraints on Mars orbiter and lander design, a convoluted history at best, with many missions ending in failure. From the earliest flyby to the latest car-size roving missions, designers have contended with building longevity into physical and electronic components, and a degree of autonomy to accommodate system faults occurring beyond short-term workarounds or fixes. Reduced solar output for energy collection, protection against hostile radiation environment, and maximising scientific return from a limited collection of instruments to understand an alien world are ongoing challenges to spacecraft engineers.

A new Mars 2.0-style engineering movement is emerging. The success of MarsCubeOne (MarsCO), the first ever deep space cubesat mission, has demonstrated that low-cost, rapid deployment missions can return useful science. The increased miniaturisation and consistency of commercial off the shelf hardware and components have opened Mars exploration to a suite of novel instruments and innovative testing that provide increased opportunities to do real science on small budgets. These reduced-scope missions and analogous testing in Mars-like environments on Earth will provide insight into a new generation of Mars missions.

The technology, missions and investigations briefly mentioned above are revealing Mars to be a member of the inner solar system family, and a unique planet with its own history. The latest analysis has ruled out the Earth-like world favoured in pre-space age times and has also dismissed the geologically dead Moon-like planet inferred from early flyby missions. Understanding of Mars geomorphology is also evolving from a single event history informed by Earth analogue studies to increasingly accounting for a distinctly Martian environment for which Earth has no equivalents. The planet we see today was influenced heavily by the presence of water, often working as part of a multi-agent process to create the landforms present today. An example is Cerberus Fossae, where lava flows triggered a very rapid release of snowmelt, erosively modifying the surrounding terrain. The Gusev Crater site is another example where water-based erosion was subsequently altered by flows of lava into a crater basin. Up to a third of the planet was likely covered by an ocean of liquid water surrounded by valley networks eroded by ice-melt in a recent "cold and wet" hypothesis. Volcanism and impact events created temporary thermal heating to generate chaotic terrain and outflow channels, often occurring simultaneously with lava flows. The ocean in the planet's northern hemisphere would have been barely above freezing, while the southern highlands remained the province of ice movements and glacial activity, evidence of which is observed today.

Mars from 1 billion years (Ga) to the present day retains a portion of its past activity shaped by wind, water, volcanism and ice. An exotic seasonal carbon dioxide (CO_2) cycle, without analogue on Earth, drives ultrafine (1–4 μm) Martian regolith to behave like a fluid, eroding V-shape channels and deposits on sand dunes and sometimes crater walls in gravity-fed flows. Winds have operated on Mars throughout its history, forming a vast sand dune sea surrounding the north pole. Wind-based erosion and deposition features throughout the planet, including sand dunes whose movements are being tracked today. Measurements in dune morphology and direction provide indications of Martian climate change, as well as the surprising ability of the thin Martian atmosphere to transport sand grain-size sediment over large distances. Global obliquity cycles of between 50 and 100 thousand years (Ka) have driven ice sublimation and deposition cycles in the mid-latitudes. Viscous flow features, sublimation thermokarsts and patterned ground are evidence of this activity at these latitudes.

Space exploration of Mars continues to be driven by the possibility of past or present life on the planet. While direct astrobiological evidence is yet to be discovered, indirect evidence suggests an environment once favourable to life. Terrestrial studies into extremophiles are defining resilience of biology and placing lower and upper environmental boundaries on potential habitability for Mars. A first-time caching activity was recently completed by the NASA Perseverance Mars Rover to drive a sample return mission, allowing access to advanced terrestrial experimentation.

The complex, multi-agent evolution of Mars as identified by decades of space exploration, suggests Earth-based analysis of returned samples may be the only method of conclusively identifying or dismissing the presence of life on Mars. Such a mission is expected to be conducted early next decade. Whatever the outcome, exploration of Mars is expected to continue as the planet remains a scientific frontier, whose explorations have application to Earth environmental and geological studies.

1.1 Pre-space-Age Mars Studies

1.1.1 Early Observations

To early astronomers, Mars appeared to be a bright star that reversed its progression across the sky every two years. Observations and analysis of the planet grew in detail in proportion to the increasing sophistication of telescopes and number of independent observers. Interpretations through astronomical sketches and even photography were made difficult by the small size of the planet and short observation windows (Fig. 1.2a–h).

In 1609 Galileo Galilei was probably the first person to identify Mars as a disk that did not always look circular with his newly invented 20× magnification refracting telescope. Because it exhibited phases as seen from Earth, this placed Mars in the category of planets rather than a stellar object (Chalmers 1985). The first distinct markings of Syrtis Major, a large basaltic volcanic region on Mars, were drawn by Christen Huygens in 1659. White patches were discovered by J. D. Cassini in 1666 and later inferred to be polar caps in 1719. The discovery and observation of

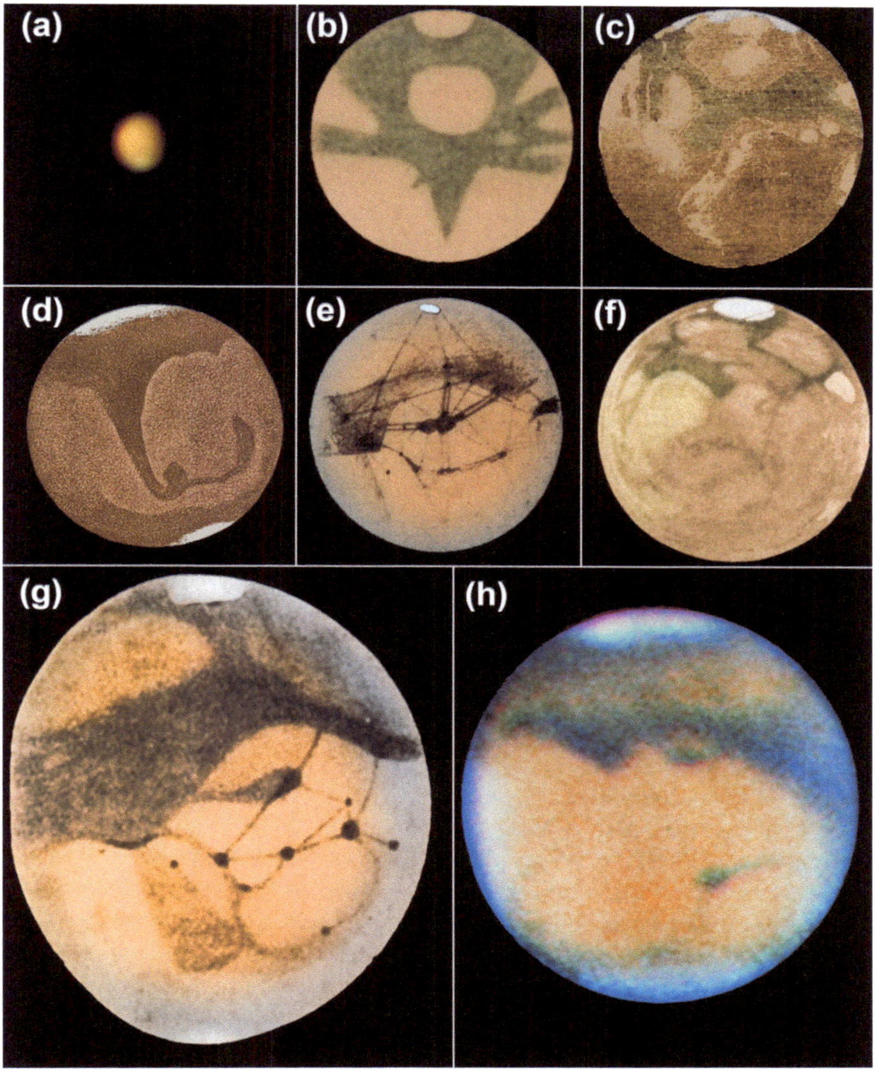

Fig. 1.2 **a** Mars as it would have appeared through a seventeenth century telescope. Its small size and 26-month opposition makes it a hard target for telescopic observations. **b** Mars as interpreted by astronomer Walter Frederick Gale in the nineteenth century, with Sirtis Major prominent (Public Domain). **c** The planet as understood by mid-nineteenth century astronomers, pre-dating Schiaparelli's interpretation of canals (Public Domain). **d** 1890s illustration of Mars showing the polar caps and Syrtis Major (Public Domain). **e** Percival Lowell sketch of Mars, showing the canals as he imagined them (Public Domain). **f** Illustration of Mars in the mid-twentieth century prior to the space age (Public Domain). **g** One of the best illustrations of Martian canals, showing a gibbous phase from 1894 (Public Domain). **h** Mid-twentieth century telescopic photography of the planet failed to conclusively identity the presence of canals (Public Domain)

permanent markings on Mars enabled estimations of the planet's rotation period in 1666, within 90 s of the modern measurements of 24.6 h.

Ongoing telescopic observations tracked changes in the light and dark markings on Mars, as well as the seasonal growth and shrinkage of the polar caps. Observations of this period to the late 1950s also seemed to identify grey to green patches contrasting with the dominant red areas. These were originally interpreted to be seas of liquid water, though more detailed observations suggested the presence of extensive vegetation that changed shape in accordance with Martian seasons. Seasonal variation was quantified by Sir William Herschel from 1777 to 1783 as being caused by a 30° tilt in the planet's pole relative to the Sun (Sheehan 1996).

1.1.2 Nineteenth Century Interpretations

The presence of an atmosphere was inferred through observations of clouds, and the realisation that polar caps required gaseous pressure to remain stable. nineteenth century computations derived from estimations of size, gravity and distance from the sun suggested a thinner atmosphere than that of Earth. Additional calculations derived from the reduced amount of solar radiation Mars received inferred a normal temperature well below the freezing point of water. This latter point was either ignored or, through interpretation of polar meltwater feeding vegetated areas, counter-argued by postulated internal heat sources raising the planet's global temperature (Young 1900).

An exceptionally close opposition of 1877 became something of a watershed moment for Mars exploration. That year, Asaph Hall discovered Mars' two moons, Phobos and Deimos. Also in that year, colourblind Italian astronomer Giovanni Schiaparelli exploited his disability plus a high-power telescope to produce a detailed map of Mars crisscrossed with a series of fine lines that he called "canali", which in Italian means channels. Unfortunately, "canali" was mistranslated as "canals", an error that sparked fascination with the Red Planet along with fanciful dreams of ancient civilisations (Fig. 1.2e, g). These dreams endured to the start of the space age. Schiaparelli's maps were also the most detailed available, showing the location of canals, and lighter areas of the planet.

The idea of canals and liquid water appeared to be supported by early spectroscopic analysis performed in 1871 by Elias Colbert (Colbert 1871). His results seemed to suggest significant amounts of water, though later measurements by more sophisticated instruments called these findings into question, particularly discoveries of strong water absorption bands in Earth's atmosphere that interfere with terrestrial telescopic observations (Owen 1982).

Early twentieth century spectral measurements by Kuiper and others discovered substantial amounts of CO_2 but failed to identify water vapour.

Percival Lowell, a self-funded Bostonian astronomer, hypothesised that the canals he observed from a purpose-built observatory were evidence for a dying Martian civilisation attempting to fight the desertification of the planet. His evidence was based on inferences of 3000+ km long, straight canals and assumptions that the green areas were vegetation. Lowell's lectures inspired science fiction author H. G. Wells, and generations of other writers, scientists, and engineers. Lowell saw a greenish wave that commenced below the north pole from the Martian Spring and fading in the following autumn. To quote Lowell, "areas of vegetation; from which it follows that Mars is very badly off for water, and that the planet is dependent on the melting of its polar snows for practically its whole supply" (Lowell 1895).

Lowell's vision of an Earth-like Mars crossed by canals was not shared by all astronomers. In 1894, using a more advanced spectroscope than Colbert, US astronomer William Wallace Campbell looked for signs of water on the Martian surface and oxygen in its atmosphere. Previous spectral "discoveries" of water and oxygen over the previous 30 years had promoted an Earth-like planet. However, Campbell's instrument showed that neither of these were present on Mars (Wright 1947). Additionally, Alfred Russell Wallace, major proponent of the development of the theory of evolution, refuted Lowell's findings and suggested Mars was uninhabitable (Wallace 1907).

Multiple observations seeking to confirm the presence of Martian canals also produced inconsistent results. Single canals seen on one day seemingly transformed into double canals on the second day. Observations of canals did not equate to the resolving power of the telescope used, with astronomers using larger instruments failing to identify features inferred from using smaller ones. Newly invented planetary photography also failed to positively confirm or refute the existence of canals. The planet's small size and limited favourable viewing conditions proved challenging for the available film emulsions of the period (e.g. Figure 1.2h). Times of atmospheric stillness required to resolve fine-scale features on Mars were simply not long enough to accommodate exposure periods needed to produce a detailed photographic image.

1.1.3 Pre-space-Age Twentieth Century

By the mid-twentieth century, French astronomer Gerard de Vaucouleurs suggested the only way Lowell and others could have inferred the presence of the canals was by imagining straight lines between features they could actually see. This went far in explaining why Lowell's maps of Mercury also showed canals despite most astronomers agreeing that the closest planet to the Sun was too hot to host any form of life.

Spectral reflectivity analysis conducted in 1935 investigated the whole of Mars. The average visible spectra approximated red sandstone. Spectroscopic analysis of the dark areas of Mars was difficult due to the small size of these areas and the limited observing times at opposition. More accurate measurements were obtained by fixing the wavelength at five discrete values between 400 and 800 μm and allowing Mars to drift through the spectrometer slit. The resulting spectra were compared with vegetation and revealed no evidence of hydrated chlorophyll, with a small peak occurring at 650 nm but no corresponding high reflectivity at visible red or near infrared. This demonstrated that the "green areas" appeared to be a colour contrast with the red hue of Mars, and, while lichens might be possible were not caused by higher orders of vegetation (de Vaucouleurs 1954).

In the USA, the legacy of Lowell continued to support the option of canals and Martian life in contrast to much of Europe's more pragmatic approach. US astronomer E. C. Slipher travelled to South Africa to take advantage of the 1954 opposition where an exceptionally good Mars opposition would favour southern skies. Many of his 20,000 photographs from the trip showed a rapid growth of the "green area" below the north pole, printed in colour in National Geographic: "… with the coming of the Martian Spring, the great green areas had begun to appear … That those dark areas are vegetation and that the change if color is somehow connected with the release of water from the melting, shrinking polar icecap is no longer seriously doubted by most observers of the mysterious planet" (Slipher 1955).

Slipher was also adamant on the existence of the Martian canals, and they were illustrated on an official map of the planet he was commissioned to contribute to by the United States Air Force. This map was the authoritative guide for the US Air Force space program and used for planning the US's first mission to the Red Planet in 1965, Mariner 4.

Just prior to the start of the space age, Mars' size, length of day and year, number of Moons, and axial tilt were known to a considerable degree of accuracy. Hard evidence also existed of Mars being much colder and drier than Earth, with a thin atmosphere lacking oxygen and appreciable amounts of water vapour. Spectroscopic analysis identified most of the planet to be similar in composition to iron-rich sandstones, with no evidence of chlorophyll-based vegetation in the so-called "green areas," or standing bodies of liquid water. Photographic evidence failed to confirm the presence of canals, with many astronomers refuting their existence, even in telescopic sketches.

In contrast, significant opinion promoted a warmer, wetter Mars where lichen-like plant or even lower forms of animal life was present. It was this view, inspired by the legacy of Lowell, that prompted the Space Science Board of the National Academy of Sciences to advocate for Mars to be prominent, if not uppermost, in plans for planetary exploration. Right up to and even during the receipt of the first images of from the Mariner 4 encounter in 1965, there was a reasonable, though not unanimous expectation that canals would be directly observed on Mars.

Figure 1.3 provides an example of the comparison between ground and space-based observations. The results of Martian telescopic mapping from the Bloemfontein Observatory between 1939 and 1941 (Fig. 1.3a) with a contemporary global map made from Viking orbiter imagery (Fig. 1.3b). Lighter and darker-toned areas are similar in position with both maps, however none of the larger geological features, such as the Tharsis volcanoes or Valles Marineris were conclusively identified prior to the early 1970s.

Terrestrial observations of Mars continue to the present day. Orbital observatories such as James Webb and Hubble (Fig. 1.4) have returned images undistorted by Earth's atmosphere, complementing data returned from orbiting spacecraft. The

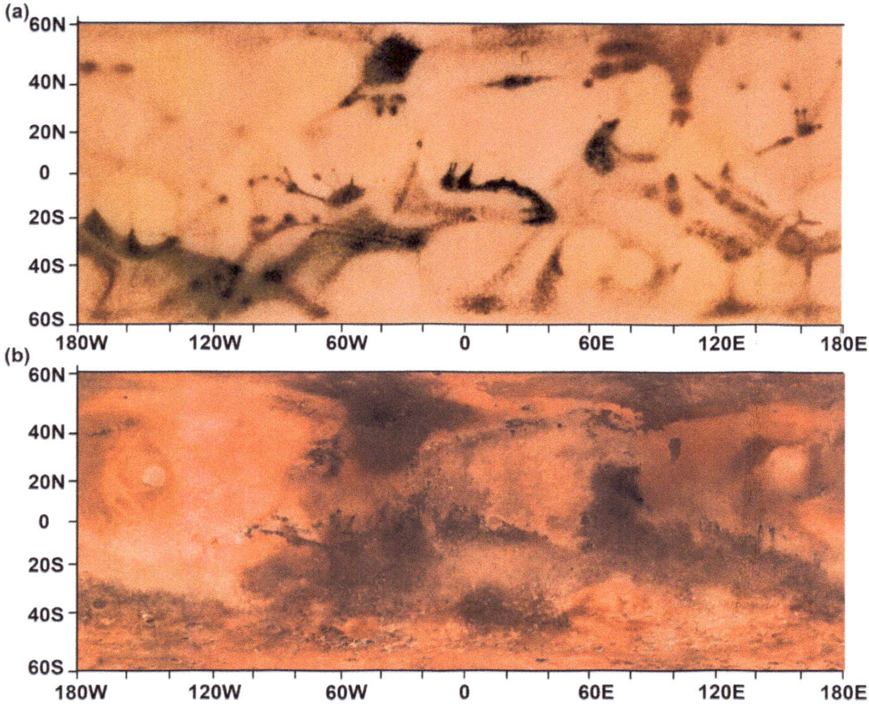

Fig. 1.3 **a** Pre-space age map of Mars produced from Bloemfontein Observatory observations between 1939 and 1941 (Public Domain). **b** Mars map produced from Viking orbiter imagery collected between 1976 and 1980 (NASA/JPL)

availability of affordable electronic sensors has also enabled near-continuous observation of changing events. Examples include occultations with other celestial bodies and dust storm monitoring, where planet-wide observations prove an advantage.

1.2 History of Mars Spacecraft Exploration

Historic and current understanding of Mars remains very dependent on the frequency and sophistication of instrumentation sent there. This is due to the absence of crewed fieldwork and total reliance on robotic space-based remote sensing and lander missions. Since the start of the space age, 50 missions have attempted to fly past, land on or orbit Mars. The first attempt was the Soviet Mars 1 probe in 1962, however it failed enroute. For brevity, only successful missions are described in this chapter, with major missions listed in Table 1.1.

The US Mariner 4 mission conducted the first successful flyby of Mars in 1965 and possessed the first camera to image the planet through an alternate series of green

Fig. 1.4 Hubble Space Telescope image of Mars captured at opposition in 2003 (opo0322i, NASA/JPL)

and orange filters at a very low resolution. The 22 returned images surprised early researchers by revealing a heavily cratered surface with a minimum of topography (Fig. 1.5a).

Additional experimentation through radio occultation and other on-board instruments identified a CO_2-rich atmosphere with a surface pressure range of 0.005–0.01 that of Earth at sea-level. The proliferation of craters plus the discovery of an ultra-thin atmosphere seemed to indicate that Mars was more like the Moon, with a geologically inert surface and absence of fluvial alteration (Leighton et al. 1965).

Mariners 6 and 7 were equipped with more sophisticated imagers, and flew past Mars' equatorial area and South pole in 3 July and 5 August 1969 respectively. The flybys did not fly over the geologically diverse terrain discovered by subsequent missions, and seemed to reinforce Mariner 4's findings (Fig. 1.5b). "Chaotic" terrain was discovered by Mariner 6 and interpreted to be younger than the surrounding cratered regions. Early inferences suggested differing geology from the Moon, with an erosive mechanism causing localised slumping and collapse of regolith (Sharp et al. 1971).

Table 1.1 Major successful Mars missions

Arrival year	Country	Name	Type
1965	USA	Mariner 4	Flyby
1969	USA	Mariner 6 and 7	Flyby
1971	USA	Mariner 9	Orbiter
1971	Soviet Union	Mars 2 and 3	Orbiter/lander
1974	Soviet Union	Mars 4, 5, 6 and 7	Orbiter
1976	USA	Viking 1 and 2	Orbiter/lander
1989	Soviet Union	Phobos 2	Orbiter
1997	USA	Pathfinder	Lander/rover
1997	USA	Mars Global Surveyor	Orbiter
2001	USA	Mars Odyssey	Orbiter
2003	Europe	Mars Express	Orbiter
2004	USA	Mars Exploration Rovers	Rover
2005	USA	Mars Reconnaissance orbiter	Orbiter
2008	USA	Phoenix	Lander
2012	USA	Curiosity	Rover
2014	USA	Mars Atmosphere and Volatile Evolution	Orbiter
2014	India	Mars Orbiter Mission	Orbiter
2016	Europe	Trace Gas Orbiter	Orbiter
2018	USA	Insight	Lander
2018	USA	MarsCO	Flyby
2021	United Arab Emirates	Hope	Orbiter
2021	China	Zhurong	Rover
2021	China	Tianwen 1	Orbiter
2021	USA	Perseverance (Ingenuity)	Rover/helicopter

Mariner 9, the first spacecraft to orbit another planet, arrived at Mars on 24 November 1971 (Fig. 1.5c), in a similar period to the arrival of the Soviet-built Mars 2 and Mars 3 orbiters and landers. The Soviet missions returned information regarding temperature, and thermal properties of the Martian surface. Darker regions were found to be 10–15 °C warmer than brighter areas, likely due to the rockier basaltic material in these regions able to retain heat when compared to lighter, dust-covered regions that cool down faster in the Martian night. The images returned from both Mars 2 and Mars 3 were adversely affected by one of the largest Martian dust storms in recorded history, and contributed little to what Mariner 9 had already discovered. The US mission was able to resume imaging operations once the Martian dust storm cleared, and the understanding of the planet underwent another paradigm shift.

Approximately 7300 images at mostly 1–3 km resolution enabled global maps to be produced, and demonstrated Mars was much more geologically diverse and

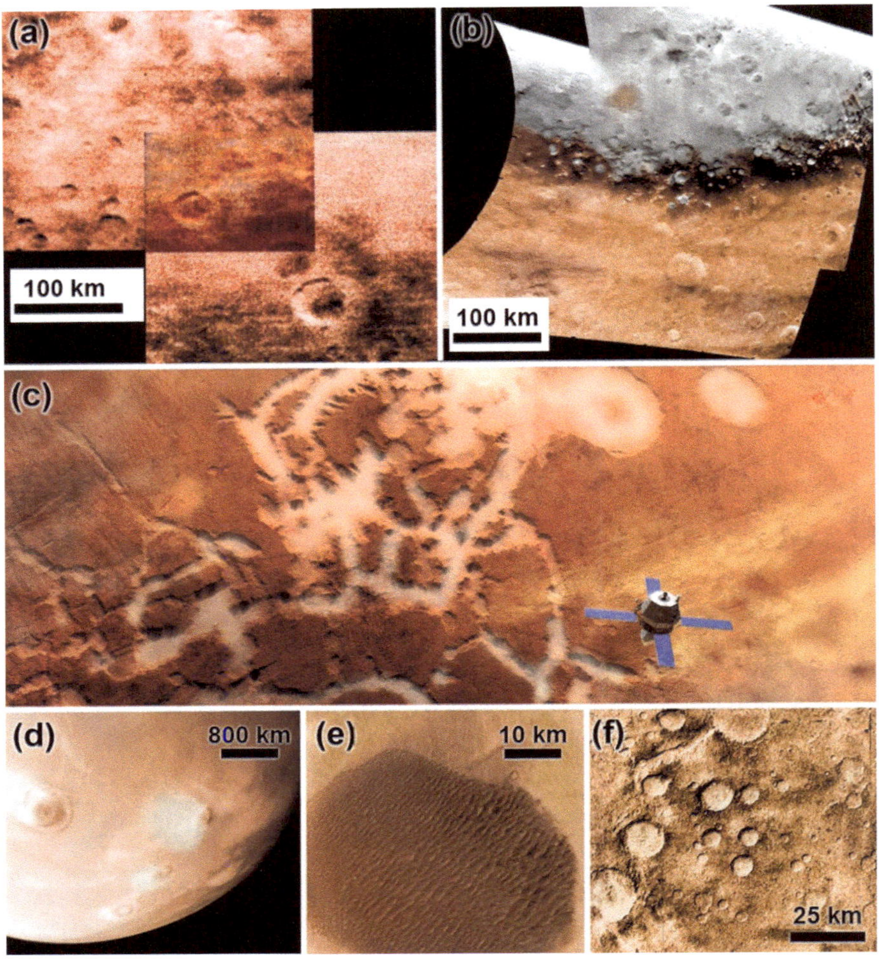

Fig. 1.5 Examples of early space-age Mars imagery. Note all images are reprocessed in colour derived from raw mission data, or from subsequent missions. **a** Mariner 4 flyby images showing cratered terrain (NASA/JPL). **b** Projected image of the planet's south pole from NASA's Mariner 7 mission (NASA/JPL). **c** Illustration of Mariner 9 in orbit over Mars, flying over Valles Marineris. **d** Tharsis volcanoes as imaged by Mariner 9 (NASA/JPL). **e** Proctor Crater dunes from the same mission (NASA/JPL). **f** A rare image from the Soviet Mars 5 mission, showing significantly eroded, flat-floored craters (NASA/JPL)

dynamic than previously thought. Features attributed to volcanism (Fig. 1.5d), wind (Fig. 1.5e), ice and water were discovered by Mariner 9. Volcanoes, including the largest in the Solar System, were discovered, with a lack of cratering indicating geologically young ages for many lava flows. Mariner 9 also discovered Valles Marineris, the largest canyon system in the Solar System and occupying 25% of the planet's circumference. Much of the current understanding of Martian geology, including significant geomorphology and elevation characteristics of the planet's northern and southern hemisphere was made possible by this mission (Masursky 1972).

The Soviet Mars 4, 5, 6 and 7 automatic stations were launched through July and August 1973, of which Mars 5 was the most successful. This mission reached the planet on 12 February 1974 and returned data, including imagery (Fig. 1.5f), for 22 orbits until an internal pressure loss ended operations (Clark 1986).

1.2.1 Additional Martian Orbiters

Post-Mariner 9, NASA sent two Viking orbiters and piggybacked landers to Mars in 1976, representing the most sophisticated and complete data collection of the planet undertaken to that time (Fig. 1.6a). Fundamental theories developed from these data remained unchanged for two decades. The twin orbiters returned 52,000 images, from which, along with other sensors, global maps, accurate albedo (fraction of sunlight reflected from the surface) and thermal emission datasets were developed. The Viking orbiter data, indicating a planet historically shaped by liquid water, were considered by some to be so complete that imaging payloads were not considered a priority on future Mars mission proposals.

The US followed up its Viking mission with Mars Global Surveyor (MGS), a replacement for a failed orbiter attempt, entering Martian orbit on 11 September 1997 (Fig. 1.6b). This orbiter was able to image the planet with resolutions two orders of magnitude greater than Viking with a maximum resolution of 1.5 m/pixel. MGS data were also used to develop very accurate topographic and mineral maps, using a thermal emission spectrometer and Mars Orbiter Laser Altimeter instrument (MOLA, Albee et al. 1998).

The Soviet Phobos 2 mission was the next (partially) successful mission to reach Mars on 28 July 1989. Its infrared (0.75–3.2 μm) spectrometer was able to obtain extensive spectra of Mars and its closest moon Phobos before mission failure. The spectroscopy data remains of interest to the present day, providing some insight into the geology of Mars' little-explored moon system (Barlow 2008).

The US Mars Odyssey orbiter and Europe's first successful Mars mission, Mars Express (carrying the Beagle 2 lander that failed on deployment), arrived at Mars on 7 April 2001 and 2 June 2003 respectively. These supplemented the results from MGS by mapping hydrogen and water content in the Martian regolith (Gamma Ray Spectrometer, GRS), providing thermophysical data (Thermal Emission Imaging System, THEMIS), stereo imagery (High Resolution Stereo Camera, HRSC), atmospheric mapping (SPICAM) hyperspectral imagery and ground penetrating radar (Mars Advanced Radar for Subsurface and Ionosphere Sounding, MARSIS, Barlow 2008).

Other NASA orbiters include the Mars Reconnaissance Orbiter (MRO), arriving on 10 March 2006 and Mars Atmosphere and Volatile Evolution (MAVEN) on 22

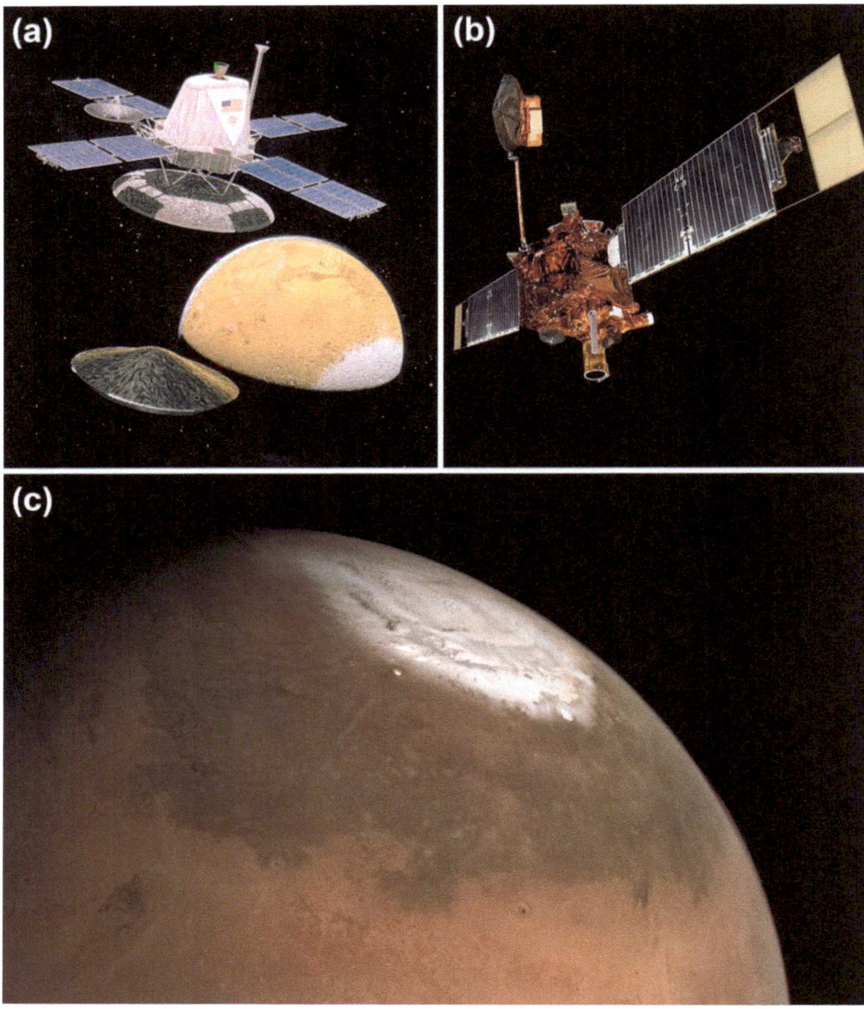

Fig. 1.6 **a** Artist impression of a Viking orbiter releasing a lander probe (NASA/JPL). **b** Mars Global Surveyor (MGS) in Mars orbit (NASA/JPL). **c** Example image of the Martian north pole taken from the UAE Hope mission (Public Domain)

September 2014. In addition to ground penetrating radar, hyperspectral and atmospheric sensors, MRO is mapping Mars at an unprecedented resolution of 30 cm per pixel, exceeding that of any sensor sent to deep space. The High-Resolution Imaging Science Experiment (HiRISE) has returned more data than all other past and present interplanetary missions combined (McEwen et al. 2023). Mars MAVEN has returned more modest amounts of data, while investigating atmospheric loss through solar wind stripping, and mapping hydrogen abundance through solar breakup of atmospheric water (Jakosky et al. 2015).

The European Space Agency (ESA) launched another orbiter that completed aerobraking in 20 February 2018, when ExoMars Trace Gas Orbiter commenced investigating the Martian atmosphere. The Nadir and Occultation for Mars Discovery (NOMAD) and Atmospheric Chemistry Suite (ACS) employed UV and infrared spectrometers to obtain the most extensive spectral coverage of the Martian atmosphere to date (López-Valverde et al. 2018).

With Mars Orbiter Mission (MOM) arriving on 24 September 2014, India became the first Asian country to successfully achieve Mars orbit. Designed as a technology demonstration, MOM remote sensors investigated atmospheric properties, as well as undertaking thermal and visible light imagery of the planet's surface (Arunan and Satish 2015). Additional countries such as United Arab Emirates and China have recently sent orbiter missions to Mars. Their Hope (example returned image, Fig. 1.6c) and Tianwen 1, reached the planet on 9 and 10 February 2021 respectively (Witze et al. 2020).

The atmospheric and surface sensing of these missions have been carried out using far UV, multispectral and thermal sensors, and in the case of Tianwen-1, hyperspectral sensing. Hope's Emirates Mars Ultraviolet Spectrometer (EMUS) measured a spectral range of 100–170 nm and investigated transient aurora occurring through solar wind proton interaction with the planet's upper atmosphere (Lootah et al. 2018). Tianwen-1's imagery and hyperspectral data, along with rover findings (below) have indicated that past liquid water activity was likely present in Utopia Planitia, the landing site for Viking 2 and that of Zhurong, the Chinese Mars Rover (Wu et al. 2022). Along with other past and present missions, these missions have contributed to the wealth of Mars' orbital data used in the chapters of this book.

Figure 1.7 provides an indication of the increased level of detail subsequent generations of Mars remote sensors have achieved, using the cratered region first imaged by Mariner 4 as a starting baseline (Fig. 1.7a). Viking (Fig. 1.7b), Mars Express (Fig. 1.7c), and MRO HiRISE (Fig. 1.7d) have vastly increased the amount and detail of imagery data available for this region and much of Mars.

1.2.2 Surface Missions

The Viking 1 and 2 landers touching down on 20 July and 3 September 1976 near Aries Vallis and Utopia Planitia respectively. They were the first to successfully return meaningful data from the Martian surface (Fig. 1.8a). Both sites were chosen as a trade between engineering constraints of reachable areas by Viking's landing systems, while also being of scientific interest. The nuclear-powered landers were able to monitor the landing sites, collecting imagery and meteorological collection data for two Mars years, with Viking 1 continuing operations until 1982.

Both Viking landers contained miniaturised experiments for searching for biological processes through regolith sample collection, the first and to-date only space

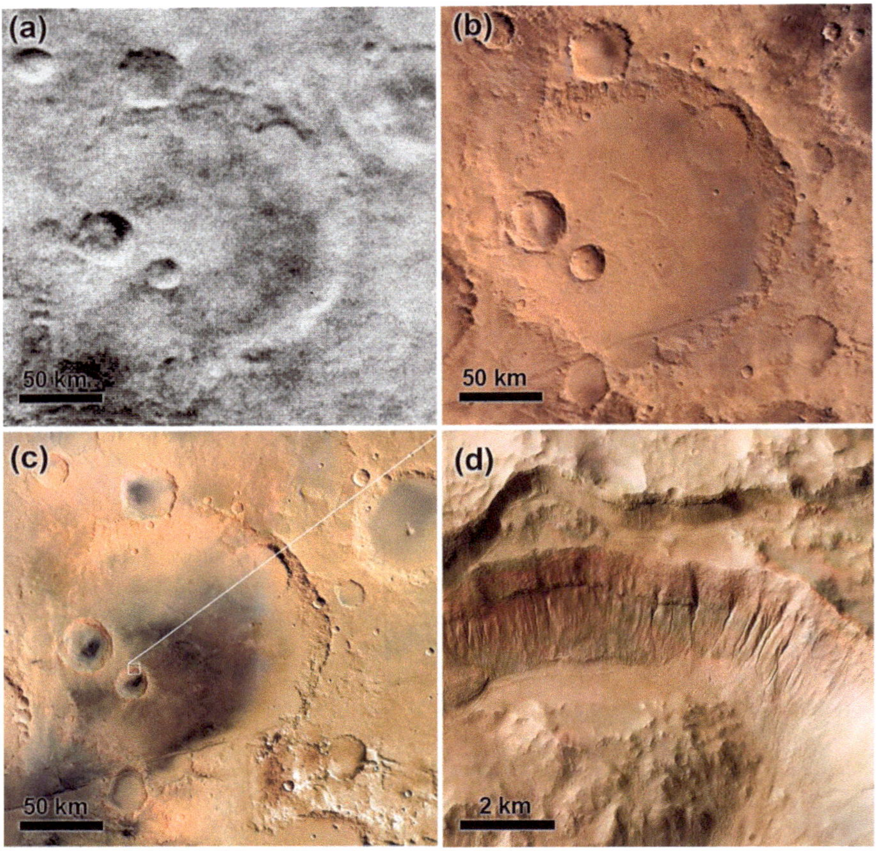

Fig. 1.7 **a** Mariner Crater first imaged by Mariner 4, with the smallest discernible features being 4 km in size (NASA/JPL). **b** The Viking orbiter returned more detailed imagery able to identify features 300 m in size (NASA/JPL). **c** Mars Express HRSC enhanced the image quality of this region (ESA/DLR/FU Berlin, CC BY-SA 3.0 IGO). **d** MRO HiRISE imagery shows the highest level of detail on Mars from orbit to date, allowing crater wall gullies to be imaged as an inset image from part c (PSP_002317_1445, NASA/JPL)

missions to do so. Experiments included Gas Exchange (investigated gasses produced from regolith to which nutrients were added), Labelled Release (injecting soil with radioactive nutrients and monitoring for metabolism activity), Pyrolytic release (simulating the Martian atmosphere using traceable radioactive compounds) plus a gas chromatograph to detect organic compounds.

The three biology experiments produced results that initially favoured a biological process, however, the gas chromatograph failed to identify the presence of any organic material in the regolith. Subsequent missions have discovered exotic chemistry, such as hyperchlorates, existing in Martian regolith, that could lead to nonbiological reactions with the experiments performed by Viking. Despite ongoing analysis of the data

Fig. 1.8 Successful Martian landers. **a** Illustration of the first mission to successfully return data from the surface, Viking 1 lander descending to Mars, narrowly avoiding a mission-ending rock. **b** Mars Pathfinder and Sojourner Rover (NASA/JPL). **c** Illustration of the Spirit Mars Rover at sunrise. **d** Curiosity Rover at Gale Crater (NASA/JPL). **e** Perseverance Rover (NASA/JPL). **f** The first successful international surface mission, Zhurong Rover, China's first lander to Mars (Public Domain). NASA has also flown static landers such as **g** Phoenix (NASA/JPL) and **h** Insight (NASA/JPL)

and laboratory simulations, the Viking search for life results remain inconclusive. Future astrobiology investigations plans favour sample return methods compared to in-situ measurements for more sophisticated searches for past or extant life (Davila et al. 2020; Davis 2021).

Pathfinder, a concept demonstrator lander/rover combination, landed at Ares Vallis on 4 July 1997. The Pathfinder lander operated a meteorology package and stereo multispectral camera while the Sojourner rover carried an Alpha Proton Spectrometer (APXS) to characterise soil and rock compositions (Fig. 1.8b; Golombek et al. 1999).

Two solar powered MER, Spirit and Opportunity explored Gusev Crater and Meridiani Planum respectively (Fig. 1.8c) from 2004. These golfcart-sized machines used airbags to cushion their landing and for the first time allowed long range exploration of their landing sites. Instrumentation, including a rock abrasion tool (RAT) returned evidence of past aqueous environments, including hot springs (Home Plate, Gusev Crater) and acid-water altered mineralogy. Spirit travelled 7.7 km over its six-year two-month operational life, and Opportunity covered over 45 km before falling silent in 2018.

NASA has landed a further two rovers, Curiosity and Perseverance, at Gale and Jezero Craters in 2012 and 2021 respectively (Fig. 1.8d, e). Perseverance also carried a 1 kg flying drone helicopter, Ingenuity, that conducted 72 flights in the thin Martian atmosphere. These latest missions represent the most sophisticated landed vehicles that have ever been sent to another planet and which operate extensive scientific payloads powered by nuclear sources. Multispectral cameras (Mastcam/Z, Hand Lens Imager, MAHLI) and spectrographs (APXS, Chemistry and Mineralogy X-Ray Diffraction, CHeMin), including those able to sample vapour produced from lasing target materials (Chemistry and Camera, CHEMCAM) are supplemented by meteorology and environmental sensors. As of publication both rovers are still operating, Perseverance recently concluding caching collected samples for a future Mars Sample Return mission (Bizony 2022).

Landing in Utopia Planitia on 14 May 2021, 1700 km from NASA's long-defunct Viking 2 lander, the Chinese Zhurong Rover represents the first successful non-US/ Soviet Union lander mission (Fig. 1.8f). The solar powered six-wheeled vehicle carried a Mars Rover Ground Penetrating Radar (RoPeR), magnetometer and meteorological sensors, plus multispectral and topography cameras. Zhurong discovered changes in Martian dune morphology in accordance with the planet's axial tilt, as well as hydrated minerals associated with the past presence of liquid saline water at the landing site. During its 356.5-day operation, the rover traversed over 1920 m from its lander prior to falling silent on 20 May 2022 (Liu et al. 2022).

The Mars Phoenix in 2008 and Insight in 2016 (Fig. 1.8g, h) were two recent static lander missions designed to investigate the immediate Martian environment. Phoenix investigated the Martian polar regions to investigate potential for local habitability and water interactions with the Martian regolith. The Phoenix mission confirmed the presence of subsurface water ice, plus likely melting of liquid water. It also discovered hyperchlorates, extremely reactive oxidising chemicals common in the

Martian regolith that probably affected the Viking search for life results (Hecht et al. 2009).

The Insight mission landed at Elsium Planitia on 26 November 2018 and placed a seismometer on the surface of the planet. During its four-year, 19-day operation it measured Marsquake activity, providing insight into the planet's interior crust, mantle and core composition (Dodson 2022; Banerdt et al. 2020).

1.3 Properties of Mars

1.3.1 Size, Mass, Distance from the Sun

Like Mercury, Venus and Earth, Mars belongs to the family of terrestrial planets orbiting the inner Solar System.

The planet is roughly half the diameter of Earth, measuring 6779 km (Fig. 1.9a) and is slightly more oblate (0.005 compared with 0.0034) than Earth. Unlike Earth, the planet does not possess an active magnetic field. Remote sensing observations of fossilised magnetic regions has provided evidence that the Martian core once generated a magnetic field that has since been lost (Langlais et al. 2004). Mars weighs 11% that of our planet, at 6.39×10^{23} kg, and has a density of 3930 kg m^{-3} (Smith et al. 1999). Despite its lighter weight, is differentiated in that it possesses a distinctive core, mantle and crust. The Martian crust lacks tectonics like Earth but instead consists of a single lithospheric plate. The thickness and rigidity of the lithosphere has allowed the growth of massive Tharsis volcanoes, compared to terrestrial volcanic chains formed through drifting tectonic plates. Deformation of the lithosphere has occurred through brittle failure caused by cooling and shrinking, volcanism-induced uplift and gravitational stresses, meteor impacts and other surface processes (Comer et al. 1985; Barlow 2008). An example includes uplifting from the Tharsis Bulge that has creased stress fractures and faulting in the surrounding terrain through gravitational loading (Comer et al. 1985).

Mars orbits between 206 and 249 million km from the sun at perihelion and aphelion respectively, with a period of 687 days compared to Earth's 365 (Fig. 1.9b). This means that variation in solar heating is more pronounced than that of Earth. Mars receiving 31% less heat at aphelion than perihelion. This is linked to seasonal changes between each hemisphere's respective summer, autumn, winter and spring (Fig. 1.9c). Present-day climate effects are linked with this solar variation, with dust storms often forming during increased heating periods, and ice clouds begin to interact with atmospheric dust during aphelion periods. Its daily rotation period is 24 h, 37 min and 22.65 s, longer than the terrestrial day.

The northern hemisphere of Mars is 1–3 km lower and has a thinner crust (average 32 km) than that of the southern hemisphere (average 58 km) and is termed the Martian Dichotomy (topographic elevation differences shown in Fig. 1.9d). It was emplaced within 500 Ma of the formation of the planet during a heightened earlier

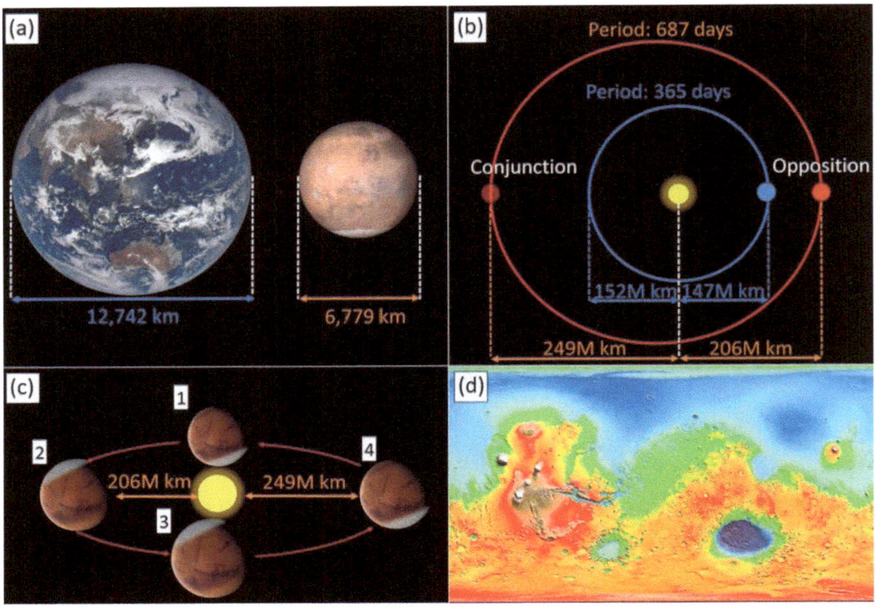

Fig. 1.9 **a** Diameter comparison between Mars and Earth (NASA/JPL). **b** Orbital distances and period comparison between Mars and Earth, including conjunction (both planets on the opposite sides of the Sun) and opposition (both planets at closest approach). The orbit of Mars has greater eccentricity when compared to that of Earth. **c** Martian seasons with (1) Southern autumn and northern spring. (2) Northern winter and southern summer occurring when the planet is closest to the sun. (3) Northern autumn and southern spring. (4) Southern winter and northern spring occurring when Mars is furthest from the Sun. **d** MOLA topography showing the northern hemisphere as significantly lower elevation (blue regions) than the southern hemisphere (yellow to red regions). The exception is the Hellas basin impact on the southern hemisphere possessing the lowest elevation regions on the planet (NASA/JPL)

period of comet and asteroid strikes. Thus, the Martian Dichotomy likely represents the results of a massive impact during this period that removed a substantial portion of the crust in this region (Golabek et al. 2011).

1.3.2 Moons

Mars' two moons Phobos and Deimos were discovered in 1877 by Asaph Hall and their size plus composition indicate that they may be captured asteroids. Remote sensing of these moons has been limited, with only one partially successful dedicated mission designed and flown. In addition to ground-based observations, most data are derived through Mars orbiter sensing, where a portion of spacecraft time was dedicated to imaging the two Martian moons. These included panchromatic or colour imagery from Mariner 9, Viking, Mars Global Surveyor, Mars Express and Mars Reconnaissance orbiter (Barlow 2008; Poggiali et al. 2022).

Phobos and Deimos orbit at 6000 km and 23, 460 km respectively from the Martian surface (Fig. 1.10a, Mars Express view, Fig. 1.10b).

Phobos, the largest of the two moons (Fig. 1.10c) measures 13.3 × 11.1 × 9.3 km in triaxial ellipsoid radii. It is losing energy to Mars through tidal interactions with each orbit, and is expected to be torn into a Martian ring or impact the surface in approximately 50 million years (Sharma 2023).

The geology of Phobos exhibits an abundance of craters, with the largest, Stickney, being 9 km in diameter. Parallel grooves and craters radiate from Stickney, being caused by secondary impacts and stress fracturing or possibly tidal flexing between it and Mars. Most of the craters on Phobos are simple bowl shapes. More complex morphology is rarer and indicative of organisation of its regolith into layers. A range

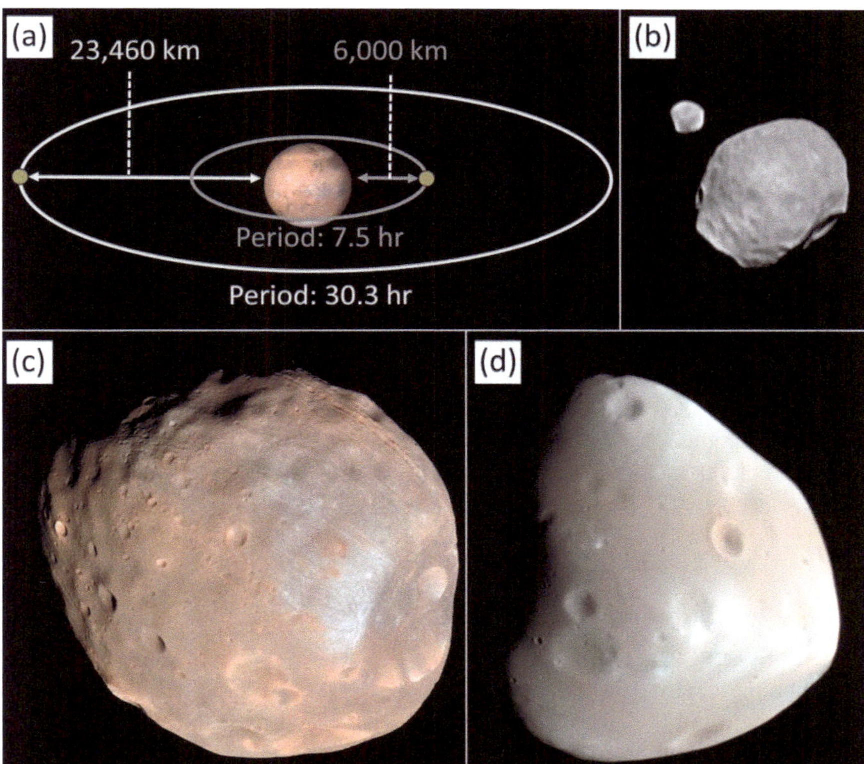

Fig. 1.10 **a** Orbits of Phobos and Deimos around Mars. **b** Both moons together as seen from Mars Express (ESA/DLR/FU Berlin, CC BY-SA 3.0 IGO). **c** Detailed view of Phobos from HiRISE, with blueish material visible near Stickney Crater (PSP_007769_9010, NASA/JL). **d** Detailed view of Deimos from HiRISE (ESP_012065_9000, NASA/JL)

of degradation is present amongst Phobos' crater population, likely caused by moonquakes generated by subsequent impacts and disturbing of the regolith, as well as micrometeorite erosion.

The Soviet Phobos 2 mission returned spectroscopic data of Phobos to infer its mineralogy. The moon's albedo is very low (0.07), with spectroscopy consistent with volatile-rich carbonaceous chondrites, similar to C-class asteroids (Hartmann 1990). A portion of material reflecting higher blue wavelengths ("blue unit") is located around Stickney Crater, suggesting it is material excavated from deeper within the moon, or a reservoir of chemically distinct material. A likely iron-rich "red unit" resides on the remainder of Phobos' surface, exhibiting Albedo variations of up to 10% suggest regolith grain size variation or impact "gardening" distributing different material on the surface (Pajola et al. 2017).

Deimos, measuring 7.6 × 6.2 × 5.4 km, features a low density similar to Phobos and suggests a porous or rubble-rich body (Fig. 1.10d). Deimos' spectra was found to be similar to the Phobos "red unit" and to "D-type" asteroids, a class of dark, reddish worlds found in the Kuiper Belt of the outer asteroid belt of the Solar System.

Deimos orbits further than Phobos from Mars at an average distance of 23,460 km and is slightly brighter than Phobos with an albedo of 0.08. Its surface is cratered but appears much smoother than the larger Martian moon, lacking grooves. This is likely due to up to the100 m thick regolith created from meteorite impacts. The abundance of fine-grained regolith compared to Phobos has been explained through the Stickney Crater impact launching much of Phobos' finer grained material into space; Deimos has not experienced an impact as large (Barlow 2008). This moon's extremely low gravity may also be the reason why crater ejecta deposits or mass wasting have not been observed on this moon. Meteor impacts would readily launch ejecta at speeds above the 5.6 m/s escape velocity for the moon.

Both Martian moons present an attractive target for future space missions, providing ideal natural orbital platforms before descending to the surface. Their location features as a staging point for telecommunications, radiation protection and landing preparation, and supports resource utilisation for future crewed missions to Mars (Deutsch et al. 2018).

1.3.3 Formation and Age of Mars

Mars, like the other terrestrial planets, is believed to have accreted through gravitational collapse of the solar protoplanetary nebular approximately 4.6 Ga. Recent theories suggest that the smaller size of Mars may have been the result of counter-interactions with the gravity of Jupiter, the Solar System's largest planet, considered to have formed at a similar period. Jupiter's orbit may have encroached on the inner Solar System, starving Mars of much of its accretion material and reducing its total mass to just 11% of Earth. This recent suggestion is used to explain the difference between our and exo-solar systems that often contain a greater abundance of larger planets than Mars (Jacobson and Morbidelli 2014).

Dating the formation and evolution of Mars is more difficult than that of the Earth. The absence of direct sampling has made crater counting the primary method for dating the planet's terrain and relies heavily on the Lunar cratering record. In general, the greater abundance of craters, the older the terrain (detailed in Chap. 8). Based on this data, as well as stratigraphy relationships, Mars's history was classified into three broad date ranges: Noachian, Hesperian and Amazonian (Fig. 1.11, example regions on Mars shown in Fig. 1.12a).

The Noachian era is the oldest of the Martian periods and is named after the heavily cratered Noachis Terra in the southern highlands. The Noachian spanned 4100–3700 Ma, during a period where the inner solar system was subjected to significant asteroid and comet collisions.

The Noachian era was the period at which Mars was at its most geologically active, with the formation of the Tharsis volcanoes and Valles Marineris. Noachian aged geology occupies 45% of Mars and features numerous craters dating from early periods of Solar System formation, and degrees of degradation indicating higher erosion rates than the present day (Fig. 1.12b; variations in crater erosion shown in part c). Large valley networks, fluvial deltas and drainage basins date from this period. The planet's denser atmosphere in this period supported standing liquid water, a likely northern ocean, rivers and outflow channels. Abundant chemical weathering occurred during this era to create hydrated minerals and clays. The Noachian was also the period in which most researchers believe that the Martian environment was most favourable for the development and sustainment of microbial life. Estimates of erosion rates on Mars show a rapid decrease from 10^{-7} to 10^{-5} Myr^{-1} in Noachian terrains (characterized by rimless, flat floored craters and valley networks) to the exceedingly slow rates of 10^{-11}–10^{-10} Myr^{-1} during the Hesperian and Amazonian periods (Golombek and Bridges 2000).

The Hesperian era (3700–3200 Ma) is a transition period between the Noachian and later Amazonian. This period was characterised by substantial volcanism and collapse of terrain in sudden and short-lived flooding events. Impact processes prevalent during the Noachian were much reduced, and the creation of extensive lava plains and volcanic features, such as wrinkle ridges and eroded volcanoes (Fig. 1.12d, e) were the dominant geologic processes of the period. Hesperian-aged plains cover approximately 30% of the planet.

The prevalent volcanism of the Hesperian era released sulphur dioxide and hydrogen sulphide into the Martian atmosphere, and chemical interactions with water produced an acidic weathering environment and creation of sulphate minerals. Associated cooling of the Martian core and mantle cooling is suggested to have frozen groundwater, forming a thick Martian cryosphere. Subsequent localised heating

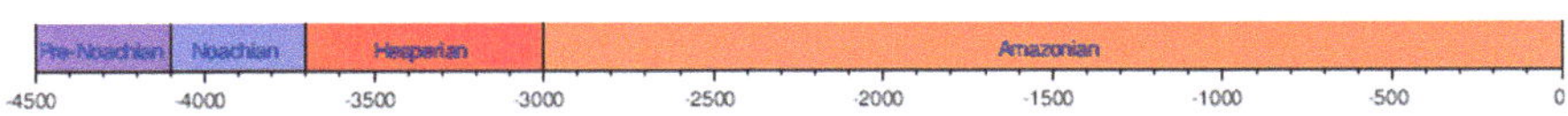

Fig. 1.11 Ages for Mars geological periods in millions of years

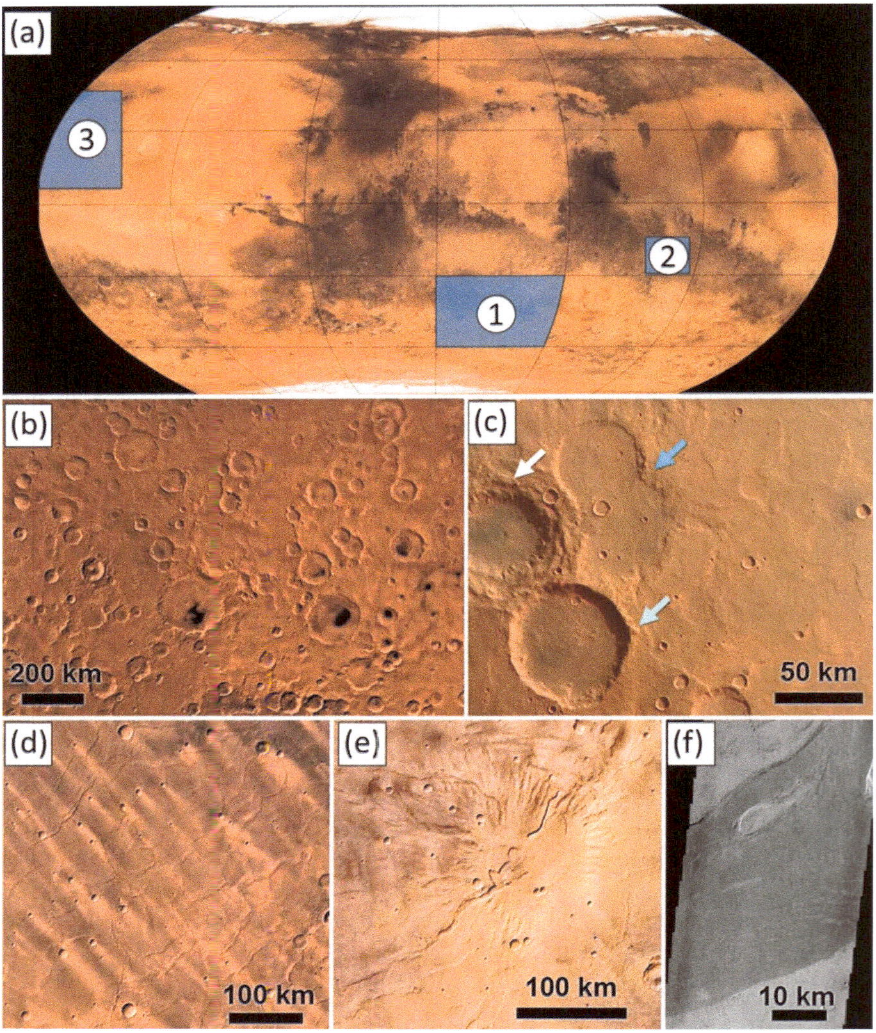

Fig. 1.12 **a** Overview locations of (1) Noachis Terra, (2) Hesperia Planum and (3) Amazonis Planitia (NASA/JPL). **b** Heavily cratered terrain typical of Noachis Terra (NASA/JPL). **c** Variations in crater shape in Noachis Terra with crater wall valley erosion marked by the white arrow, flattening of the crater floor showed by the light blue arrow, and virtual infilling and erasure of two craters marked by the darker blue arrow (NASA/JPL). **d** Wrinkle ridges within Hesperia Planum (NASA/JPL). **e** Tyrrhenus Mons: a heavily eroded, Hesperian-aged volcano (NASA/JPL). **f** Marte Vallis is a representative Amazonian-aged feature (PIA13249, NASA/JPL)

through volcanic activity melted ice-rich materials to cause the catastrophic flooding described above (Carr and Head 2010).

The Amazonian period is the longest and most recent Martian age, commencing from 3200 Ma to the present day. This era is characterised by low crater impacts, a thin atmosphere and hyper-arid, aeolian-driven conditions observed at the present time. Younger lava flows, and outflow features, such as Marte Vallis (Fig. 1.12f), may be present within this terrain (Barlow 2008).

Remote sensing has changed the current understanding of the Amazonian era. Mars was initially thought as virtually geologically inactive by this period, but new findings, such as isolated volcanic events occurring less than 5 Ma, small-scale fluvial activity and flows of ice-rich material suggest a more dynamic planet than once thought.

Although Mars is not the pre-space-age world once imagined by astronomers, the robotic missions described in this chapter have revealed compelling evidence of a dynamic and evolving planet. Mars is an alien world with a thin CO_2 atmosphere, a heavily cratered surface, and vast canyon systems, yet its landscape has been shaped over time by volcanism, wind, ice, and ancient liquid water. These same processes continue to operate on Earth today, highlighting important geological parallels between the two planets despite their very different environments.

The flyby, orbital and surface missions described above are responsible for making these discoveries and rely on robust engineering practices and technological developments. The Martian environment is hostile to electronics and subsystems; and many attempts to reach Mars have failed. Environmental considerations and engineering practices will be discussed in the next chapter.

References

Albee AL, Palluconi FD, Arvidson RE (1998) Mars global surveyor mission: overview and status. Science 279:1671–1672. https://doi.org/10.1126/science.279.5357.1671

Arunan S, Satish R (2015) Mars orbiter mission spacecraft and its challenges. Curr Sci 1061–1069

Banerdt WB, Smrekar SE, Banfield D, Giardini D, Golombek M, Johnson CL, Lognonné P, Spiga A, Spohn T, Perrin C, Stähler SC (2020) Initial results from the InSight mission on Mars. Nat Geosci 13:183–189. https://doi.org/10.1038/s41561-020-0544-y

Barlow N (2008) Mars: an introduction to its interior, surface and atmosphere. Cambridge University Press, New York, 264 pp

Bizony P (2022) NASA missions to Mars: a visual history of our quest to explore the red planet. Motorbooks International, USA

Carr MH, Head JW III (2010) Geologic history of Mars. Earth Planet Sci Lett 294:185–203. https://doi.org/10.1016/j.epsl.2009.06.042

Chalmers A (1985) Galileo's telescopic observations of Venus and Mars. Br J Philos Sci 36:175–184

Clark PS (1986) The Soviet Mars programme. J Br Interplanet Soc 39:2–18

Colbert E (1871) Star studies: what we know of the universe outside the Earth. The Western News Company, Chicago, 115 pp

Comer RP, Solomon SC, Head JW (1985) Mars: thickness of the lithosphere from the tectonic response to volcanic loads. Rev Geophys 23:61–92

Davila A, Kahre MA, Quinn R, Marais DJD (2020) The biological potential of present-day Mars. Planet Astrobiol 16:169

Davis B (2021) Did the 1976 Mars Viking LR biology experiments detect life from Earth. J Astrobiol 7:1–5

de Vaucouleurs G (1954) Physics of the planet Mars. Faber and Faber, London, 365 pp

Deutsch AN, Head JW, Ramsley KR, Pieters CM, Potter RW, Palumbo AM, Bramble MS, Cassanelli JP, Jawin ER, Jozwiak LM, Kaplan HH (2018) Science exploration architecture for Phobos and Deimos: the role of Phobos and Deimos in the future exploration of Mars. Adv Space Res 62:2174–2186. https://doi.org/10.1016/j.asr.2017.12.017

Dodson G (2022) NASA retires InSight Mars Lander mission after years of science. NASA. Accessed 5 Dec 2023

Golabek GJ, Keller T, Gerya TV, Zhu G, Tackley PJ, Connolly JA (2011) Origin of the Martian dichotomy and Tharsis from a giant impact causing massive magmatism. Icarus 215:346–357

Golombek MP, Bridges NT (2000) Erosion rates on Mars and implications for climate change: constraints from the Pathfinder landing site. J Geophys Res Planets 105:1841–1853

Golombek MP, Anderson RC, Barnes JR, Bell JF III, Bridges NT, Britt DT, Brückner J, Cook RA, Crisp D, Crisp JA, Economou T (1999) Overview of the Mars Pathfinder Mission: launch through landing, surface operations, data sets, and science results. J Geophys Res Planets 104:8523–8553

Hartmann WK (1990) Additional evidence about an early intense flux of C asteroids and the origin of Phobos. Icarus 87:236–240

Hecht MH, Kounaves SP, Quinn RC, West SJ, Young SM, Ming DW, Catling DC, Clark BC, Boynton WV, Hoffman J, DeFlores LP (2009) Detection of perchlorate and the soluble chemistry of Martian soil at the Phoenix lander site. Science 325:64–67

Jacobson SA, Morbidelli A (2014) Lunar and terrestrial planet formation in the Grand Tack scenario. Philos Trans R Soc A Math Phys Eng Sci 372:20130174

Jakosky BM, Lin RP, Grebowsky JM, Luhmann JG, Mitchell DF, Beutelschies G, Priser T, Acuna M, Andersson L, Baird D, Baker D (2015) The Mars atmosphere and volatile evolution (MAVEN) mission. Space Sci Rev 195:3–48

Langlais B, Purucker ME, Mandea M (2004) Crustal magnetic field of Mars. J Geophy Res Planets 109. https://doi.org/10.1029/2003JE002048

Leighton RB, Murray BC, Sharp RP, Denton AJ, Sloan RK (1965) Mariner IV photography of Mars: initial results. Science 149. https://doi.org/10.1126/science.149.3684.627

Liu J, Li C, Zhang R, Rao W, Cui X, Geng Y, Jia Y, Huang H, Ren X, Yan W, Zeng X (2022) Geomorphic contexts and science focus of the Zhurong landing site on Mars. Nat Astron 6:65–71. https://doi.org/10.1029/2022EA002624

Lootah F, Almatroushi H, Holsclaw G, Deighan J, Chaffin M, Lillis R, Fillingim M, England S, AlMheiri S, Reed H (2018) Scientific payload of the Emirates Mars Mission: Emirates Mars ultraviolet spectrometer (EMUS) overview. In: European planetary science congress, Sept 2018, pp EPSC2018-1006

López-Valverde MA, Gerard JC, González-Galindo F, Vandaele AC, Thomas I, Korablev O, Ignatiev N, Fedorova A, Montmessin F, Määttänen A, Guilbon S (2018) Investigations of the Mars upper atmosphere with ExoMars trace gas orbiter. Space Sci Rev 214:1–45

Lowell P (1895) MARS. Houghton, Mittlin, 228 pp

Masursky H (1972) An overview of the geological results from Mariner 9. J Geophys Res 78:4009–4030

McEwen AS, Byrne S, Hansen C, Daubar IJ, Sutton S, Dundas CM, Bardabelias N, Baugh N, Bergstrom J, Beyer R, Block KM (2023) The high-resolution imaging science experiment (HiRISE) in the MRO extended science phases (2009–2023). Icarus 115795. https://doi.org/10.1016/j.icarus.2023.115795

Owen T (1982) The composition of the Martian atmosphere. Adv Space Res 2:75–80. https://doi.org/10.1016/0273-1177(82)90107-7

Pajola M, Simioni E, Lucchetti A, Re C, Cremonese G, Thomas N, Pommerol A, Tornabene L (2017) Refining the boundary between the Phobos blue/red spectral units with the ExoMars-CaSSIS imagery. In: European planetary science congress, vol 11

Poggiali G, Matsuoka M, Barucci MA, Brucato JR, Beck P, Fornasier S, Doressoundiram A, Merlin F, Alberini A (2022) Phobos and Deimos surface composition: search for spectroscopic analogues. Mon Not R Astron Soc 516:465–476. https://doi.org/10.1093/mnras/stac2226f

Sharma BK (2023) Comparative study of tidal evolution of Mars-Phobos-Deimos based on kinematic model and based on seismic model. J Earth Environ Sci Res 190:2–12. https://doi.org/10.47363/JEESR/2023

Sharp RP, Soderblom LA, Murray BC, Cutts JA (1971) The surface of Mars: 2, uncratered terrains. J Geophys Res Planets 76:331–342

Sheehan W (1996) The planet Mars: a history of observation & discovery. University of Arizona Press, 270 pp

Slipher EC (1955) New light on the changing face of Mars. National Geographic, Sept 1955

Smith DE et al (1999) The global topography of Mars and implications for surface evolution. Science 284:1495–1503. https://doi.org/10.1126/science.284.5419.1495

Wallace AR (1907) Is Mars habitable?: a critical examination of Professor Percival Lowell's book "Mars and its canals," with an alternative explanation. MacMillan & Co, London, 110 pp

Witze A, Mallapaty S, Gibney E (2020) All aboard to Mars. Nature 583:184–188

Wright WH (1947) Biographical memoir of William Wallace Campbell, vol xxv third memoir. National Academy of Sciences of The United States of America, Biographical Memoirs, 74 pp

Wu B, Dong J, Wang Y, Rao W, Sun Z, Li Z, Tan Z, Chen Z, Wang C, Liu WC, Chen L (2022) Landing site selection and characterization of Tianwen-1 (Zhurong rover) on Mars. J Geophys Res Planets 127. https://doi.org/10.1029/2021JE007137

Young CA (1900) A textbook of general astronomy for colleges and scientific schools. The Athenarum Press, London, 630 pp

Chapter 2
Space-Age Exploration and Technologies

2.1 Space Engineering Considerations for Mars

Mars mission development is typically undertaken to fill a gap in knowledge about the Red Planet because an orbiter or lander spacecraft return data to gain new understanding. At the start of the space age, research questions were put forward to enhance our Mars understanding. However, they were wide in scope. In this sense, two early examples were confirming at a broad scale the composition of the planet's atmosphere and the imaging of surface features at a better resolution rather than the previous telescopic observations. In addition, early missions, being constrained by technology available at the time were able to provide huge order-of-magnitude increases in understanding of Mars with relatively rudimentary instruments (Chap. 1). Over the following decades, as more missions returned data, scientific questions have become much more refined, requiring more tailored engineering designs to answer them.

Every mission flown to Mars has had to contend with finite resources including financial cost, human workforce resources, and schedule. The 26-month opposition cycle where Mars and Earth make their closest approach in their orbits represents the most energy efficient method for launching missions and places a hard constraint on flight schedules. Missions not ready in time for an upcoming launch window are delayed until the next launch opportunity, causing significant impacts on operational and workforce costs, as well as delays in returned returning scientific data. Mission budget is another constraint. This includes development costs of design, analysis, and test of low NASA technology readiness levels (TRL, Mankins 1995). TRL concepts are applied to engineering models, production and testing of flight subsystems; and operation and sustainment of the flown space mission. Other costs to be factored into system design include project management and engineering staffing costs, redundancy against failure (e.g. Woellert et al. 2011; Selva and Krejci 2012) as well as technology limitations in operating in a hostile environment.

Balancing the constraints with scientific return, used for every flown Mars mission, involves conducting design and engineering trades. These make evidence-based

S. Hobbs, *Sensing Mars: Remote Sensing to Understand the Red Planet*,
https://doi.org/10.1007/978-3-032-24416-1_2

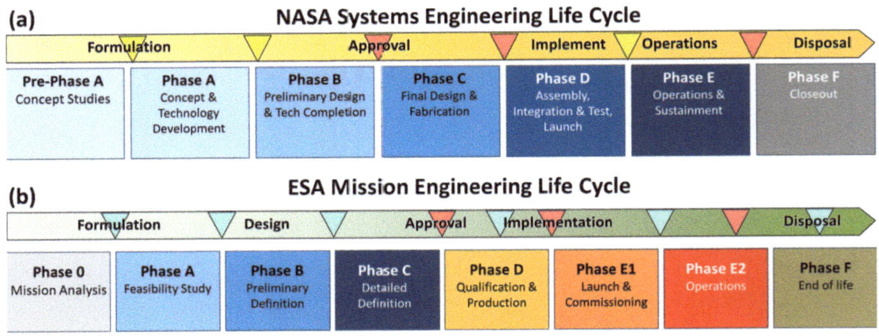

Fig. 2.1 Systems engineering lifecycle showing primary phases and decision points. **a** Process overview as used by NASA. Yellow triangles denote reviews that occur prior to proceeding to the next phase, red triangles denote major reviews. **b** The equivalent mission engineering lifecycle as used by ESA with process reviews (blue triangles) and milestone reviews (red triangles)

options development achievable within a schedule and budget, while still returning useful scientific data.

Space engineering and operations follow a lifecycle that help mission planners formulate, design, operate, sustain, and end missions in a planned and logical manner (Ridolfi et al. 2009; Hirshorn et al. 2017). One of the most popular is the NASA Space Systems Engineering framework that splits the life of a program into seven key phases, linked by critical decision points that need to be made prior to progressing to the next phase (Fig. 2.1a). ESA has equivalent mission program phases (Fig. 2.1b). These processes are tailored to the scope and complexity of the mission. Design engineers may elect to undertake a condensed version of the engineering cycle, a key decision taken early during concept development phase. Primary phases for NASA missions include Pre-Phase A, and Phases A through F (Fig. 2.1a). These map with the ESA definitions (Fig. 2.1b) and will be discussed in more detail.

2.1.1 NASA Engineering Phases

Pre-Phase A is the highest level of planning and concept design for spacecraft missions. High-level goals are defined in this pre-phase, such as better understanding the history of a surface site, or even to demonstrate the operation of novel technology on Mars. A broad range of concepts are developed that will need further development and analysis to refine in terms of TRL, cost and schedule. A mission concept review concludes this phase, where the high-level options are formally accepted for consideration and progression to the next phase. Planetary protection measures and policy compliance measures commence from this phase and continue throughout the lifecycle. Planetary protection operates to minimize the possibility of terrestrial biological contamination of Martian sites through sterilizing space hardware below pre-determined bacterial spore counts. Failure to sterilize equipment could otherwise degrade or destroy whatever native biology that may be present, or introduce foreign biota to an alien environment (e.g. Olsson-Francis et al. 2023).

Phase A takes the high-level concepts from the previous phase and develops them into more detailed architecture able to satisfy identified mission needs and provide reasonable estimates of costs and schedule. During this phase, the high-level mission needs are broken down into more detailed goals and requirements from which expectations of performance and utility can be measured. Examples may include defining volume and weight constraints, the need to distinguish between surface mineral types, or imaging the Martian surface at a certain level of detail. In Phase A, engineering models may be developed as concept demonstrators and to help reduce mission risk.

Engineering trade studies to critically analyse and provide evidence for specific concept options are conducted in this phase to identify increased utility for reduced cost. As an example, sensors with high resolution are typically larger due to long focal length optics, requiring more payload volume to accommodate (Wertz et al. 2011). They are generally data intensive, requiring wide communication bandwidth to return large volumes of information that compromise the ability of other sensors in the same spacecraft to operate. Such a design is useful for highly detailed geomorphic (analysis of formation and evolution of surface landforms) investigations of the Martian surface, and is used by the HiRISE instrument on MRO. This payload is not appropriate for small-scale missions, or where analysis of larger parts of the planet are required, such as dust storm or weather monitoring (Ebben et al. 2007). Another trade consideration may entail a choice between designing a new instrument or payload from scratch (high risk) or incorporating previously flown technology that, while not novel, might be suitable to satisfy the mission requirements.

Results of the trade studies refine and develop a suite of payload instruments that satisfy developed requirements and guide space vehicle design and systems. A systems requirements review and mission design review, where identified requirements and proposed mission design are formally accepted, conclude this phase.

Phase B refines the mission in enough detail to provide a baseline for meeting the project's objectives. In this phase, technology and prototype development, as well as hardware and software trade studies are conducted with a high level of detail. Concept demonstrations and risk reduction activities are completed, with results used to break down project requirements to detailed system and subsystem requirements levels. A preliminary design is also developed during Phase B that provides a substantially detailed concept spacecraft where payloads and subsystems are identified with performance metrics. Operational planning is later conducted to a substantial level of detail to demonstrate in detail how the mission will operate and achieve its goals.

As Phase B progresses, the preliminary design of the mission is refined, with fundamental or large-scale changes becoming less likely. A series of preliminary design reviews may be undertaken to align design proposals to requirements. A final preliminary design review formally accepts the requirement to design approach, the result being a presentation of clearly-defined design specifications. It should be noted that significant design changes required after the conclusion of Phase B typically result in major cost increases for the planned mission.

Phase C develops the final detailed design of the mission and commences fabrication. The concept of the spacecraft is developed to provide an end-to-end design that is validated using a series of complex trade studies. Prototypes developed during Phase C will closely resemble the final hardware. Testing is planned to mimic actual operating conditions as much as possible, generating real metrics used to verify designs and mission operations. The last step prior to fabrication is a critical design review. This is the final assessment before committing designs to physical hardware and is conducted during Phase C.

Tailoring of this phase for suitable missions may entail splitting up Phase C into two sub-phases: a final design closeout that concludes with the critical design review, and the second sub-phase entailing fabrication. During the fabrication phase, hardware and software are developed in accordance with the mission plans approved from the critical design review. Changes or alterations made during this phase become prohibitively expensive as hardware will need to be modified, re-engineered, or discarded and rebuilt from scratch. As a mitigation strategy, outputs are carefully monitored so that undesirable trends or faults are detected and rectified early.

Phase D takes the fabrication product developed in the previous phase and conducts assembly, integration, validation, and launch. Flight readiness and mission readiness reviews provide formal assurance that the built spacecraft will function as planned to satisfy requirements agreed to in Phase A, and will survive the launch into space. Operation manuals and procedures will have been developed by Phase D and final go/no go decisions to fly will be made at the conclusion of this phase.

Phase E conducts the primary mission through launch, primary operations, and sustainment. The primary needs and goals are met through operation of the payloads. As an example, Phase E for the Perseverance Rover covered the vehicle launch, cruise to Mars, entry, descent, and landing, as well as its day-to-day operations. Mission operations planning is conducted prior to each Martian sol (one Martian rotation period) consisting of payload operations scheduling and synchronisation of communications passes (Farley et al. 2020). Modifications conducted during Phase E may entail upload of new software during operation, amendments to in-flight hardware of course being cost-prohibitive. This phase also includes sending mission data to Earth for analysis.

Phase F details the closeout and disposal of the mission. All scientific and engineering data are downloaded for the final time and archived. In the context of Mars missions, Phase F is commenced upon failure of a mission, where Phase E operations are no longer sustainable. For on-orbit missions, Phase F may occur following exhaustion of attitude control propellant or failure of a spacecraft to respond to commands (Zielinski 2007). Successful surface missions have predominantly succumbed to the harsh Martian conditions, where either hardware failure or power loss forced an end of operations (e.g. Callas 2015).

2.1.2 Deep Space Operational Impacts on Spacecraft Engineering

The space environment is much more hostile on all equipment intended to operate there, with vacuum, temperature extremes and high energy radiation being significant hazards. Being further from the Sun than Earth, the environment around Mars poses its unique challenges. These challenges shape the development of a Mars mission throughout the engineering life-cycle, from design and build (pre-Phase A to D), mission operations (Phase E) and disposal (Phase F). Failure to design, build and test these missions with sufficiently robust components and operating procedures often leads to mission failure. Specific implications for spacecraft engineering are described below.

Variable Background Radiation

Deep-space probes operate well beyond the Earth's magnetic field and are subject to increased background radiation. The radiation in the space environment originates from the Sun, as well as from interstellar and intergalactic sources. The Sun outputs ionising and non-ionising radiation, including UV radiation known to degrade spacecraft components such as solar substrates, as well as charged particles from the solar wind. Transient solar events such as coronal mass ejections and solar flares can massively increase the amount of radiation present in the deep space environment. Results returned from the Radiation Assessment Detector on board the Curiosity Rover has helped characterize the Martian radiation environment (Guo et al. 2021). Further from the Sun, interplanetary missions become increasingly affected by interstellar and intergalactic radiation (Table 2.1; Garrett et al. 2008).

Jupiter and Earth's orbits possess significant amounts of radiation, while the interplanetary space environment is lower (Table 2.1). Aluminium shielding of 2.54 mm would be sufficient to reduce radiation dosage to less than 1 krad/yr, with a hypothetical three-year mission receiving a maximum dose of 3 krad. Many existing space mission components are tested to verify toleration of up to 20 krad of radiation, providing off-the-shelf options for mission consideration (Crowley 2018).

Increased Communication Distance

For spacecraft in low-Earth orbit, size and power constrictions on communications equipment are not as severe as those required for deep-space missions. The proximity of a spacecraft to Earth has allowed the use of small transmitters with modest power outputs and near-instantaneous communications. The varying distances between Mars and Earth make this form of communication impossible, with speed of light one-way delays ranging between 4.3 and 21 min, depending on where the two planets are in their orbits (Love and Reagan 2013). Missions to Mars also require higher power transmitters (e.g. 20 W Viking orbiter, 100 W MRO transmitter, Koktas and Basar 2024) and often large parabolic, high gain antennas. Dishes of 3 m and 70 m diameter at the Deep Space Network can receive and transmit data to spacecraft (Fig. 2.2).

Table 2.1 Representative radiation environments

Mission	Dominant species	Total dose (krad/yr)		Total SEU risk (relative)	SFU susceptibility (bit-flip/Gbits/ye)[a]			
		With 100 mil Al	With 400 mil Al		Trapped protons	Cosmic rays[b]	Solar protons[c]	
							With 100 mil Al	With 400 mil Al
LEO	p+	< 1	< 1	Low	+	10^{-2}	0	0
LEO-Polar	p+	6	1	Moderate	+	10^{-2}	10^{-2}	10^{-5}
MEO	e− (soft)	200–2000	10–100	Severe	++	10^{-2}	10^{-3}	10^{-4}
GEO	e− (soft)	< 100	< 1	Moderate	0	10^{-2}	10^{-3}	10^{-4}
Jup-Galileo	e− (hard)	250	25	Severe	++	10^{-2}	0	0
Jup-Europa	e− (hard)	1500	800	Severe	++	10^{-2}	0	0
Interplanetary	p+	< 1	< 1	Low	0	10^{-2}	10^{-1}	10^{-3}
DOD/Nuclear	e− (soft)	50–10,000	1–100	Severe	N/A	N/A	N/A	N/A

[a] For devices with LET = 10 MeV cm^2/mg
[b] For Adams' 90% worst case galactic cosmic rays
[c] For 99th percentile solar flare events

The Deep Space Network consists of three sites places 120° apart to ensure continuous coverage of space missions regardless of the Earth's rotation. The sites are at Goldstone, near Barstow, California; Madrid Deep Space Communications Complex, Spain; and Tidbinbilla, near Canberra, Australia.

Where high bandwidth is required, such as for imaging payloads, mission designers use X-band (~ 8 GHz), where higher frequencies allow transmission of vast volumes of data (e.g. MRO: 2.6–3.2 Mb/s). Lower frequency systems with low gain antennas are commonly used to transmit spacecraft telemetry data at low bandwidths. The typically omnidirectional nature of low gain antennas has the advantage of not needing accurate pointing to Earth. This assists during a mission anomaly, where critical commands can be sent and received if the spacecraft has lost the ability to correctly orient itself (Vacchione 2006).

Surface missions, typically having greater size, mass and weight constraints, may use a lower-powered communications system and employ orbiters as relay stations to Earth. In this application, the transceiver needs only needs to be powerful enough to produce a usable signal to reach Mars orbit. More powerful orbiter transceivers then relay the information to the much more distant Deep Space Network receivers on Earth. Operation of the relay method uses standardised communications protocols, such as Proximity-1, jointly established between NASA, ESA, Russia, and Japan (Koktas and Basar 2024). This also allows for better collaboration between international efforts. An additional constraint with surface to orbit communications is the requirement to wait for the chosen orbiter to rise in the sky with respect to the

Fig. 2.2 70 m dish at Tidbinbilla, one of the tracking stations in the Deep Space Network (Public Domain)

lander for such operations to be possible. The communication period is dictated by the time in which the orbiter remains overhead with respect to the lander, as well as scheduling with orbiter payload operation periods. During surface and orbital missions, line-of-sight transmission windows are monitored closely and affect not only operational planning but also priority of data transfer (Edwards 2007).

Power Availability

The electrical power system (EPS) of spacecraft needs to be efficient, flexible and reliable.

The EPS is a mission critical subsystem on which all other functions of the spacecraft depend. It must meet the power requirements of the spacecraft for the duration of the mission. At its basic level, the EPS collects energy from the Sun for solar-powered vehicles or nuclear sources as in the case of the Curiosity or Perseverance rovers, converts the energy to electrical power and regulates it to voltages the spacecraft electronics can manage. Excess energy is stored in a battery while the remainder is switched out and distributed to the rest of the spacecraft. A representative solar energy flow is illustrated in Fig. 2.3 (after Patel 2005).

While the battery can supplement high power requirements for short periods, the spacecraft must be able to collect and convert more energy than it uses over its mission to replenish the energy storage. Solar arrays on Mars operate under reduced solar energy availability, given Mars' increased distance from the sun. In addition, lower operating temperatures, altered solar spectrum due to suspended atmospheric

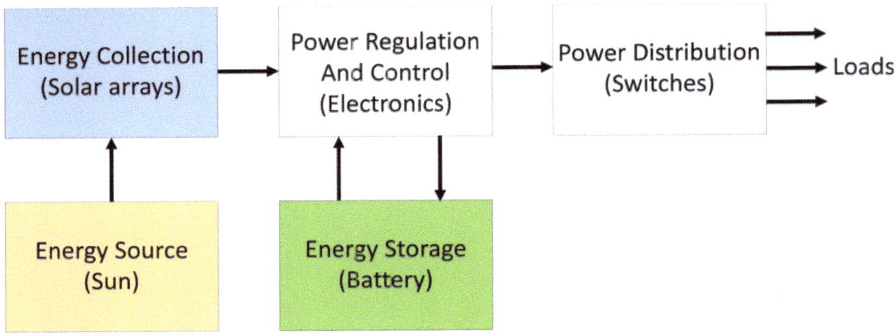

Fig. 2.3 Typical electrical power system (EPS) structure

dust and deposition of dust onto horizontal surfaces constrain the output of Martian solar arrays (Mazumder et al. 2006). Triple-junction cells constructed from layers of Gallium Indium Phosphide, Gallium Arsenide and Germanium provided MER with increased energy-gathering capacity in the redder end of the spectrum that characterizes the Martian operating environment. The solar arrays were constructed of lightweight material that minimised payload mass while also being able to survive the g forces of launch, entry and landing (Landis et al. 2004).

Nuclear sources are an alternative to solar panels and have flown on the Viking Mars landers, as well as the more recent Curiosity and Perseverance Rovers. Radioisotope thermoelectric generators (RTG) or multi-mission RTG (MMRTG) use thermocouples of dissimilar metals to convert heat into electrical power. The thermal energy is provided through the radioactive decay of Plutonium-238 that is produced in a ceramic form to reduce accidental spread in case of a mission accident. RTGs so far flown to Mars are not efficient, with Curiosity only able to convert 120 W of electrical power from 2000+ W of thermal energy. The most recent RTG flown to Mars on the Perseverance Rover only has 6.3% efficiency, and its application is to charge chemical battery banks better equipped for higher power demands (Holgate et al. 2015). Despite the difficulty in manufacturing Plutonium-238, increased weight and reduced efficiency when compared to solar arrays, RTGs are able to provide consistent power day or night over decades (Chen et al. 2025).

Thermal Considerations

The space environment negatively impacts the lifespan of a spacecraft and its subsystems. Thermal cycling is a significant concern beyond Earth orbit. Temperatures changes in orbit or on the surface causes damaging expansions and contractions in solar cells and electrical connections. In deep-space missions, thermal cycling damage was inferred as a contributing factor for the mission ending events on two surface missions on Mars. The Mars Pathfinder Mission for example was believed to have failed following a thermal induced open circuit to its power system (Mishkin

2003). Temperature regulation is thus required for subsystems and electronic components to remain within their manufactured tolerances or risk malfunction or failure (e.g. Garzon 2012).

Nuclear heating sources have also been used to compliment or as an alternative to battery-powered heaters. These use the natural decay of radioactive materials, such as Plutonium, to produce heat and in some cases supply power over a period of years, reducing power demands from a traditional solar powered energy system (e.g. Schmidt et al. 2011). Extra engineering is required to protect electronics and sensors from the radiation these sources emit, by either increasing the distance between the heater and such components, or by providing additional shielding. In addition, the procurement and handling of such radioactive devices, including host country nuclear policies, add additional constraints to the use of these devices (Wertz et al. 2011).

2.2 Travel Between Earth and Mars

Mars and Earth make their closest approach in their orbits every 760 days. Use of this orbital cycle enables transfer of spacecraft between the planets using the minimum possible rocket energy (Fig. 2.4a). As Mars' orbit is more eccentric than that of Earth, the lowest energy periods occur roughly every 15 years, and require half the rocket propulsion energy of the peak transfers (Fig. 2.4b).

A Hohmann transfer orbit represents the path for the lowest energy travel between Earth and Mars. This elliptical orbit (Fig. 2.4a) has its perigee (lowest point) at the location of the lower orbit (Earth) and the apogee (highest point) at the more distant orbit (Mars). A Hohmann transfer spacecraft would take 259 days to reach Mars with a change in velocity (delta-v) of 3.9 km/s, and would be launched when Earth and Mars approach opposition, the point at which they are closest to each other in their orbits (Stinner and Begoray 2005). In the case of a return from Mars to Earth, a burn to slow a spacecraft's velocity would allow it to follow a reverse Hohmann transfer orbit, "falling" back to Earth over another nine-month period (Stinner and Begoray 2005; Burke et al. 2010).

2.2.1 Navigation, Propulsion and Attitude Control

Accurately positioning a space vehicle and its payloads are critical for remote sensing applications. This includes being able to predict where a spacecraft is at any given time, as well as its attitude and orientation in space. Increased accuracy in these attributes increases the chance for scientific return from the mission. Satellites in Earth orbit, while still being exposed to the space environment, can take advantage of measures to assist with navigation. Position, Navigation and Timing (PNT) systems, commonly known as geo-positioning systems (GPS), assist with geolocation and navigation (Parkinson et al. 2006). GPS consists of a constellation of mid-Earth

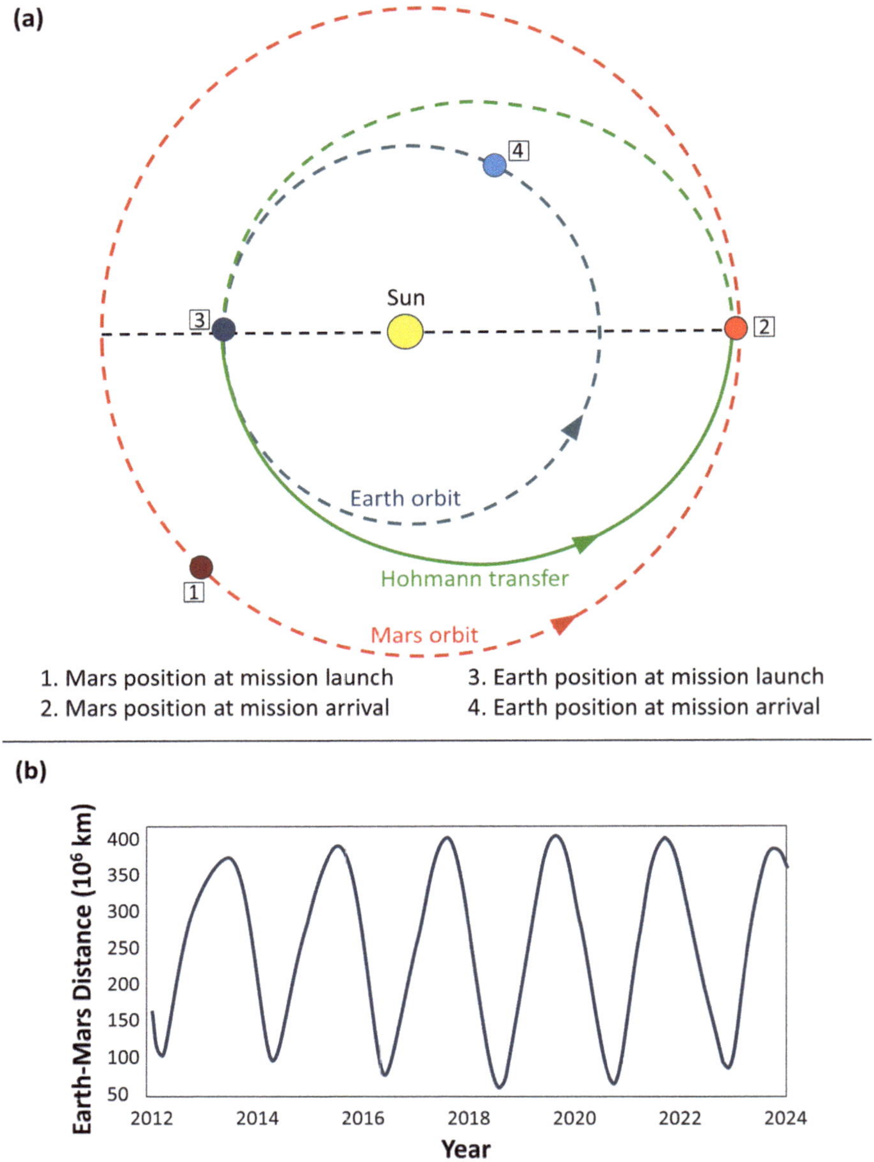

Fig. 2.4 **a** Illustration of a Hohmann transfer between Earth and Mars, where a space mission is targeted to "catch up" with the planet in its orbit. **b** Historic variations of distance between Earth and Mars. Shorter distances allow for greater payload mass and save on fuel

orbit satellites that use precise signal timing to geolocate receivers mounted on in-orbit spacecraft. The Earth's magnetic field is also available to assist with navigating and attitude control. Magnetorquer devices, using energized coils along orthogonal axes, react to the force of the Earth's magnetic field (Li et al. 2013). Deep-space missions, operating far from Earth orbit, are unable to make use of either GPS or the Earth's magnetic field. Spacecraft positioning in Mars orbit instead relies on trajectory calculations based on communicator wave propagation. This method operates though transmission of precisely timed radio transmissions sent from Earth to the spacecraft and measuring the time it takes for the signals to be received and retransmitted. Differences in the expected times as calculated by trajectory predictions and the actual returned signal allow for precise course corrections to be made (Wertz et al. 2011).

Spacecraft orientation can also be provided through celestial navigation methods, using Sun and/or star tracking systems as well as reaction wheels (Overview, Fig. 2.5). Star trackers detect the brightness of target stars, providing information on spacecraft attitude that can then be held using the rotational inertia of reaction wheels placed at 90° angles to each other. This method assists the pointing of imagers or other remote sensing instruments at specific portions of the planet's surface. Early spacecraft relied on notional pointing, introducing errors in registration of different datasets. High-resolution sensors, such as Context Camera (CTX) and HiRISE, require far greater pointing accuracy than lower resolution instruments. At these sensitivities, determination of physical forces acting on the space vehicle, such as solar wind pressure, thermal effects and gravitational anomalies are paramount. MRO underwent star tracker to HiRISE calibration activities in October and December 2005 using stellar clusters to determine to high accuracy the boresight of the high-resolution imager. Additional accuracy was exploited through overlapping charged coupled device (CCD) with known ground spatial distances for deriving precise geometric corrections to collected imagery (Johnston et al. 2005; McEwan et al. 2007).

2.2.2 *Propulsion*

Terrestrial satellites, once achieving orbit, do not usually require significant additional propulsion, and can use attitude control to conduct their mission. Propulsion energy is a concern for deep-space counterparts needed to reach their planetary targets, either as part of a primary mission or as part of a hosted payload (Martin-Mur et al. 2016). The velocity required to reach Mars is typically provided by the launch vehicle. Following departure from the Earth's surface, the upper stage then generates the velocity to escape Earth's gravity and reach Mars. The liquid hydrogen and oxygen fueled Centaur upper stage is a popular choice for Mars missions. This stage can be restarted during flight, initially for the spacecraft to achieve Earth orbit and a second time to accelerate the Mars mission away from Earth and to its destination. In the case of India's MOM, a less powerful Polar Satellite Launch Vehicle

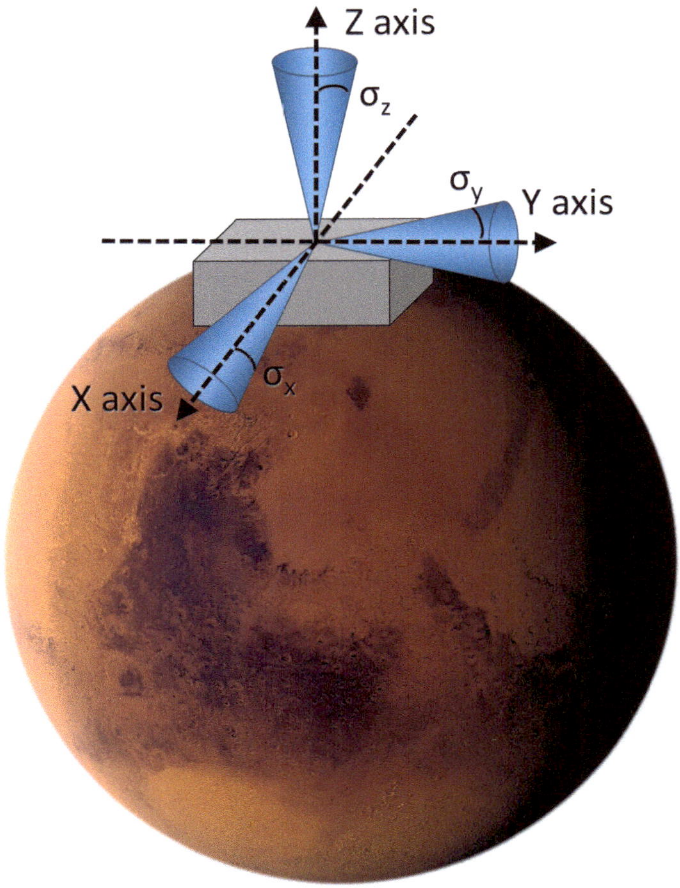

Fig. 2.5 Overview diagram of a star tracker in operation (grey box), to orient a spacecraft along three axes (X, Y, Z). Pointing errors, represented by sigma "σ" cones reduce remote sensing accuracy, with high-resolution sensors requiring low sigma values

was the only rocket deemed suitable at the time. This meant that the MOM spacecraft itself was forced to use thrusters to increase its perigee altitude until achieving trans-Mars injection (Helfrich et al. 2015).

A "secondary payload" concept is intended to reduce the requirement for smaller spacecraft to carry large propellants for deep-space transit. In this scenario, the smaller mission, such as a cubesat, is transported to its destination via a larger spacecraft, as was the case of the Mars Insight lander that carried MarsCO (Hodges et al.

2017). Small-scale thrusters or electric propulsion would enable the final maneuvering once the secondary payload is released from the main spacecraft (Thangavelautham et al. 2017). The disadvantages of this method are that it requires the availability of a host mission, as well as increasing the complexity of the piggybacked mission's operations.

2.2.3 Presence of Significant Perturbation Forces

Deep space exerts external forces that affect the trajectory of deep-space spacecraft, including gravity fields from celestial bodies and solar activity. Direct sunlight exerts pressure on material exposed to its rays which over time will change a spacecraft's velocity. Differing surfaces of a spacecraft reflect different amounts of light, resulting in a rotational force, or solar radiation pressure (SRP) torques acting on the vehicle. SRP is of greater consideration to larger spacecraft, and smaller cubesat-style vehicles are less affected by its influence (Wertz et al. 2011).

Outside of the Earth's magnetic field, the solar wind exerts greater pressure on spacecraft operating in deep space. This natural space weather phenomenon consists of charged particles streaming radially outwards from the Sun at supersonic speeds. The solar wind decreases in force with an increase in distance from the Sun, before slowing down and finally ceasing on encountering winds from the interstellar region (Cummings and Odenwald 2008). This pressure needs to be accounted for during trajectory and Mars transit planning.

2.2.4 Arriving at Mars

Spacecraft must decelerate to achieve orbit or land on Mars. For example, Mariner 9 and Viking both used large braking rockets to decrease their interplanetary velocity, allowing them to be captured by Mars' gravity field (Soffen and Snyder 1976). While effective in achieving their desired orbit, these spacecrafts required sacrificing payload mass to carry sufficient fuel to slow down. Later orbiter missions such as MGS and MRO, employed smaller rockets to achieve a highly elliptical Martian orbit, then used the planet's atmosphere to execute a manoeuvre called aerobraking. This entailed sending the vehicle through the upper portions of the atmosphere and using the resulting friction forces to lower the apogee over successive cycle, until a circular orbit was achieved. This minimised fuel use at the expense of taking longer to achieve the orbit required for conducting Phase E of the mission (Lyons et al. 1999).

Orbital Considerations

For most Mars spacecraft, the type of orbit and distance from the planet has a direct bearing on what remote sensing can be achieved.

The velocity V of a satellite in an elliptical orbit (Fig. 2.6a) is provided by the vis-viva equation (Rauschenbakh et al. 2003):

$$V = (GP(2/r - 1/a))^{0.5} \quad (2.1)$$

where GP is the Standard Gravitational Parameter (for Mars, is $4.2828 \times 10^{13}\ m^3\ s^{-2}$), "r" is the distance between the satellite and the centre of Mars, and "a" is the semi-major axis (the longest radius of the orbit ellipse).

As an example, MRO orbits between 255 km (minimum altitude above the surface) and 320 km (maximum altitude above the surface).

Converting to orbital heights above Mars centre (in metres) provides:

- 3,389,500 (Mars radius) + 255,000 = 3,644,500 m. This value is used for "r" at periapsis (closest point in orbit to Mars).
- 3,389,500 (Mars radius) + 320,000 = 3,709,500 m. This value is used for "r" at apoapsis (furthest point in orbit from Mars).

"a" is calculated as the average of 3,644,500 and 3,709,500, equalling 3,677,000 m.

Using these values for formula 2.1 for periapsis provides:

- $V = (4.2828 \times 10^{13} \times (2/3{,}644{,}500 - 1/3{,}677{,}000))^{0.5}$;
- V (periapsis) = 3443 m/s (approx.).

Using these values for formula 2.1 for apoapsis provides:

- $V = (4.2828 \times 10^{13}(2/3{,}709{,}500 - 1/3{,}677{,}000))^{0.5}$;
- V (apoapsis) = 3383 m/s (approx.).

The Hope spacecraft orbits much higher at 20,000 and 43,000 km above the Martian surface.

Converting to orbital heights above Mars centre (in metres) provides:

- 3,389,500 (Mars radius) + 20,000,000 (minimum altitude above the surface) = 23,389,500 m. This value is used for "r" at periapsis.
- 3,389,500 (Mars radius) + 43,000,000 (maximum altitude above the surface) = 46,389,500 m. This value is used for "r" at apoapsis.

Calculating for "a" as the average of "a" is calculated as the average of 23,389,500 and 46,389,500, equalling 34,889,500 m.

Similar to the above, using these values for formula 2.1 for periapsis provides:

- $V = (4.2828 \times 10^{13}(2/23{,}389{,}500 - 1/34{,}889{,}500))^{0.5}$;
- V (periapsis) = 1560 m/s (approx.).

Using these values for formula 2.1 for apoapsis provides:

- $V = (4.2828 \times 10^{13}(2/46{,}389{,}500 - 1/34{,}889{,}500))^{0.5}$;
- V (apoapsis) = 787 m/s (approx.).

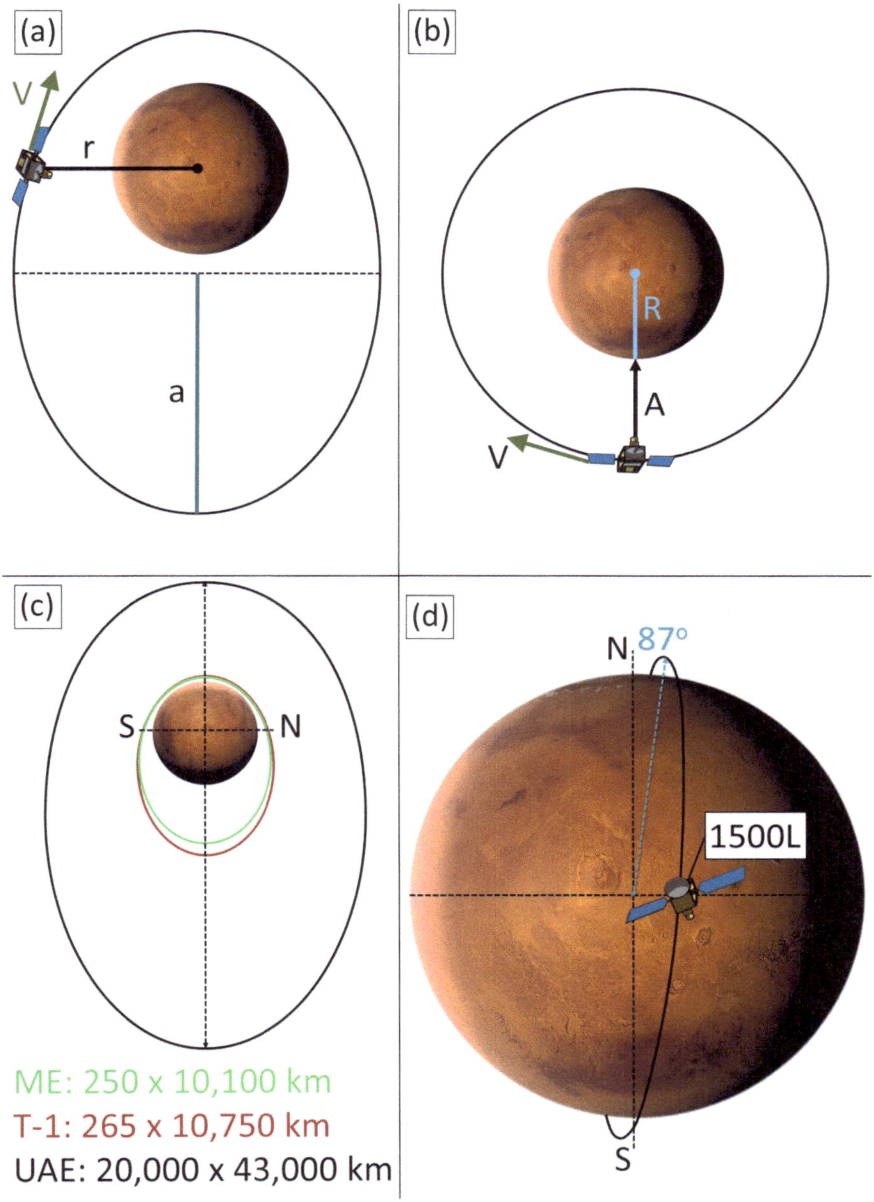

Fig. 2.6 **a** Principle components of a spacecraft in an elliptical planetary orbit showing "a", the semimajor axis, "r", the distance between the centre of Mars and the spacecraft and "v", its velocity. **b** Equivalent overview to part (a) but for a circular orbit. **c** Examples of elliptical orbits used by Mars Express ("ME", green), Chinese Tianwen-1 orbiter ("T-1", dark red), and UAE Hope spacecraft ("UAE", black). **d** MRO HiRISE sun-synchronous orbit where the spacecraft, inclined 87° to the poles, crosses the planet's equator at 1500 local time

For a circular orbit the formula becomes (Fig. 2.6b):

$$V = (GP/(A + R))^{0.5} \tag{2.2}$$

where

- R is the radius of the body, and
- A is the altitude of the spacecraft above the surface of the body.

For example, assuming a circular orbit for the Mars Odyssey orbiter of 400 km (and converting to metres):

A + R is 400,000 + 3,389,500 = 3,789,500 m;
GP is 4.2828×10^{13} m^3 s^{-2}.

Using these values for formula 2.2 provides:

- $V = (4.2828 \times 10^{13}/3{,}789{,}500)^{0.5}$;
- V = 3361 m/s.

Both formulae demonstrate that satellite velocity increases as the orbital height is reduced. From a remote sensing perspective, the higher orbital velocity places greater constraints on sensor design, because faster frame rates are required to avoid smear during exposures.

As with terrestrial remote sensing missions, choosing the preferred Martian orbit is subject to engineering trade-offs for scientific return within cost and technology constraints, underpinned by the foundational needs of the missions. An example of an engineering trade off occurred with the Indian MOM, where the choice of launch system constrained the amount of manoeuvring fuel the spacecraft carried. This meant that, following arrival at Mars, MOM was able to achieve only a highly elliptical orbit of 420 × 80,000 km, higher than most recently flown Mars missions (Sundararajan 2013).

Elliptical orbits have been favoured for Martian missions, including the UAE Emirates Hope probe (worked example above, represented as black ellipse, Fig. 2.6c; Amiri et al. 2022). A near-polar, elliptical orbit of 265 × 10,750 km was also used by the Chinese Tianwen-1 mission that arrived at Mars on 10 February 2021 (represented as dark red ellipse, Fig. 2.6c; Wang et al. 2025). These orbits allow for longer dwell times over regions of the planet and communication windows with Earth as they approach apoapsis, and a reduced time of being in Mars' shadow, hence reducing power collection constraints (Wertz et al. 2011). They also allow for greater mission flexibility. As an example, the elliptical orbit of Mars Express at 250 × 10,100 km provided an advantage of multiple scientific investigations, at the cost of variable resolutions during its low altitude mapping phase (represented as green ellipse, Fig. 2.6c). The higher apoapsis (furthest distance from the planet's surface) allowed for global measurements to be made by its atmospheric and spectral sensors, and sub-500 km altitude used for HSRC, MARSIS and high-resolution hyperspectral captures (Pischel and Zegers 2009).

Additional Martian orbit types used or considered by mission planners include: sun-synchronous, critical inclination, frozen orbits and aerostationary (Mars-specific) orbits. Low altitude orbits are desirable for many terrestrial and interplanetary mapping missions, where the closer proximity to the surface maximises the ability to identify small surface details of limited aperture sensors. A type of low-altitude orbit used by MRO is sun-synchronous, where the spacecraft crosses the Martian equator at the same time (1500 local time) each sol (Fig. 2.6d). This provides consistent lighting for the remote sensors, minimising shadows due to higher solar grazing angles near the equator and avoiding morning mists that would otherwise obscure terrain. It also provides a consistent sun angle, which makes for constant hill shading and easier recognition of landforms. The sensor height above the surface has little variation, meaning that pixel size over the surface is consistent during planetary mapping. Disadvantages of these orbit types are that they require more manoeuvring to achieve, and are more greatly affected by variations in Mars' gravity and atmosphere than higher orbits. Lower orbits also subject a spacecraft to increased eclipse time, where the bulk of the planet blocks the Sun, reducing total power and thermal energy available to a spacecraft (Wertz et al. 2011).

Aerostationary orbits for Mars exist 17,032 km above the planet's equator, and operate in the same fashion to geostationary orbits around Earth. A satellite placed in this type of orbit maintains its position over the same part of the surface: its period equates to the planet's rotation of 24.62 h. Such orbits are ideal for communication and navigation, though no flown mission has yet taken advantage of this type of orbit (Liu et al. 2010).

Frozen orbits allow for long term predictability through stability of eccentricity (maintaining the orbit's shape over time) and perigee argument (maintaining the orientation of the orbit angle between where the spacecraft crosses the Martian equator to its closest point to the planet's surface). Mars Express was placed in such a frozen elliptical orbit (Hechler and Yáñez 2003). Frozen orbits require a minimum of manoeuvring fuel to maintain, accommodating irregularities cause by the planet's shape and gravity field.

2.2.5 *Reaching the Surface*

Surface mission operations use different localisation and navigation methods to operate on the planet. For direct-to-surface missions, employed for all NASA post-Viking landers, celestial navigation and trajectory control are still required to transfer these missions from Earth to Mars using a cruise module that is discarded prior to landing. In the case of piggybacked orbiter/lander systems such as the US Viking and Chinese Tianwen-1 missions, celestial navigation is performed by the orbital craft (Soffen and Snyder 1976; Li et al. 2021).

Entry, descent and landing (EDL) on Mars presents significant systems engineering challenges that are unique to the planet. Unlike the Moon, Mars possesses an atmosphere capable of causing substantial heating on re-entry, though not thick enough (6–7 mBar) to substantially reduce terminal velocity to a safe landing. High-speed parachutes are required to provide additional deceleration, but are unable to slow a lander to survivable speeds due to the thin Martian atmosphere (Korzun et al. 2010). Higher elevations are traditionally avoided for EDL considerations owing to the lack of atmosphere to safely slow a spacecraft at these altitudes. Wind and weather modelling are also crucial steps for successful EDL. Prevailing winds may be shaped by surface topography, causing turbulence or unfavourable atmospheric currents affecting parachute deployment. Increased dust loads may also alter atmospheric density and restrict surface visibility required for obstacle avoidance for precision landing (Braun and Manning 2012).

Surface conditions are also carefully assessed during engineering planning, with ideal terrain possessing a minimum of surface dust, hazardous rocks or outcrops and low slopes to allow for safe landing. Additional considerations for surface trafficability are required for enabling roving vehicles, such as Curiosity and Perseverance, to traverse to differing geological sites (e.g. Nelessen et al. 2019).

Landing on Mars has used a variety of technologies. For the first phase of landing, aeroshells provide the initial deceleration from interplanetary space to atmospheric entry, following further deceleration by parachutes. Terminal descent methods have depended on mission type and included descent rockets, air bags, and a combination of both. Landing accuracy has also increased significantly since the first successful soft landing by Mars 3 (Marov and Petrov 1973). Better understanding of atmospheric profiles, plus the use of sophisticated landing systems such as the sky crane have produced a significant reduction in the landing error ellipse size between subsequent missions to the most recent Perseverance Rover that landed in Jezero Crater (Fig. 2.7; Dwyer Cianciolo and Powell 2017).

The Perseverance Rover employed real time imaging software to map its position by correlating landmarks previously identified in orbital imagery to what the descent camera was imaging. Additional calculations based on remaining fuel and other constraints identified and targeted the rover to a safe landing area (Johnson et al. 2022).

Once a lander is on the ground, triangulation using imaged surface features is a dominant method for precisely determining position. Topographic features such as crater rims or hills within the landers field of view are imaged, and positions compared to orbital imagery. For the latest generation of Mars rovers, this process is significantly enhanced with the ability of HiRISE to detect objects 25 cm in size, allowing the vehicles themselves to be directly imaged. This direct imaging, plus on-orbit topographic mapping compliments rover on-board navigation systems to plan future traverses and scientific investigations (Wan et al. 2021).

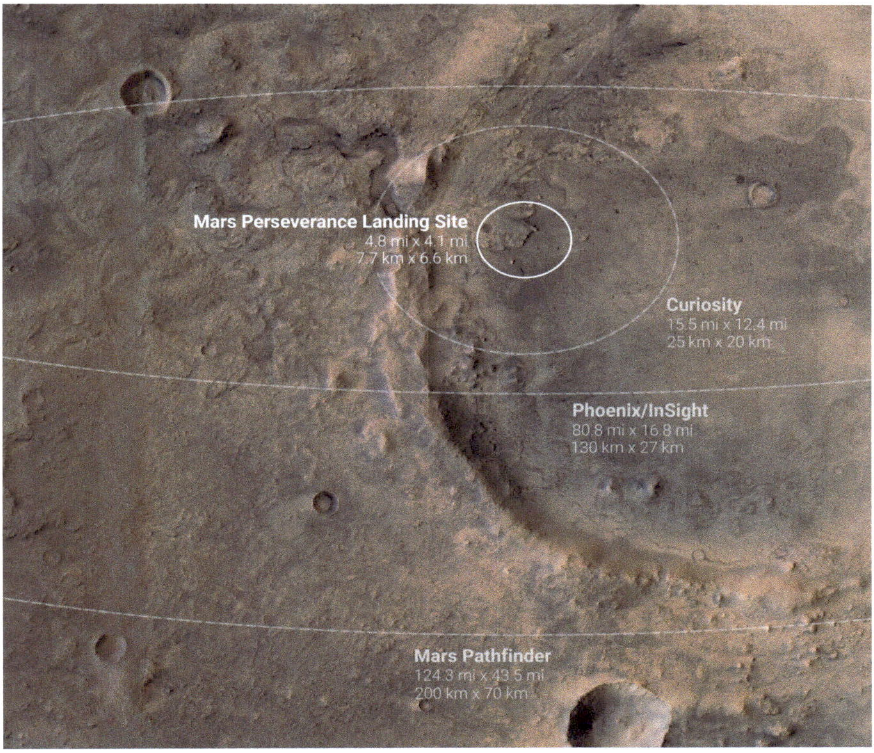

Fig. 2.7 Landing ellipse sizes of Mars missions showing the increase in accuracy with improved landing technology (PIA24377, NASA/JPL)

2.3 The Influence of Space 2.0

The invention of the cubesat standard, along with subsequent development of miniaturised components, has drastically lowered the cost to put payloads into Earth orbit. The ability to plan, design, build and launch a cubesat mission is well within the budget of small businesses, universities and even schools. Researchers are increasingly taking advantage of reduced scope, "Space 2.0" missions for deep space and Mars exploration (Asundi and Fitz-Coy 2013). As well as savings in cost, additional advantages include less wait time between concept design and return of scientific results and the potential to piggyback on larger missions. This latter benefit has the advantage of using excess mass allowance that would otherwise be wasted on ballast, increasing the frequency of missions able to be sent to deep-space (Alanazi and Straub 2019).

Concurrent with the development of small satellites has been the reduction in size and weight, and increased affordability of remote sensors. These developments are making Mars Space 2.0 missions more achievable and attractive to the scientific community (Walker et al. 2018).

MarsCO was the first successful cubesat mission to operate near Mars and consisted of two spacecraft, flown in conjunction with the InSight mission in 2018 (Fig. 2.8a, b). The twin 6U spacecraft operated 3500 km above the Martian surface, relaying InSight radio data back to Earth. MarsCO demonstrated that low-cost missions consisting of off-the-shelf components can contribute to Mars exploration (Schoolcraft et al. 2017).

At the time of publication, additional cubesat missions to Mars are being considered. Example proposals include cubesats profiling the Martian atmospheric and polar properties using constellation or hyperspectral technologies (Kleinboehl et al. 2022; Pagano et al. 2022), as well as a mission to the Martian moons (Nallapu et al. 2020).

The design and scope of these Space 2.0 missions are necessarily constrained by cubesat limitations of payload space, power, propulsion and communication. The small size of the cubesat standard limits the power collection available, constraining payload operations. In addition, propulsion systems are generally limited to operation near Mars, with Earth-Mars transfer dependant on a primary payload, as was the case with MarsCO (Schoolcraft et al. 2017).

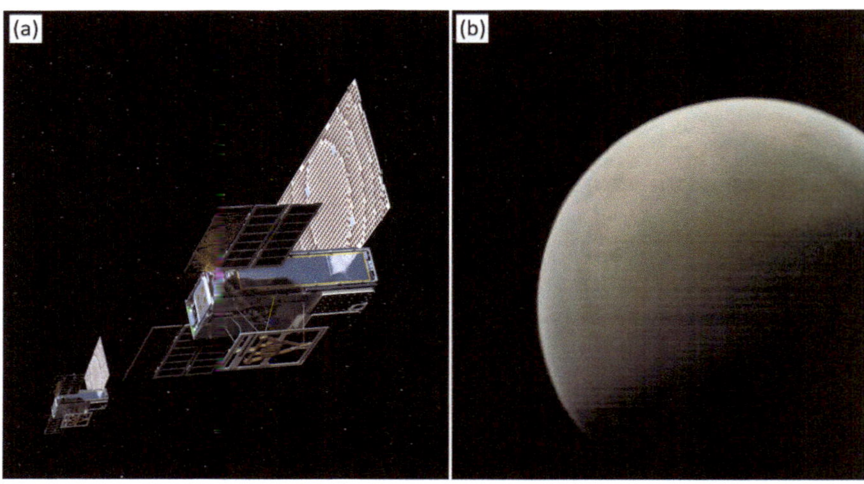

Fig. 2.8 **a** Illustration of MarsCO cubesats in operation near Mars (NASA/JPL). **b** Image returned from the MarsCO mission of Mars (NASA/JPL)

2.3.1 Trade Study Analysis

Historic missions to Mars have incorporated custom design of sensors as part of their mission. The prohibitive cost and schedule for this, plus the increased availability of commercial off the shelf scientific-grade sensors has made this option less attractive for mission designers. Instead, for Space 2.0, engineering trade studies underpinned by mission requirements and operational constraints help drive optimal design and choice of available sensors within agreed-to constraints. This involves researching and obtaining a list of component options that represent possible solutions to the chosen mission requirements (Fig. 2.9). Utilities such as sensor performance, resolution and sensitivity, along with constraints including mass, volume and power draw are then evaluated across the alternative options until an informed decision can be made on choosing the optimal component.

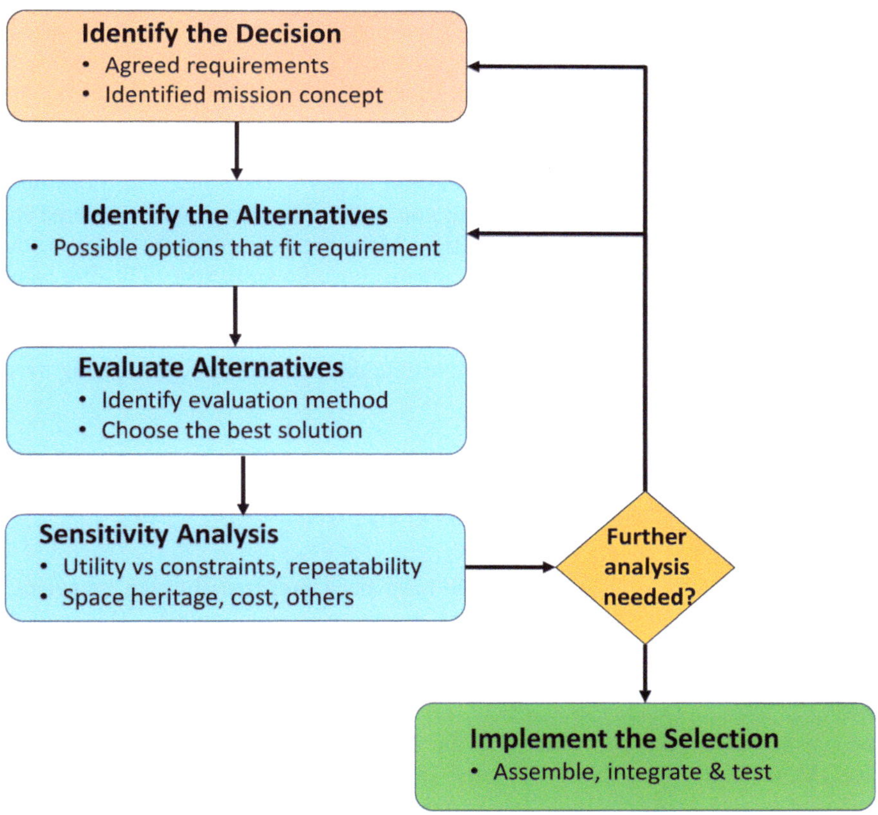

Fig. 2.9 Example alternative options workflow. A clear understanding of the mission requirements and constraints are essential for identifying suitable options to trade. Constraints such as costs, power usage and payload volume are popular trade options

If some attributes, such as space heritage, mass or cost are considered more important to mission designers, these can be weighted to increase their influence in the trade analysis.

In the case of evaluation scores being very similar, sensitivity analysis involving changing weights or statistical analysis, can be performed (Fig. 2.9).

2.3.2 *Commercial-Grade Components*

For cost-constrained missions, such as those represented by institutions wishing to fly cubesats, purchase of science-grade instruments may be out of the reach. Low-cost imaging sensors have driven a revolution in science, technology, mathematics and education (STEM) and citizen science applications, particularly when used in conjunction with unmanned aerial vehicles (Shailesh et al. 2015).

Early in the space race, these types of sensors were perceived to be low quality and not robust enough for assured operation at Mars. Recent research has tested this belief, driven by cost and schedule described above. On orbit cubesats have successfully demonstrated space heritage for what had previously been considered to be hobby electronics. Ardusat was launched using the popular Arduino micro-controller, plus low-cost sensors to Earth orbit in 2014. The concept demonstrator operated successfully throughout its mission until it deorbited and re-entered the Earth's atmosphere (Geeroms et al. 2015). More recently cubesats based on the popular Raspberry Pi ARM board and cameras have flown to space. Testing using a NASA radiation facility has quantified the survivability of these low-cost devices by exposing them to increasing amounts of radiation expected to be found in space (Violette 2014). Favourable results demonstrate the feasibility of considering these applications for Mars missions.

2.3.3 *Mars Analogue Environment Application for Space 2.0*

Many flown Mars exploration missions have been prepared through knowledge gained by undertaking analogous research in terrestrial environments. This reduces risk of flying unproven hardware to Mars as well as characterises the performance of an instrument in a similar Earth-bound setting. An example is the testing of the Mars Viking cameras in Death Valley, USA. This region was chosen to approximate what the sensor was expected to observe at the Ares Vallis and Utopia Planitia landing sites (Mutch and Viking Lander Imaging Team 1978).

More recently, NASA's Field Integrated Design and Operations (FIDO) rover, a precursor to the Spirit and Opportunity Rovers, was tested, along with its sensors, near Flagstaff, Arizona, USA. Field trial results characterised the performance of demonstrator technology in a Mars-like setting. Rover operators were able to appreciate the quality and limitations of interpreting returned imagery and data in the

Mars-like setting of Arizona. These could then be checked against actual samples collected from on-site, a method impossible on Mars. Calibration and field validation of the sensors using geological formations similar to those expected at the Spirit and Opportunity Mars landing sites. These field operations shaped the flown final rover mission operations and science data collection methods (Anderson et al. 2006).

Similarly, an engineering version of the Perseverance Mastcam-Z camera, the Mastcam-Z Analog Spectral Imager (MASI), was trialled in a field setting at Hawthorne, Nevada, USA. The results from the analogue trial supplemented laboratory calibration data to provide useful predictions of Mastcam-Z performance on the surface of Mars (Hayes et al. 2021).

2.3.4 *Space 2.0 Case Study: Identifying the Utility of Low-Cost Remote Sensing for High Altitude Ballooning Applications*

Development and trade analysis of low-cost sensors described above, along with the usefulness of field testing in Mars analogue environments has opened pathways for flying commercial-grade components in space missions (e.g. Geeroms et al. 2015). This case study trials commercially available sensors in an analogue environment to baseline performance in Mars-like conditions. High altitude balloon (HAB) flight testing provides a useful analogue environment to space. The environment above 30 km altitude experiences conditions similar to the Martian surface, with temperatures dropping below − 40 °C and over 99% of Earth's atmosphere is below the balloon. Atmospheric absorption and distortion at these altitudes also approximates that for Earth Observation satellites, providing an analogue for on orbit image calibration. This environment also exposes payloads to increased high-energy radiation environments, though not as high as would be found outside Earth's atmosphere.

The advantages of HAB flights are they provide a more accessible, lower-cost method of testing in low-pressure environments. Multiple HAB flights can be conducted to provide additional testing opportunities, assisting in reducing the risk of failure in space environments. The ability to accommodate spectral absorption is a valuable test, when compared to ground-level of low altitude drone imagery such as may be provided by an Ingenuity-style Mars helicopter (Balaram et al. 2021).

Balloon testing has limitations such as the inability to completely mimic the space environment such as launch stresses. Balloon platform stability is a further consideration for launches of this type; the balloon launch provider used in this work had developed a passive system to reduce unwanted yaw and pitch motions during flight. In addition, while HAB exposes payloads to increased high-energy radiation, flights are still well within the Earth's protective magnetic field. In contrast, deep space payloads are subjected to solar and galactic radiation, required hardened components or shielding to protect them from degradation. Despite these limitations,

HAB flights remain an attractive option for supplementing space testing of remote sensors.

HAB flights have also been used to assist in placing constraints in survivability for astrobiology considerations. Fungal spores of *Aspergillus niger* and bacteria cells of *Salinisphaera shabanesis*, *Staphyloccus capitis* subsp. *capitis* and *Buttiauxella* sp. were flown to altitudes of up to 38 km altitude for a period of five hours. The advantage as compared with studies in surface analog environments such as the Atacama Desert and Antarctic Dry Valleys (ADV) are that the conditions more closely approximate those found on Mars, being many kilometres above free oxygen and moisture sources. The HAB flight showed that *Aspergillus niger* and *Salinisphaera shabanesis* were able to survive these conditions, not only demonstrating the possibility of life surviving on Mars, but also having implications for spacecraft sterility methods (Cortesão et al. 2021).

Recently, a low-cost RGB sensor was flown on a HAB in West Wyalong, NSW, Australia, to trial its utility in a near-space environment. This class of sensors is useful for visible/near infra-red (VIS/NIR) remote sensing sensitive to wavelengths between 400 and 1000 nm. They have extensive terrestrial application in low-cost vegetation analysis. As much of the discovered Mars mineralogy exhibits spectral phenomenology in this range, VIS/NIR sensors have found application in Mars orbit or lander missions (detailed in Chap. 4).

To ensure accessibility in cost constrained mission designs, most HAB flights occur over agricultural or vegetated areas, precluding access to Mars-analogue environments and associated mineralogy. Useful analysis for utility in Mars applications can be achieved, however, by taken advantage of the highly reflective properties in vegetation's NIR portion of the electromagnetic spectrum (Price 1992) and comparing results gained in flight with on-orbit Earth observation sensors such as Landsat imaged at similar periods to the flights.

For this case study, a single-filter approach was chosen to simplify sensor design. Single-filter sensing has also been trialed for vegetation applications; in these cases, the ability of coloured filters to separate visible and NIR across different channels of a Bayer-filtered (red, green and blue filters overlaid onto an imaging sensor to produce colour images) camera (e.g. Glenn and Tabb 2019). The use of a bandpass filter provided two-channel imagery Additional channels could be made available through using a filter wheel, such as with Mastcam on Curiosity, but this would require time to position the filters over the camera between exposures. As such this method was not deemed practical for a dynamic activity such as HAB, where position and heights are not constant throughout the flight.

The use of the Normalized Difference Vegetation Index (NDVI) is also distinctively advantageous from an engineering perspective since the harsh stratospheric environment required a degree of hardware robustness and redundancy. NDVI, is a popular band ratio calculation for environmental applications such as terrestrial satellite interpretations. Typically providing values between -1 and 1, NDVI takes advantage of the high NIR and low red (RED) reflectance of vegetation with pixel-by-pixel processing of:

$$\mathrm{NDVI} = (\mathrm{NIR} - \mathrm{RED})/(\mathrm{NIR} + \mathrm{RED}) \tag{2.3}$$

NDVI has shown utility in vegetation analysis, and used in previous multispectral experiments (Wang et al. 2020).

For example, measuring vegetation to obtain a RED value of 0.12 (12% reflectance in the sensor's red band) and NIR of 0.48 (48% reflectance in the sensor's NIR band), NDVI could be calculated as:

- NDVI = (0.48 − 0.12)/0.48 + 0.12) = 0.36/0.6
- NDVI = 0.6, signifying healthy, dense vegetation.

Two payloads were built to provide redundancy, with the main difference being the inclusion of different sensors as the primary imaging source. Baseline testing for space applications was further developed by using a cubesat format was used for the design of the payloads (Twiggs 2003).

The payload consisted of an RGB imager fitted with A. Jaegers Optics #62A red photographic filters to trial the utility of single-filter NDVI in near-space applications. This filter operated by segregating visible information to the red Bayer filter and NIR to the blue Bayer filter, with results from the green Bayer filter not being used. The balloon flights were timed to ensure maximum altitude was attained within two hours of local noon. This ensured minimum shadowing, and enhanced the fidelity of returned NDVI results.

As in-flight calibration target imaging was not practical for HAB flight, ground-based testing was performed to provide an initial characterization of an IMX219 second sensor with a lens of f-2 aperture and 3.04 mm focal length. This readily available, commercial off the shelf (COTS) sensor represented low-cost analogues to the sensor used in Mastcam-Z on the Perseverance rover. Ground-based analysis entailed of imaging a calibration target illuminated by direct sunlight in an analogous fashion to the mineral tests conducted for Mars cameras (Chap. 4). Soil samples from the HAB flight areas were obtained together with representative vegetation for ground truth of captured results. Colour swatches and a white Lambertian reflector were also used for sensor calibration. Outputs were compared with an SR-3500 laboratory-grade spectro-radiometer for consistency. Analysis showed high correlation of values for the colour swatches in red and blue channels of the IMX219 sensor.

HAB flight analysis was conducted by using the image at the payload's maximum altitude just prior to balloon burst for the camera to identify NDVI performance above 99% of Earth's atmosphere. The image was geo-referenced using QGIS, an open-source geospatial analysis platform (Menke et al. 2016) and regions of interest representing soil and vegetation were extracted (Fig. 2.10a). Results were compared with the equivalent sites obtained from a Landsat 7 Earth reflectance image (bands 3 and 4, red and NIR respectively) acquired within two weeks of the HAB flights, providing a comparison between near-space and low-Earth orbit remote sensing (Vermote et al. 2016).

Additional balloon images from closer to the ground were sampled of vegetation and soil and built environment where the balloons were at lower (< 5 km) altitudes (Fig. 2.10b). This was to provide a comparison of the camera's performance at light

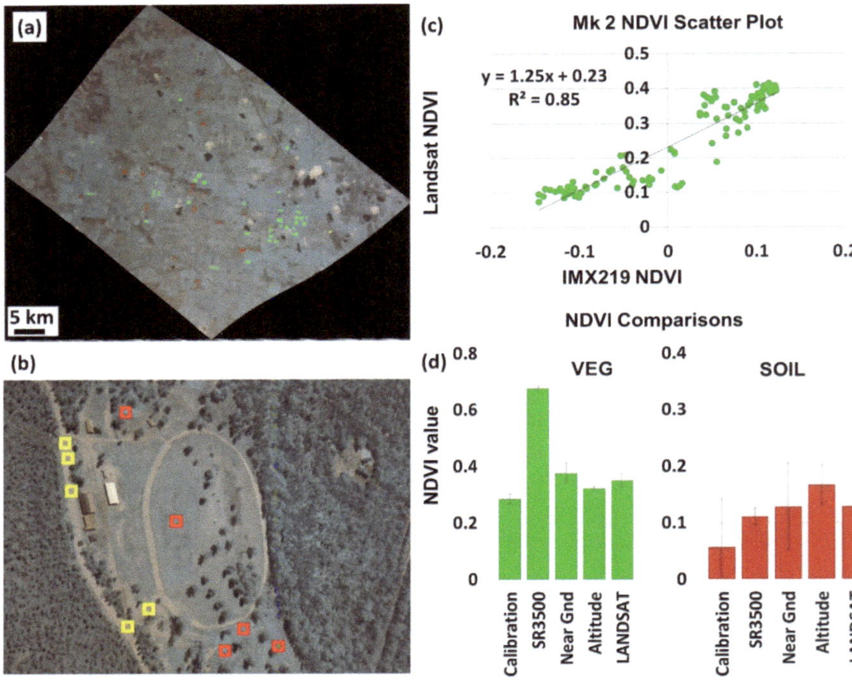

Fig. 2.10 **a** Custom sensor (IMX219) two-band image at the maximum flight altitude of 32 km showing the soil (red) and vegetation (green) sample polygons. **b** Low-altitude (~ 750 m) red-filter image showing example locations of sampled vegetation and soil. **c** Goodness of fit (R^2) analysis between IMX219 camera and Landsat at high altitude. **d** Comparison of average NDVI values for vegetation (green) and soil (red) for the IMX219 camera in calibration, low and high-altitude settings with SR3500 and Landsat values. Error bars denote statistical standard deviation between individual values

aircraft height and determine the consistency of NDVI ratio from the cameras. The single-filtered camera was able to return imagery throughout the flight to a maximum altitude of 34 km.

Comparison with Satellite imagery showed a close agreement using a goodness of fit value of R^2 (R^2 of 0.85, Fig. 2.10c), showing the low-cost sensor can produce scientifically useful information in the near-space environment. Comparison of NDVI values of vegetation and soil for ground imagery, low and maximum altitude HAB imagery also showed a consistency of results (Fig. 2.10d). NDVI results for the soil calibration samples and balloon flight (Fig. 2.10d) exhibited more variance than for the vegetation samples (average standard deviation of 0.036 for vegetation, 0.51 for soils, Fig. 2.10d). The goodness of fit formulae-applied soil NDVIs as reported by the sensor (average 0.12, Fig. 2.10d) were comparable to the Landsat or SR-3500 results, both near the ground and at altitude (Landsat 0.13, SR-3500 0.11, Fig. 2.10d).

Temperature, pressure and G-forces measured during the HAB flight demonstrated the Mars-like environmental conditions acting on the payload (Fig. 2.11a–c, red arrow denotes maximum altitude). Temperature variation throughout the flight was 72 °C, with minimum being − 44 °C. Pressure reduced to approximately 1% sea level, while G-forces acting on the payload increased substantially post-balloon burst (red arrow, Fig. 2.11c).

These results demonstrated the utility of using HAB and low-cost sensor for remote sensing in Mars-like conditions. Ground-based calibration testing indicated that the camera was able to capture useful for remote sensing data with use of a single filter. This sensor was thus able to return scientifically useful results, indicating their utility for low-cost space missions. These results have assisted in reducing inherent risks associated with spaceflight by raising the TRL of the sensors. Additionally, the performance of the sensor was favourably compared to the Landsat on-orbit results.

Analogue activities such as the case study described above highlight the importance of designing and testing to maximise survivability and science return in the Martian environment. The 760-day period between favourable launch conditions from Earth to Mars require robust systems engineering techniques, where NASA's

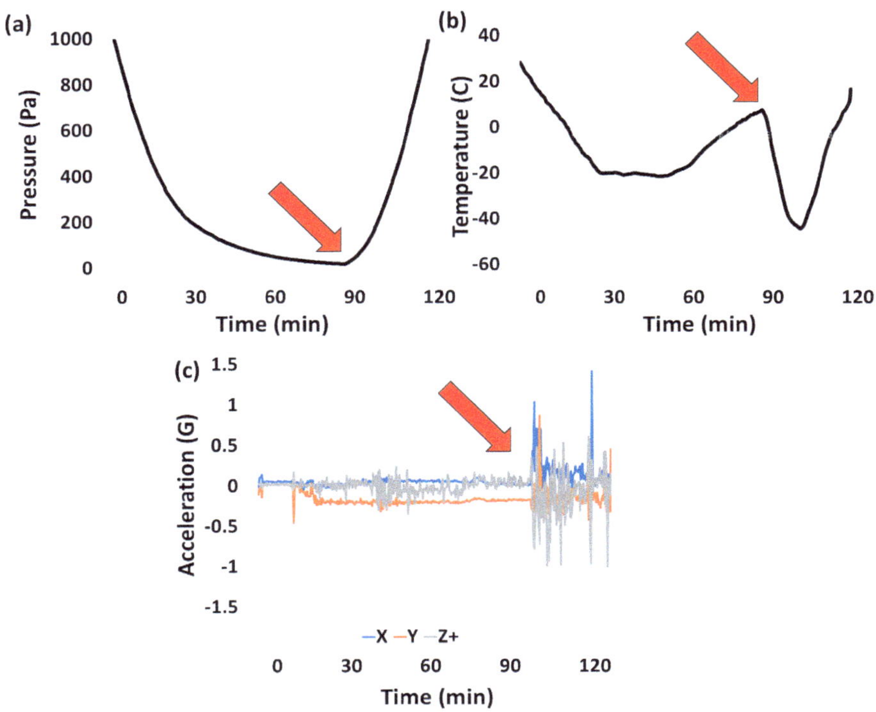

Fig. 2.11 Environmental measurements of the HAB payload during flight. Red arrow denotes maximum altitude and balloon burst. **a** Pressure. **b** Temperature. **c** G-force acceleration in three dimensions

phased approach accommodates mission scope and design changes early in the process before final designs are committed to construction and operation.

The advent of Space 2.0 is streamlining systems engineering practices and driving down mission costs through reducing mission scope and drawing on reducing component costs. The first cubesat-format Mars mission has already successfully flown, with more missions being designed. Space 2.0 also increases the opportunity for smaller-budget agencies and institutions to explore Mars.

References

Alanazi A, Straub J (2019) Engineering methodology for student-based cubesats. Aerospace 6. https://doi.org/10.3390/aerospace6050004

Amiri HES et al (2022) The Emirates Mars mission. Space Sci Rev 218. https://doi.org/10.1007/s11214-021-00868-x

Anderson AC, Haldemann AFC, Dohm J, Huntsberger TA (2006) Dress rehearsal for the 2003 Mars exploration rovers, vol 11. Mars Analog Research, American Astronomical Society, Long Beach CA, USA, pp117–128

Asundi SA, Fitz-Coy NG (2013) Cubesat mission design based on a system engineering approach. In: IEEE aerospace conference proceedings, paper, 2180. https://doi.org/10.1109/AERO.2013.6496900

Balaram J, Aung M, Golombek MP (2021) The ingenuity helicopter on the perseverance rover. Space Sci Rev 217:56. https://doi.org/10.1007/s11214-021-00815-w

Braun RD, Manning RM (2012) Mars exploration entry, descent, and landing challenges. J Spacecr Rockets 44. https://doi.org/10.2514/1.25116

Burke, LM, Falck RD, McGuire ML (2010) Interplanetary mission design handbook: Earth-to-Mars mission opportunities 2026 to 2045. Cleveland, USA

Callas JL (2015) Mars exploration rover spirit end of mission report (No. JPL-Publ-16-2)

Chen S, Ma X, Gou J, Gao G, Gong C (2025) An efficient multi-layer TEG with thermal-electrical interfaces for Mars rover RTGs. Appl Therm Eng 280. https://doi.org/10.1016/j.applthermaleng.2025.128383

Cortesão M, Siems K, Koch S, Beblo-Vranesevic K, Rabbow E, Berger T, Lane M, James L, Johnson P, Waters SM, Verma SD (2021) MARSBOx: fungal and bacterial endurance from a balloon-flown analog mission in the stratosphere. Front Microbiol 12:177. https://doi.org/10.3389/fmicb.2021.601713

Crowley K (2018) Exploring the concept of a deep-space solar-powered small spacecraft. Master's thesis, The Faculty of California Polytechnic State University, San Luis Obispo

Cummings A, Odenwald S (2008) Issue 62: where is the heliopause? In: NASA sun-earth day: 2008—space weather around the world. https://sunearthday.nasa.gov/2008/TTT/62heliopause.php

Dwyer Cianciolo, A., Powell, R.W., 2017. Entry, descent, and landing guidance and control approaches to satisfy Mars human mission landing criteria. In AAS/AIAA Space Flight Mechanics Meeting (No. AAS Paper 17–254).

Ebben TH, Bergstrom J, Spuhler P, Delamere A, Gallagher D (2007) Mission to Mars: the HiRISE camera on-board MRO. In: Focal plane arrays for space telescopes III, Sept 2007, vol 6690, pp 102–123. https://doi.org/10.1117/12.739930

Edwards C (2007) Relay communications for Mars exploration. Int J Satell Commun Netw 25:111–145. https://doi.org/10.1002/sat.871

Farley KA, Williford KH, Stack KM, Bhartia R, Chen A, de la Torre M, Hand K, Goreva Y, Herd CD, Hueso R, Liu Y (2020) Mars 2020 mission overview. Space Sci Rev 216:1–41. https://doi.org/10.1007/s11214-020-00762-y

Garrett HB, Shapiro AA, Yang JY (2008) Interstellar space missions: ultra reliability and engineering issues—part II. In: 46th AIAA aerospace sciences meeting and exhibit. https://doi.org/10.2514/6.2007-501

Garzon MM (2012) Development and analysis of the thermal design for the OSIRIS-3U CubeSat. Master's thesis, Pennsylvania State University, Pennsylvania

Geeroms D, Bertho S, De Roeve M, Lempens R, Ordies M, Prooth J (2015) ARDUSAT, an Arduino-based cubesat providing students with the opportunity to create their own satellite experiment and collect real-world space data. ESA Publications Division C/O ESTEC

Glenn DM, Tabb A (2019) Evaluation of five methods to measure normalized difference vegetation index (NDVI) in apple and citrus. Int J Fruit Sci 19:191–210. https://doi.org/10.1080/15538362.2018.1502720

Guo J, Zeitlin C, Wimmer-Schweingruber RF, Hassler DM, Ehresmann B, Rafkin S, Freiherr von Forstner JL, Khaksarighiri S, Liu W, Wang Y (2021) Radiation environment for future human exploration on the surface of Mars: the current understanding based on MSL/RAD dose measurements. Astron Astrophys Rev 29:1–81. https://doi.org/10.1007/s00159-021-00136-5

Hayes AG, Corlies P, Tate C, Barrington M, Bell JF, Maki JN, Caplinger M, Ravine M, Kinch KM, Herkenhoff K, Horgan B (2021) Pre-flight calibration of the Mars 2020 Rover Mastcam Zoom (Mastcam-Z) multispectral, stereoscopic imager. Space Sci Rev 217:1–95. https://doi.org/10.1007/s11214-021-00795-x

Hechler M, Yáñez A (2003) Mars express orbit design. Acta Astronaut 53:497–507. https://doi.org/10.1016/S0094-5765(03)80010-3

Helfrich C, Berry DS, Bhat R, Border J, Graat E, Halsell A, Kruizinga G, Lau E, Mottinger N, Rush B, Wilson R (2015) A journey with MOM. In: 25th international symposium on space flight dynamics, Oct 2015, Munich, Germany

Hirshorn SR, Voss LD, Bromley LK (2017) NASA systems engineering handbook (No. HQ-E-DAA-TN38707)

Hodges R, Chahat N, Hoppe D, Vacchione J (2017) A deployable high-gain antenna bound for Mars: developing a new folded-panel reflectarray for the first CubeSat mission to Mars. IEEE Antennas Propag Mag. https://doi.org/10.1109/MAP.2017.2655561

Holgate TC, Bennett R, Hammel T, Caillat T, Keyser S, Sievers B (2015) Increasing the efficiency of the multi-mission radioisotope thermoelectric generator. J Electron Mater 44:1814–182115. https://doi.org/10.1007/s11664-014-3564-9

Johnson AE, Aaron SB, Ansari H, Bergh C, Bourdu H, Butler J, Chang J, Cheng R, Cheng Y, Clark K, Clouse D (2022) Mars 2020 lander vision system flight performance. In: AIAA SciTech 2022 forum, p 1214. https://doi.org/10.2514/6.2022-1214

Johnston MD, Graf JE, Zurek RW, Eisen HJ, Jai B (2005) The Mars reconnaissance orbiter mission. In: 2005 IEEE aerospace conference, Mar 2005. IEEE, pp 447–464

Kleinboehl A, Kass D, Piqueux S, Greybush S, Kenyon M (2022) Mars Climate CubeSat Constellation (MC3)—a low-cost orbital constellation for atmospheric profiling, polar science and surface thermophysics. In: Proceedings of the low-cost science mission concepts for Mars exploration, Pasadena, CA, USA, pp 29–31

Koktas E, Basar E (2024) Communications for the planet mars: past, present, and future. IEEE Aerosp Electron Syst Mag 1–35

Korzun AM, Dubos GF, Iwata CK, Stahl BA, Quicksall JJ (2010) A concept for the entry, descent, and landing of high-mass payloads at Mars. Acta Astronaut 66:1146–1159. https://doi.org/10.1016/j.actaastro.2009.10.003

Landis GA, Kerslake TW, Jenkins PP, Scheiman DA (2004) Mars solar power. In: NASA technical conference publication, second international energy conversion engineering conference, American Institute of Aeronautics and Astronautics, Providence, Rhode Island, NASA/TM—2004-213367

Li J, Post M, Wright T, Lee R (2013) Design of attitude control systems for CubeSat-class nanosatellite. J Control Sci Eng. https://doi.org/10.1155/2013/657182

Li C, Zhang R, Yu D, Dong G, Liu J, Geng Y, Sun Z, Yan W, Ren X, Su Y, Zuo W (2021) China's Mars exploration mission and science investigation. Space Sci Rev 217. https://doi.org/10.1007/s11214-021-00832-9

Liu X, Baoyin H, Ma X (2010) Five special types of orbits around Mars. J Guid Control Dyn 33. https://doi.org/10.2514/1.48706

Love SG, Reagan ML (2013) Delayed voice communication. Acta Astronaut 91:89–95. https://doi.org/10.1016/j.actaastro.2013.05.003

Lyons DT, Beerer JG, Esposito P, Johnston MD, Willcockson WH (1999) Mars global surveyor: aerobraking mission overview. J Spacecr Rocket 36:307–313

Mankins JC (1995) Technology readiness levels. White Pap 6(1995):1995

Marov MY, Petrov GI (1973) Investigations of Mars from the Soviet automatic stations Mars 2 and 3. Icarus 19:163–179. https://doi.org/10.1016/0019-1035(73)90120-6

Martin-Mur T, Gustafson E, Young B (2016) Interplanetary cubesat navigational challenges. In: International symposium space flight dynamics, Munich, Germany

Mazumder MK, Biris AS, Johnson CE, Yurteri CU, Sims RA, Sharma R, Pruessner K, Trigwell S, Clements JS (2006) Solar panel obscuration by dust and dust mitigation in the Martian atmosphere. Part Surf Detect Adhes Removal 9:167–195

McEwan AS et al (2007) Mars reconnaissance orbiter's high resolution imaging science experiment (HiRISE). J Geophys Res 112. https://doi.org/10.1029/2005JE002605

Menke K, Smith Jr R, Pirelli L, Van Hoesen J (2016) Mastering QGIS. Packt Publishing Ltd., Birmingham, UK. ISBN: 978-1-78646-037-0

Mishkin A (2003) Sojourner. Berkley Publishing Group, New York, 334 pp

Mutch TA, Viking Lander Imaging Team (1978) The Martian landscape. NASA Special Publication SP-425, Scientific and Technical Information Office, National Aeronautics and Space Administration

Nallapu RT, Dektor G, Kenia N, Uglietta J, Ichikawa S, Herreras-Martinez M, Choudhari A, Chandra A, Schwartz S, Asphaug E, Thangavelautham J (2020) Trajectory design of Perseus: a cubesat mission concept to Phobos. Aerospace 7. https://doi.org/10.3390/aerospace7120179

Nelessen A, Sackier C, Clark I, Brugarolas P, Villar G, Chen A, Stehura A, Otero R, Stilley E, Way D, Edquist K (2019) Mars 2020 entry, descent, and landing system overview. In: 2019 IEEE aerospace conference. IEEE, pp 1–20. https://doi.org/10.1109/AERO.2019.8742167

Olsson-Francis K, Doran PT, Ilyin V, Raulin F, Rettberg P, Kminek G, Mier MPZ, Coustenis A, Hedman N, Al Shehhi O, Ammannito E (2023) The COSPAR Planetary Protection Policy for robotic missions to Mars: a review of current scientific knowledge and future perspectives. Life Sci Space Res 36:27–35. https://doi.org/10.1016/j.lssr.2022.12.001

Pagano T, Mischna M, Liuzzi G (2022) Hyperspectral infrared temperature sounding of Mars' atmosphere in a CubeSat. LPI Contrib 2655:5019

Parkinson KJ, Dempster AG, Mumford PJ, Rizos C (2006) FPGA based GPS receiver design considerations. J Glob Posit Syst 5:70–75

Patel MR (2005) Spacecraft power systems. CRC Press, Boca Raton, FL. https://doi.org/10.1201/9781420038217

Pischel R, Zegers T (2009) Mars Express science planning and operations. In: Mars Express: the scientific investigations, ESA, SP-1291, pp 249–256

Price JC (1992) Estimating vegetation amount from visible and near infrared reflectances. Remote Sens Environ 41:29–34. https://doi.org/10.1016/0034-4257(92)90058-R

Rauschenbakh BV, Ovchinnikov MY, McKenna-Lawlor S (2003) Qualitative analysis of the properties of orbits. In: Essential spaceflight dynamics and magnetospherics, pp 17–24. https://doi.org/10.1007/0-306-48027-1

Ridolfi G, Mooij E, Corpino S (2009) A system engineering tool for the design of satellite subsystems. In: AIAA modeling and simulation technologies conference, Aug 2009, p 6037

Schmidt G, Dudzinski L, Sutliff T (2011) Radioisotope power: a key technology for deep-space exploration. In: 6th international energy conversion engineering conference (IECEC), p 5640. https://doi.org/10.2514/6.2008-5640

Schoolcraft J, Klesh A, Werne T (2017) MarCO: interplanetary mission development on a CubeSat scale. In: Space operations: contributions from the global community, pp 221–231. https://doi.org/10.2514/6.2016-2491

Selva D, Krejci D (2012) A survey and assessment of the capabilities of cubesats for Earth observation. Acta Astronaut 74:50–68. https://doi.org/10.1016/j.actaastro.2011.12.014

Shailesh KR, Kurian CP, Kini SG (2015) Auto-calibration of emission spectra of light sources captured using camera spectrometer. In: Proceedings of the 2015 international conference on smart sensors and systems (IC-SSS), Bangalore, India, 21–23 Dec 2015. Institute of Electrical and Electronics Engineers (IEEE), Piscataway Township, NJ, USA. https://doi.org/10.1109/SMATSENS.2015.7873614

Soffen GA, Snyder CW (1976) The first Viking mission to Mars. Science 193:759–766. https://doi.org/10.1126/science.193.4255.759

Stinner A, Begoray J (2005) Journey to Mars: the physics of travelling to the red planet. Phys Educ 40:35–45

Sundararajan V (2013) Mangalyaan—overview and technical architecture of India's first interplanetary mission to Mars. In: AIAA space 2013 conference and exposition, AIAA 2013-5503

Thangavelautham J, Asphaug E, Dektor G, Kenia N, Uglietta J, Ichikawa S, Choudhari A, Herreras-Martinez M, Schwartz S (2017) An interplanetary cubesat mission to Phobos. In: 48th annual lunar and planetary science conference, Abstract 1707, The Woodlands, Texas. Lunar and Planetary Institute, Houston, TX

Twiggs JPS (2003) Cubesat design specification document. Californian Polytechnic State University

Vacchione J (2006) The Mars missions. Spaceborne Antennas for Planetary Exploration, Wiley, pp 157–211

Vermote E, Justice C, Claverie M, Franch B (2016) Preliminary analysis of the performance of the Landsat 8/OLI land surface reflectance product. Remote Sens Environ 185:46–56. https://doi.org/10.1016/j.rse.2016.04.008

Violette DP (2014) Arduino-Raspberry Pi: hobbyist hardware and radiation total dose degradation. Electrical, Electronic and Electromechanical (EEE) Parts for Small Missions, USA (No. GSFC-E-DAA-TN17893)

Walker R, Binns D, Bramanti C, Casasco M, Concari P, Izzo D, Feili D, Fernandez P, Fernandez JG, Hager P, Koschny D, Pesquita V, Wallace N (2018) Deep-space cubesats; thinking inside the box. Astron Geophys 59:26–30. https://doi.org/10.1093/astrogeo/aty232

Wan W, Yu T, Di K, Wang J, Liu Z, Li L, Liu B, Wang Y, Peng M, Bo Z, Ye L (2021) Visual localization of the Tianwen-1 lander using orbital, descent and rover Images. Remote Sens 13:3439. https://doi.org/10.3390/rs13173439

Wang L, Jin J, Song Z, Wang J, Zhang L, Rehman TU, Ma D, Carpenter NR, Tuinstra MR (2020) LeafSpec: an accurate and portable hyperspectral corn leaf imager. Comput Electron Agric 169:1–11. https://doi.org/10.1016/j.compag.2019.105209

Wang C, Yan J, Lu S, Sun S, Yang W, Cao J, Qiu D, Li X, Barriot JP (2025) Evaluating the contributions of Tianwen-1 orbiter tracking data to the Mars gravity field modelling. Astrophys J 992. https://doi.org/10.3847/1538-4357/ae0e22

Wertz J, Everett DF, Puschell JJ (2011) Space mission engineering: the new SMAD. Microcosm Press, CA, 1029 pp

Woellert K, Ehrenfreund P, Ricco AJ, Hertzfeld H (2011) Cubesats: cost-effective science and technology platforms for emerging and developing nations. Adv Space Res 47:663–684. https://doi.org/10.1016/j.asr.2010.10.009

Zielinski S (2007) Computer failure caused loss of Mars spacecraft. EOS Trans Am Geophys Union 88:192–192. https://doi.org/10.1029/2007EO170007

Chapter 3
Remote Sensing Considerations for Mars

Remote sensing has shaped terrestrial environmental policy, rural planning and military strategy for decades. For much of Solar system exploration, where direct field observations are not yet possible, remote sensing is the only method available to researchers. To date, 10 missions have successfully operated on the Martian surface, though these have only explored a tiny fraction of the planet. This leaves orbital remote sensing as the dominant method for large scale exploration (Greeley 2013).

In its broadest definition, remote sensing collects data about an object or area of interest without the sensor or collector being in direct contact. This describes the operation of imaging cameras, spectrometers, laser range finding systems and other technologies. Other payloads that do not fall into the remote sensing category but have operated on the Martian surface include those that have directly interacted with Martian material. Examples include the search for life experiments using Martian regolith, the Phoenix lander sampling ovens, and the drilling tools used with the Curiosity and Perseverance Mars rovers. These supplement and provide ground truth analysis for orbital or other remotely sensed data.

3.1 Electromagnetic Energy and Influences

3.1.1 The Electromagnetic Spectrum

The electromagnetic spectrum (EMS) consists of energy travelling in wavelengths spanning long radio frequencies below 1 Hz to gamma rays exceeding 10^{25} Hz (Fig. 3.1a; Fröhlich 2006).

The relationship between velocity (speed of light in a vacuum, c = 299,792,458 m/s), frequency, f and wavelength λ is given by:

$$C = f\lambda \tag{3.1}$$

S. Hobbs, *Sensing Mars: Remote Sensing to Understand the Red Planet*,
https://doi.org/10.1007/978-3-032-24416-1_3

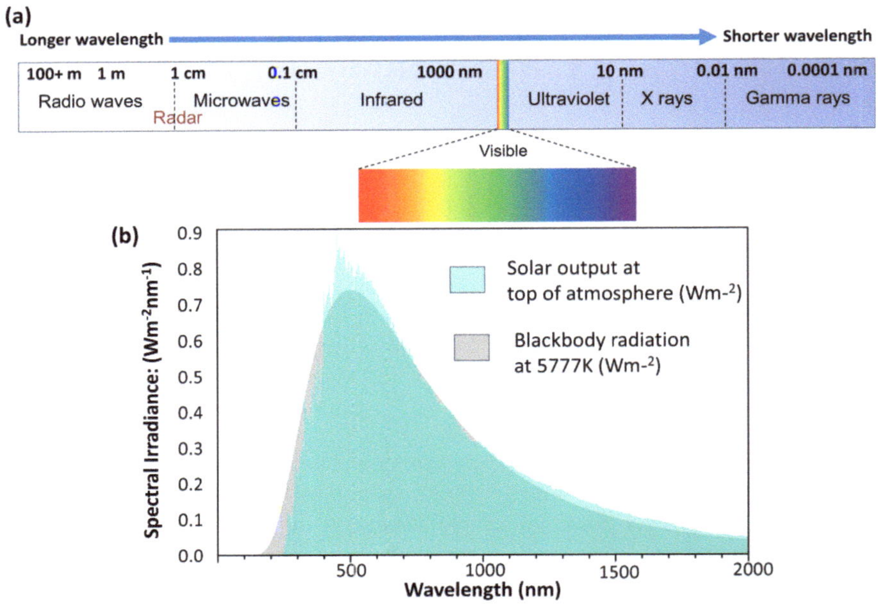

Fig. 3.1 **a** The electromagnetic spectrum, with visible wavelengths highlighted. **b** Solar irradiance at the top of the Martian atmosphere (and terrestrial atmosphere) compared to the ideal blackbody radiation curve at 5777 K

Electromagnetic energy is prevalent in the universe, emitted by objects with temperatures above absolute zero according to the Stefan–Boltzmann law (De Lima and Santos 1995):

$$M = \sigma T^4 \tag{3.2}$$

where M is the radiated energy from an object's surface, σ is the Stefan–Boltzmann constant of 5.6697×10^{-8} W/m^2/K and T is the absolute temperature. As shown by Eq. 3.2, increases in temperature equate to significant increases in radiated energy. In our Solar System, the dominant radiative emitter is the Sun.

As illustrated in the solar spectrum illustrated in Fig. 3.1b, not all frequencies are emitted equally, with output increasing dramatically to a peak of ~ 580 nm. Approximately half of the Solar output is in the visible wavelengths of 400–700 nm, before declining to much lower levels past 1200 nm (Iqbal 1983). This attribute is common across all Solar System bodies. The EMS emitted by the Sun interacts with Solar System bodies, including surfaces and atmospheres. EMS may be reflected, or absorbed by a body, converted to thermal energy and re-emitted to space. It is these interactions that passive remote sensors operate to detect and characterise (Curran 1985).

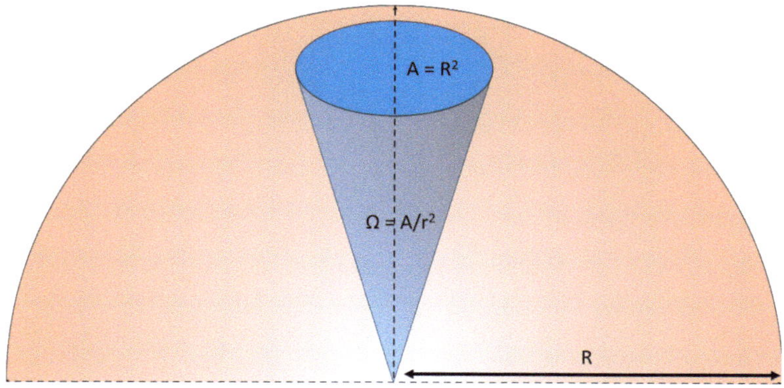

Fig. 3.2 Solid angle, Ω, showing relationship with the area of a sphere and radius

The amount of solar flux available to the sensor affects remote sensing. Solar flux available on the Martian surface is approximately 585.2 W/m^2. The energy available to the sensor that is reflected from a body is described by the solid angle, and decreases proportional to the square of the distance (Fig. 3.2; Granger 2013).

A Mars orbiter circling hundreds of kilometres above the planet receives much less reflected energy from an object on the ground than a surface mission within metres of the same object. The available radiation available per pixel of the sensor is also dependent on the area each pixel covers on the target as influenced by the focal length of the objective optics (the first portion of the sensor that receives the incident radiation) and the amount of time to which the pixel is exposed to that energy, defined as the integration time. The formula used to evaluate reflected energy (E_r) with distance is (McCluney 1994; Granger 2013):

$$E_r = S_f t P_a (O_a/4)/D^2 \tag{3.3}$$

where

- E_r = Reflected energy available at sensor (W)
- S_f = Incident solar flux (W/m^2)
- t = Integration time (s)
- P_a = Area of target area a given pixel covers (m^2)
- O_a = Objective area (m^2)
- D = distance from object (m).

As an example, assume an imager equivalent to the MRO CTX has the following attributes (Malin et al. 2007):

- An objective diameter of 108 mm,
- t of 100 ms,
- P_a of 36 m^2 based on pixel resolution of 6 m per pixel, and
- D of 300 km (within MRO periapsis and apoapsis distances).

Formula 3.3 would provide (converting to metres):

- O_a is πr^2 or $\pi \times 0.054^2 = 0.009161\ m^2$
- $S_f = 590\ W/m^2$
- $E_r = 590 \times 0.1 \times 36 \times (0.009161/4)/(300{,}000)^2$
- $E_r = 5.40 \times 10^{-11}$ W (approx.).

Passive and Active Sensors

Remote sensors fall into two broad categories: active and passive. Active remote sensors emit energy that interacts with the target object and measurements are taken from reflected or re-emitted energy. An example active sensor sent to Mars is the MOLA instrument that fired laser pulses from orbit. Measuring the time difference between fired pulse and reflected return enabled creation of highly detailed topographic information of the planet (Kreslavsky and Head 2000). Additional examples of active sensors are the ground penetrating radar systems flown on orbiters and recent rovers. Described in more detail in Chap. 5, these systems send electromagnetic energy at discrete wavelengths into the Martian surface, where conductivity properties of underlying geology produce returns that can be analysed and interpreted.

Passive sensors obtain data from energy that is reflected or emitted from a target object or area. For Mars, most of this energy originates from the Sun. Electromagnetic radiation interacts with the Martian surface, producing reflected and/or emitted radiation in accordance with the formula:

$$\mathrm{Tt} = \mathrm{Rl} + \mathrm{Al} + \mathrm{Tl} \tag{3.4}$$

where Tt, the total solar flux consists of reflected (Rl), absorbed (Al) or radiation transmitted through an atmosphere or the vacuum of space (Tl, Curran 1985).

Thermal sensors are examples of detectors able to pick up emitted radiation. They exploit temperature changes of geology resulting from absorption and re-emission of wavelengths at ~ 3–35 μm. Unlike visible remote sensing that is conducted during the local Martian day, thermal emissions may also be detected during the Martian night.

Imaging cameras and reflectance spectrometer instruments are popular examples of passive reflective sensors. These may measure solar reflection from a target in wavelengths equating to UV, visible, near and shortwave infrared parts of the electromagnetic spectrum (Fig. 3.1a). Capturing multiple wavelengths and suitably calibrating them to remove effects of the sensor and radiation source can produce spectral reflectance curves that show characteristics of the material being examined. Different atmospheric and surface minerals exhibit distinct spectral curves, allowing identification of material type.

In the case of surface materials, absorption features in parts of the spectrum are caused by electron orbit transitions and vibrations (Burns 1993). Many of these absorption features occur in the infrared portion of the spectrum (0.7–2.6 μm), being influenced by the presence of transition metal ions such as Fe^{2+}, Fe^{3+}, Mg^{2+} common in the molecular structure of Martian geology (Schade et al. 2004).

The discovery of hematite using reflectance spectroscopy at Meridiani Planum caused researchers to choose this site to land the Opportunity Mars Rover. The presence of this mineral in hydrated environments on Earth suggested that its abundance at Meridiani may have been an indicator for past liquid water at this Martian site (Barlow 2008).

3.1.2 The Influence of the Martian Atmosphere

Incident solar radiation must pass through a planet's atmosphere prior to illuminating a surface feature of interest and then back again to be collected by a remote sensor. Through the process, the radiation will suffer a degree of absorption and scattering, incurring a loss of total radiation. The terrestrial atmosphere is dominated by water and oxygen absorption bands, where "windows" of 0.4–1.3, 1.5–1.8, 2–2.5, 3.5–5.5 and 8–14 μm respectively allow transmission of radiation (Fig. 3.3a). These windows are exploited by orbital remote sensors to maximise reflected signal return. Spectral sensitivities that cross into regions not covered by these windows will suffer severe attenuation and are generally avoided by sensor designers.

The different composition of the Martian atmosphere poses unique challenges to sensors operating at the planet. Besides receiving reduced incident radiation

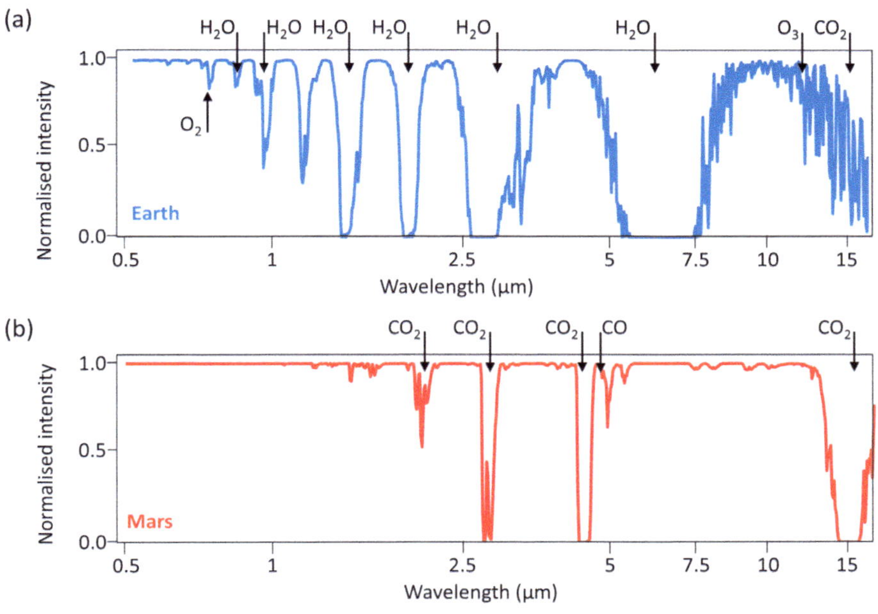

Fig. 3.3 Solar absorbance spectra of the atmospheres of **a** Earth and **b** Mars as generated by NASA's Planetary Spectrum Generator. Major absorbance bands are labelled (NASA/JPL)

(Fig. 3.1a), the thinner atmosphere of Mars reduces the amount of Rayleigh scattering (incident photons scattered by gas molecules that gives Earth's sky its characteristic blue colour). Vastly higher amounts of fine dust compared to the terrestrial atmosphere significantly increases Mie scattering (incident photons scattered by particles larger than the photon wavelength, such as suspended dust). Longer wavelengths are scattered more uniformly through this process, so the daytime atmosphere on Mars appears a salmon pink.

The Martian atmosphere is composed primarily of CO_2, leading to strong solar absorption bands at 2, 2.6, 4 and 15 μm, rendering detection in these regions difficult (Fig. 3.3b). In addition, the atmosphere is much drier than that of Earth, minimizing attenuation by water vapour and allowing wider transmission "windows" when compared to the terrestrial atmosphere (Fig. 3.3a, b).

3.2 Sensor Resolution, Processing and Calibration Considerations

3.2.1 Spatial Resolution

Returned raster sensor data is collected in the form of a two-dimensional array of picture elements, or pixels. Each pixel, arranged in rows and columns, represents a location in space and time and stores data as digital numbers (DN), representing intensity of received radiation such as visible light reflectance. Increasing the number of pixels (and hence the number of rows and columns in a sensor) and decreasing the distance between the sensor and the object to be investigated increases spatial resolution, the area of the Martian surface covered by the pixel. Ground sample distance (GSD) is the measure between the centres of two adjacent pixels of an image sensor as measured on the ground and is used to characterise spatial resolution. The relation between sensor pixel resolution of a detector array and GSD is given by (Fig. 3.4):

$$GSD_h = AS_h/FlI_h \tag{3.5}$$

$$GSD_w = AS_w/FlI_w \tag{3.6}$$

where A is the sensor altitude above the surface, S_h and S_w are the physical sensor heights and widths respectively, Fl is the remote sensor focal length, I_h and I_w the image height and width in pixels respectively. Where the detector is square, the heights and widths are the same.

As an example, the UAE Hope Emirates Exploration Imager orbiting at a 23,000 km altitude phase of its orbit, S_h and S_w of 16.9 and 22.5 mm respectively,

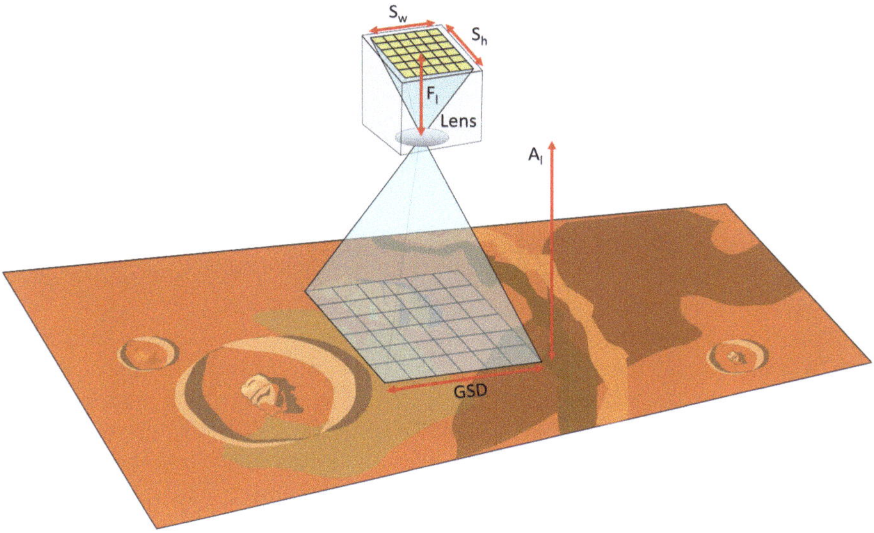

Fig. 3.4 Hypothetical raster sensor in orbit over Mars showing the relationship between GSD, altitude (A_l), lens focal length (F_l), sensor width and sensor height (S_w, S_h)

Fl of 51 mm and I_w and I_h of 4096 × 3072 pixels (Sharaf et al. 2017) would provide (from Eqs. 3.5 and 3.6):

- $GSD_w = (23{,}000{,}000 \times 0.0225)/(0.051 \times 4096)$
- $GSD_w = 2.477$ km/pixel.
- $GSD_h = (23{,}000{,}000 \times 0.0169)/(0.051 \times 3072)$
- $GSD_h = 2.481$ km/pixel.

In the above example of the Hope mission these GSD values would change as the spacecraft varies its altitude from the Martian surface in its elliptical orbit.

Higher spatial resolution in imagery means more pixels per unit area, enhancing the interpretation of features and patterns and hence more accurate interpretations of underlying geology or Martian phenomenology. This relationship with spatial resolution also applies to elevation data, where discovery and interpretation of finer scale landscape slope processes are only possible with higher resolution elevation data (Pain 2005). Increased spatial resolution comes at the cost of increased data volumes that must be stored and transmitted back to Earth.

3.2.2 Spectral Resolution

Spectral resolution determines the ability of the sensor to distinguish different portions of the electromagnetic spectrum. Panchromatic imagers, returning high-resolution monochrome images, possess low spectral resolution. Returned data consist of a single band over a range of wavelengths. These sensors are popular for many Earth observation satellites, with on-orbit vehicles needing to return a single channel of information. In addition, panchromatic sensors typically have higher spatial resolution than multi-band counterparts through trading spatial resolution at the expense of spectral resolution (e.g. Švab and Oštir 2006).

Multispectral imagers, being sensitive to three or more bands, have been flown since Mariner 9 and have higher spectral resolution than panchromatic sensors (Barlow 2008). Hyperspectral instruments such as Observatoire pour la Minéralogie, l'Eau, les Glaces et l'Activité (OMEGA) and Compact reconnaissance imaging spectrometer for Mars (CRISM) capture hundreds of spectral bands and are among the highest spectral resolution sensors sent to Mars orbit (Bibring et al. 2004).

3.2.3 Temporal Resolution

Temporal resolution refers to the revisit rate of an orbiter to return to the same observation area. The greater the frequency of revisit, the higher the temporal resolution. Applications for high temporal resolution include monitoring transient Martian phenomena, such as sand dune movement and other aeolian transport, frost cycles and contemporary impact cratering. Orbiters with wider sensor coverage facilitate higher temporal resolution but is often at the expense of spatial resolution (Gebhardt and Abuelgasim 2019). As an example, Mars Express was designed to revisit the same point over Mars every three days on its ascent past the equator in its orbit (Hechler and Yáñez 2003).

3.2.4 Optical Designs

For most passive sensors, refractive or reflective optical elements are required to direct and focus collected energy to the sensor for detection. Choosing the most appropriate design for these systems, including elements, supporting optics and sensors, is critical for remote sensing, particularly that of Mars. Operating on or around Mars presents additional challenges, particularly reduced light levels available to the sensor and colder thermal environments. One such consideration is resolving power, the minimum distance at which two objects imaged by the sensor can be determined. In addition to the number of pixels manufactured on the remote sensor, optical design has a direct impact on sensor performance, including GSD.

It is important to note that sensor design considerations entail engineering trades and constraints (Chap. 2). An example for trade is the design of the optical system, which is normally refractive or reflective in nature. The choice between using refractive or reflective optics (or a combination of both) is governed by the desired light collecting capacity, size, weight and cost constraints. Material composition also effects the utility of optics for applications. Optical glass is commonly used for imaging lenses, but will absorb UV and thermal radiation, making it unsuitable for these applications. Other lens substrates need to be used for these wavelengths, such as silicon for thermal applications (e.g. Fonollosa et al. 2008) or quartz, calcium fluoride or lithium fluoride lenses for UV sensing (Ballman et al. 1968; Brixner and Winkler 1981). In these applications, reflective elements may be a better engineering choice for the sensor.

Optical components are also designed to collect as much incident radiation as possible within the space constraints available, and produce a minimum of distortion. Common forms of distortion that must be corrected for include chromatic and spherical aberration, coma, field curvature and vignetting (e.g. Hawkes 2012). Chromatic aberration occurs when different wavelengths focus at different points of the focal plane; spherical aberration is caused by the shape of the optical element and can be corrected for through making the element parabolic in shape. While these distortions can occur in any optical system, they are more likely with short focal length refractive optics, such as those required to fit inside cubesat or landed rover payload spaces.

Corrections for lens distortions and aberrations require the development of more complex optical systems. Solutions typically include developing aspherical lenses to counter spherical aberration and using achromat doublet lenses of differing dispersion values to correct for chromatic aberration. Each of these techniques increases complexity and weight of refractive optical systems (Salva 2018). Flown refractive designs used for optical systems include the lens triplet (Fig. 3.5a) used in the MER Pancams (Bell et al. 2003) and combinations of moving and static lens triplets and subassemblies to create distortion free images at multiple focal lengths (Mastcam-Z, Fig. 3.5b; Bell et al. 2021).

Reflective designs flown or considered for Mars applications include the three mirror astigmat (TMA), where curved mirrors reduce optical aberration and maximise the field of view (Fig. 3.5c). TMAs have employed through-hole optics, where radiation from the secondary element is reflected through a hole in the primary mirror such as used in HiRISE (HiRISE Fig. 3.5d; McEwen et al. 2007) or where the three elements are offset (Fig. 3.5c). The latter configuration maximises incident radiation collection by not placing an obstruction at the centre of the primary element, but require more complex design and layout, and potentially occupy more space.

Even with a perfect optical system, impossible to actually construct, the theoretical limit to its performance is still constrained by its diffraction limit. Diffraction produces a pattern of elements that reduce in size from increased distance from the centre. The diffraction shape for a circular aperture, the most popular shape for flown optical Mars sensors, produces an Airy disk, or point spread function and not a perfect

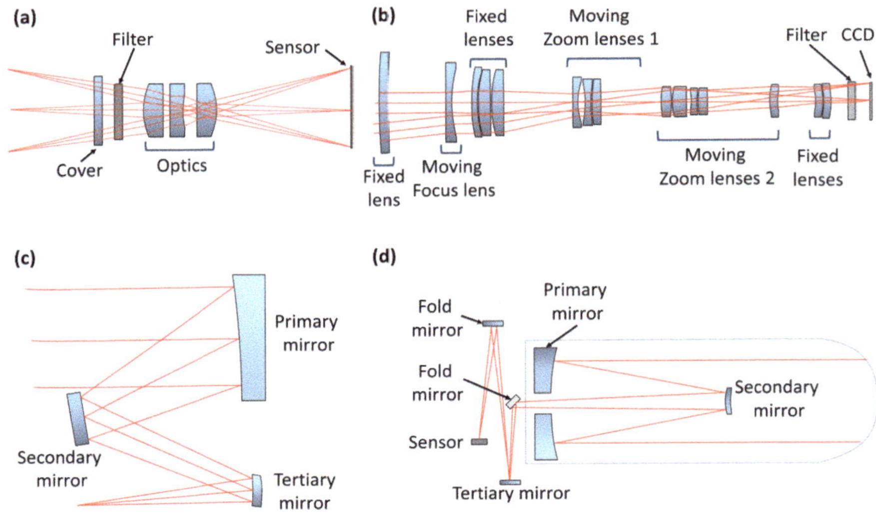

Fig. 3.5 **a** MER optical layout showing the lens triplet used to minimise aberration. **b** Mastcam-Z optical layout revealing a more complex array of lens elements required to facilitate zoom and focus. **c** Three mirror astigmat optical layout using reflective elements to negate distortion. **d** HiRISE camera optical layout incorporating a three mirror astigmat plus folding mirrors

point. For these systems, diffraction limits constrain the maximum resolution, X, by (Wertz et al. 2011):

$$X = (h\lambda)/D \tag{3.7}$$

where h is the distance between the sensor and the target, λ is the wavelength, D is the aperture diameter of the optical instrument.

As an example, using the green (550 nm) wavelength of the 0.5 m aperture HiRISE, orbiting 300 km from Mars would provide a theoretical diffraction limit of (converting to metres):

- $X = (300{,}000 \times 0.00000055)/0.5$
- $X = 0.33$ m.

This theoretical value is in close agreement with stated HiRISE GSD of 0.3 m for this orbit height (McEwen et al. 2007). This equation also highlights the dependency of diffraction on the wavelength. For thermal imaging systems, where wavelengths are up to tens of microns, diffraction becomes a major limiting factor on the spatial resolution of these sensors, making it more difficult for high resolution to be achieved.

3.2.5 Sensor Data Processing

Processing of remotely sensed data is essential for obtaining meaningful information regarding the target object. The increased diversity of remote sensing technologies used on orbit or landed on Mars has led to the capture of large quantities of data with distinctive properties. Consequently, remote sensing image processing and analysis has evolved to extract as much meaningful information as possible from returned data. Common image processing and analysis methods include pre-processing, calibration, enhancements, classification and feature extraction.

Consideration for obtaining this data commences with instrument optical design to ensure the maximum energy is available at the sensor for detection, while at the same time fitting into a compact space and minimizing aberrations or distortions. Mission operation parameters also ensure appropriate solar illumination angles to reduce shadowing or dust obscuration, and also maximizing survival in the space environment. This might involve maintaining constant sensor temperature as even small movement in optical assemblies from expansion or contraction of materials are detrimental to the quality of the end product.

Pre-processing

The detection element of a remote sensing system responds in a consistent manner to incident radiation directed by the optical system, allowing characterisation of the phenomenon being investigated. These sensors are imperfect, with manufactured pixels returning small, but different voltages to otherwise uniform input. Sensor response to different wavelengths may also not be uniform. An example is that the standard sensor used in many imaging applications have their highest sensitivities around the green (550 nm) portion of the spectrum with sensitivities falling off towards the NIR (e.g. Poletto et al. 1997).

Optical or focussing elements may also leave an artificially brighter region in the centre of an image, with intensity falloff towards the edges known as vignetting (Fig. 3.6a). Vignettes may also vary with changes in focal length, such as the Mastcam-Z. Detector uniformity may also change over time, as radiation and other environmental effects degrade the optical or electronic components of the sensor and producing vignetting. Variations are not limited to optical imagers, temperature sensitivity variations across the sensor, and pixel noise can also be present in thermal and other detectors (e.g. Edwards et al. 2011).

Flat fielding, a technique used in commercial photography applications, is used to identify and correct for these variations in sensor brightness. This process may include imaging a Lambertian (surface has the same reflectance from any viewing angle) reflective target or integrating sphere, ensuring uniform illumination of the sensor (Soderblom et al. 2008). The use of a calibration target can also assist with flat fielding the sensor. Examples include a specialised flat field lamp carried on board, as was used with the Hope Emirates Mars UV Spectrometer (Holsclaw et al. 2021), or THEMIS calibration flag that was frequently imaged to enable calibration (Christensen et al. 2004).

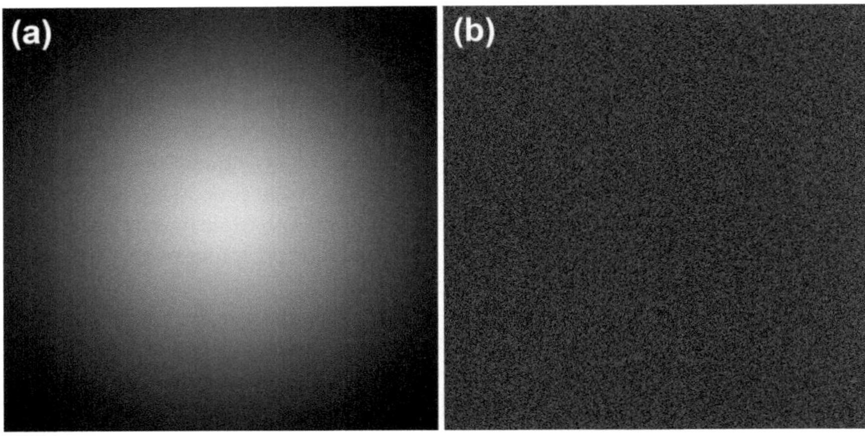

Fig. 3.6 Sensor attribute examples. **a** Image flat field example, showing exposure differences between different portions of the frame. **b** Image noise example

Sensor Noise

The ability of remote sensing to distinguish phenomena is also influenced by additional limitations of the sensor itself. Noise, undesired pixel values generated independently of a received signal, affect detectors and originate from thermal effects, where an increase in temperature increases the amount of random noise generated by the sensor (Fig. 3.6b; Irie et al. 2008). Thermal noise can particularly degrade the sensor signal during long exposures, and methods to manage sensor temperatures are of benefit to reduce this effect (e.g. McEwen et al. 2007). An additional noise source originates from the reading of sensor voltages themselves. Readout noise results from variations within pre-amplification circuitry for each pixel and manifests during conversion of electron charge to voltage and increases with frame rate (Gach et al. 2003). For fast frame rates, such as when rapid exposures during a hyperspectral scan of a bright object are captured, readout noise is of greater consideration in terms of corrupting pixel signals.

Another noise source is shot noise. This results from statistical variations in photon arrival rate, efficiency of the sensor in capturing the photons, and their position in striking the sensor. Large photon counts in higher exposures drastically reduce the significance of these statistical variations, and accompanying shot noise. This type of noise is a greater risk in low light conditions, where photon counts are low and statistical variations more significant (Reibel et al. 2003).

The strength of the incident signal must exceed the "noise floor" of the sensor in order to be registered as a meaningful result. Signals below this threshold can't be distinguished to obtain useful information. This is particularly of interest in the Martian environment, where solar energy levels are much less than those of Earth. Signal to noise ratios (SNR) are commonly used to characterize the performance of a given sensor. It is a ratio of the measured signal to the overall measured noise of each pixel; higher SNR are the desired result for engineering trade studies (Howell 2006). SNR is given by (Fellers and Davidson 2020):

$$\mathrm{SNR} = IQE\mathrm{t}/\left(IQE\mathrm{t} + N_{\mathrm{d}}\mathrm{t} + N_{\mathrm{r}}^{2}\right)^{1/2} \tag{3.8}$$

where

- I = Photon flux (photons/pixel/s)
- QE = Quantum efficiency (photo carriers/photon)
- t = Integration time (s)
- N_d = Dark current, including dark current shot noise plus dark current non-uniformity (carriers/pixel/s)
- N_r = Readout noise (carriers).

As an example, using the following sensor values:

- $I = 50{,}000$
- $QE = 0.4$
- $t = 0.1$
- $N_d = 5$ electrons/pixel/s
- $N_r = 4$ electrons.

Calculating signal electrons provides:

- $IQE\mathrm{t} = 50{,}000 \times 0.4 \times 0.1 = 2000$ electrons.

Calculating dark current provides:

- $N_d\,\mathrm{t} = 5 \times 0.1 = 5$ electrons.

Calculating SNR provides:

- $\mathrm{SNR} = (2000)/(2000 + 0.5 + 4^2)^{0.5}$
- $\mathrm{SNR} = (2000)/(44.90546)$
- $\mathrm{SNR} = 44.5$, which is within the range of mapping missions.

Noise Mitigation

Cooling the sensor is a method for increasing SNR. Other methods may include increasing collection of incident radiation through larger objective optics, increasing exposure times or combining charges of adjacent pixels through "binning" (Nasibov et al. 2010; He et al. 2021). Binning provides an increased sensor signal through averaging statistical variations, but at the expense of overall resolution. This method is more applicable with sensors possessing high pixel resolutions for their greater availability of pixels to bin, than those of resolutions of 1 Mpixel or less. Pixel binning is commonly employed with Martian remote sensing, particularly with spectral instruments such as CRISM and Tianwen-1 MMS (Murchie et al. 2007). Binning may also be used to provide a regional-level context understanding of remotely sensed information to identify potential data trends, such as mapping aerosol trends in the

Martian atmosphere (Douté et al. 2014). Such binning may be performed as part of an exploitation phase on Earth after data is transmitted from the spacecraft.

Noise characterisation and mitigation are performed pre-flight to understand unwanted sensor noise behaviour in response to changes in exposure, gain and temperature. For example, dark current management may artificially increase sensor levels (sensor bias) to avoid undersaturation (zero DN values) and noise impacts as described above. Dark current can be measured during ground-based sensor testing, where images are captured in zero incident radiation conditions. In the case of optical imagers, this is conducted in the absence of light. A series of frames are captured under varying exposure and thermal conditions to baseline the performance response of the sensor. During flight and operations, additional measurements may be taken to measure changes in sensor performance over the life of a Mars mission (Bell et al. 2009, 2013).

Once flat field and dark currents are characterised, data processing pipelines can be used to apply corrections to raw imagery for scientific use. Additional processing may be applied, such as characterisation and post-process removal of dust effects on the imaging optics. The presence of this dust causes artefacts in the captured data and often accumulates over time. A new flat field can be obtained, and used to process out the artefacts provided the dust contamination consistently appears in the same place of the sensor (Metzger et al. 1999).

3.2.6 *Radiometric Calibration*

Following characterisation and remediation of sensor biases and artefacts, such as vignetting and noise described above, a series of additional processes are required to remove unwanted influences of the external environment and extract meaningful information from the target being studied.

As on Earth, reflected radiation from Mars is affected by a variety of phenomena, shown in Fig. 3.7. Assuming the Sun is the primary light source for passive sensors collecting reflected energy, the solar spectrum dictates the power and output of EMR at various wavelengths. At the top of the Martian atmosphere, Solar output and absorption spectra will be identical for Earth, though the amount of available incident energy is less (585.2 W/m^2/compared with 1300 W/m^2 for Earth). This latter attribute constrains the amount of energy available for sensors, requiring considerations for optics to increase light gathering capacity and lower-noise sensors to increase SNR.

For Mars orbiters, radiation output from the Sun passes through the atmosphere to the ground region of interest, then back up through the atmosphere before finally being recorded by the sensor. As shown in Fig. 3.3a, b, the Martian atmosphere differs significantly from Earth's, with dust, CO_2 as the dominant gas, plus and its decreased density absorbing different parts of solar. Energy. On its path from the sun to the sensor, a portion of the solar radiation is: reflected back into space by the atmosphere (Fig. 3.7 [1]), scattered by the Martian atmosphere (Fig. 3.7 [2]), absorbed by the surface (Fig. 3.7 [3]), scattered or absorbed by the atmosphere on

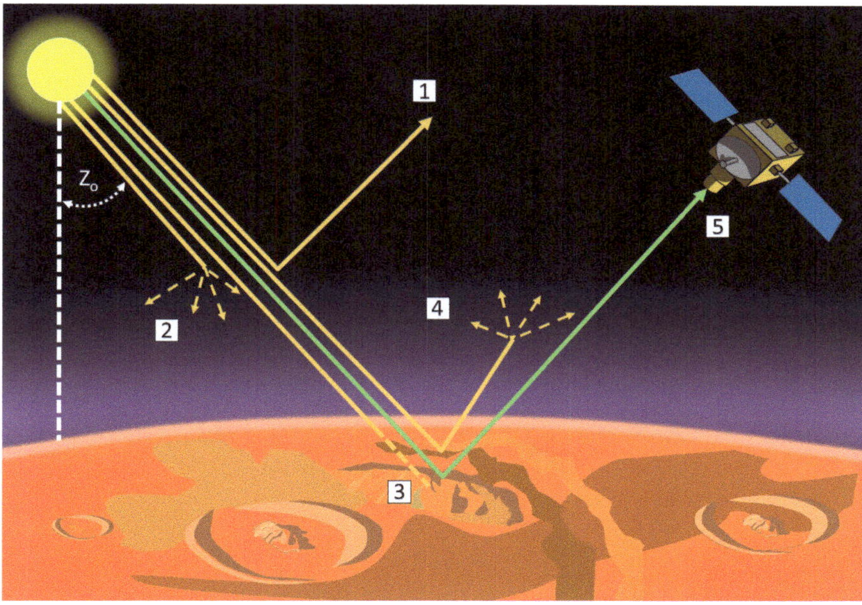

Fig. 3.7 Illustration of the Solar energy budget for a passive remote sensor. Z_o denotes the solar angle from zenith. [1] Reflection from the Martian atmosphere. [2] Atmospheric scattering. [3] Absorption by the surface. [4] Reflected energy scattered by the Martian atmosphere. [5] Reflected energy reaching the spacecraft sensor

its path back into space (Fig. 3.7 [4]). Only a portion of the incident solar energy is reflected from the surface and reaches the sensor detector (Fig. 3.7 [5]). In the case of surface missions, the radiation travel distance is halved, travelling from the Sun through the atmosphere to the target object and reflected to the sensor. These factors, along with the sensor limitations described above, must be accounted for before true reflectance values can be obtained.

The position and elevation of the Sun varies the amount of illumination available to the scene. The orbit of Mars means that its distance from the Sun of 25.2% eccentricity is much greater than the 3.4% value of Earth. Intensity varies by the square of the distance from the Sun, meaning that increased eccentricity will alter solar radiation availability. At the top of the Martian atmosphere (free of any atmospheric effects), normalised solar irradiance E is given by (Lillesand and Kiefer 2000):

$$E = E_o \cos Z_o / d^2 \tag{3.9}$$

where

E_o is the solar irradiance at 1 astronomical unit (AU), approximately 1361 W/m^2
Z_o is the sun's angle from the zenith (30°)
d is the Mars-sun distance in AU (average 1.524).

For example, if the sun is 30° from zenith, formula 3.9 would provide:

- $E = (1361 \times \cos(30))/(1.524)^2$
- $E = (1361 \times 0.866)/(2.323)$
- $E = 507.5\ W/m^2$.

The effects of the Marian atmosphere act to attenuate incident and reflected radiation through scattering and absorption. It also as a reflector, adding scattered "path radiance" into the signal detected by the sensor. Atmospheric influence, which is dependent on wavelength, is given by (Lillesand and Kiefer 2000):

$$R_t = R_o ET/\pi + L_p \tag{3.10}$$

where

- R_t is the total spectral radiance (object plus atmosphere) measured by the sensor
- R_o is the actual reflectance of the object (desired value)
- E is the irradiance on the object
- T is the transmission of the atmosphere (minus the scattering away from the sensor and absorption, Fig. 3.7 [2], [3] and [4])
- L_p is the path radiance (Fig. 3.7 [1]).

As an example, assume the following values for 600 nm wavelength of sunlight to calculate path radiance using formula 3.10:

- $R_o = 0.25$ (surface reflectance value).
- $E = 400\ Wm^{-2}\ \mu m^{-1}$ (Solar irradiance for this wavelength at Mars).
- $T = 0.95$ (Clear Martian atmosphere transmittance—may vary for Martian dust storms).
- $L_p = 1.5\ Wm^{-2}\ sr^{-1}\ \mu m^{-1}$ (Path radiance from dust or haze).
- $\pi = 3.1416$.

Calculating for R_t:

- $R_t = 1.5 + (0.25 \times 400 \times 0.95)/\pi$
- $R_t = (95/\pi) + 1.5$
- $R_t = 31.74\ Wm^{-2}\ sr^{-1}\ \mu m^{-1}$.

Solar spectrum, illumination angle and atmospheric effects all need to be accounted for through using the equations above (Eqs. 3.2–3.4, 3.9, 3.10) before R_o can be calculated and provide meaningful information. The extant dustiness of the Martian atmosphere has been modelled using radiative transfer models, to account for its deposition and effects on reflectance calibration (Kinch et al. 2020). These models, while useful, are unable to account for short-term changes in atmospheric dust, and as mitigation, surface-based images have employed physical calibration targets imaged by sensors throughout the mission. The targets consist of colour and greyscale swatches of materials capable of resisting degradation in the Martian environment, such as ceramics or space-grade pigments. In later Mars landers, such as Perseverance, permanent magnets were installed beneath the calibration targets to isolate magnetic Martian dust from the targets. This reduced the contamination of dust coating on these targets and subsequent obscuration of the spectral properties (Buz et al. 2019).

The spectra of the swatches are obtained using laboratory-grade instruments, such as spectro-radiometers and captured under varying illumination conditions to completely characterise their response. Images of the targets on Mars as returned from sensors can then be used to derive conversion values to transform DNs to radiance values by removing the effects of the Martian atmosphere.

Following capture of the calibration targets and subsequent flat field and dark current processing, a conversion from DN to radiance can be achieved. Ideal sensor detectors generally provide a linear response to increases in radiation levels. Exposures are carefully managed to avoid portions of undersaturation or oversaturation in cases where responses are not linear (Janesick et al. 1987).

For Mars orbiter calibration, direct imaging of on-board calibration targets has sometimes been impractical. Stellar objects, such as targeting Earth's moon during the outward journey or stars with well-characterised spectral properties allow comparison between sensor spectral performance and published values (McEwen et al. 2007). In the case of HiRISE, this process is conducted as part of a pipeline to reduce processing steps for exploiting this dataset (Hou et al. 2016).

The considerations and processes described in this chapter are extensively used for a variety of remote sensor technologies flown to Mars. The effectiveness of remote sensing depends on the electromagnetic spectrum, solar irradiance, sensor–target geometry, and interactions between radiation, the Martian surface, and its CO_2-dominated, dust-laden atmosphere, which strongly influences absorption and scattering. Optical design choices, diffraction limits, and environmental constraints further govern achievable resolution, while careful calibration and processing—such as flat fielding, noise reduction, and correction for optical distortions—are essential to convert raw data into reliable scientific information.

For context, an overview of the diversity of major remote sensing instruments successfully operated in Mars orbit and on the surface is listed in Appendix A. These also include optical sensors which will be discussed in detail in the next chapter.

References

Ballman AA, Dodd DM, Kuebler NA, Laudise RA, Wood DL, Rudd DW (1968) Synthetic quartz with high ultraviolet transmission. Appl Opt 7:1387–1390

Barlow N (2008) Mars: an introduction to its interior, surface and atmosphere. Cambridge University Press, New York, 264 pp

Bell III JF, Squyres SW, Herkenhoff KE, Maki JN, Arneson HM, Brown D, Collins SA, Dingizian A, Elliot ST, Hagerott EC, Hayes AG (2003) Mars exploration rover Athena panoramic camera (Pancam) investigation. J Geophys Res Planets 108, https://doi.org/10.1029/2003JE002070

Bell III JF, Wolff MJ, Malin MC, Calvin WM, Cantor BA, Caplinger MA, Clancy RT, Edgett KS, Edwards LJ, Fahle J, Ghaemi F (2009) Mars reconnaissance orbiter Mars color imager

(MARCI): instrument description, calibration, and performance. J Geophys Res Planets 114. https://doi.org/10.1029/2008JE003315

Bell III J, Malin M, Caplinger M, Fahle J, Wolff M, Cantor B, James P, Ghaemi F, Posiolova L, Ravine M, Supulver K, Calvin W, Clancy R, Edgett K, Edwards L, Haberle R, Hale A, Lee S, Rice M, Williams R (2013) Calibration and performance of the Mars reconnaissance orbiter context camera (CTX). Mars 8. https://doi.org/10.1555/mars.2013.0001

Bell JF, Maki JN, Mehall GL, Ravine MA, Caplinger MA, Bailey ZJ, Brylow S, Schaffner JA, Kinch KM, Madsen MB, Winhold A (2021) The Mars 2020 perseverance rover mast camera zoom (Mastcam-Z) multispectral, stereoscopic imaging investigation. Space Sci Rev 217. https://doi.org/10.1007/s11214-020-00755-x

Bibring JP, Soufflot A, Berthé M, Langevin Y, Gondet B, Drossart P, Bouyé M, Combes M, Puget P, Semery A (2004) Mars Express: the scientific payload. ESA Spec Publ 1240:37–58

Brixner B, Winkler MA (1981) Ultraviolet spectrograph lens. In: Los Alamos conference on optics '81, Dec 1981, vol 288. SPIE, pp 12–14

Burns RG (1993) Mineralogical applications of crystal field theory. Cambridge University Press (CUP), New York, USA, pp 460–470

Buz J, Ehlmann BL, Kinch K, Madsen MB, Johnson JR, Rice MS, Maki J, Bell JF (2019) Photometric characterization of Lucideon and Avian Technologies color standards including application for calibration of the Mastcam-Z instrument on the Mars 2020 rover. Opt Eng 58:027108. https://doi.org/10.1117/1.OE.58.2.027108

Christensen PR, Jakosky BM, Kieffer HH, Malin MC, McSween HY, Nealson K, Mehall GL, Silverman SH, Ferry S, Caplinger M, Ravine M (2004) The thermal emission imaging system (THEMIS) for the Mars 2001 Odyssey Mission. Space Sci Rev 110:85–130

Curran PJ (1985) Principles of remote sensing. Longman Group Ltd, Hong Kong, 282 pp

De Lima JAS, Santos J (1995) Generalized Stefan–Boltzmann law. Int J Theor Phys 34:127–134

Douté S, OMEGA Team (2014) Monitoring atmospheric dust spring activity at high southern latitudes on Mars using OMEGA. Planet Space Sci 96:1–21. https://doi.org/10.1016/j.pss.2013.12.017

Edwards CS, Nowicki KJ, Christensen PR, Hill J, Gorelick N, Murray K (2011) Mosaicking of global planetary image datasets: 1. Techniques and data processing for thermal emission imaging system (THEMIS) multi-spectral data. J Geophys Res 116. https://doi.org/10.1029/2010JE003755

Fellers TJ, Davidson MW (2020) CCD noise sources and signal-to-noise ratio. In: Concepts in digital imaging technology. www.hamamatsu.magnet.fsu.edu/articles/ccdsnr.html. Accessed 27 Apr 2020

Fonollosa J, Rubio R, Hartwig S, Marco S, Santander J, Fonseca L, Wöllenstein J, Moreno M (2008) Design and fabrication of silicon-based mid infrared multi-lenses for gas sensing applications. Sens Actuators B Chem 132:498–507. https://doi.org/10.1016/j.snb.2007.11.014

Fröhlich C (2006) Solar irradiance variability since 1978: revision of the PMOD composite during solar cycle 21. Space Sci Rev 125(1–4):53–65. https://doi.org/10.1007/s11214-006-9046-5

Gach J, Darson D, Guillaumé C, Goillandeau M, Cavadore C, Balard P, Boissin O, Boulesteix J (2003) A new digital CCD readout technique for ultra–low-noise CCDs. Publ Astron Soc Pac 115:1068–1071. https://doi.org/10.1086/377082

Gebhardt C, Abuelgasim A (2019) Scientific innovations from the Mars aerosol optical depth based on satellite data with a temporal resolution of hours. Prog Earth Planet Sci 6:1–10. https://doi.org/10.1186/s40645-019-0276-z

Granger RG (2013) An atmospheric radiative transfer primer. CRC Press, London, 205 pp

Greeley R (2013) Introduction to planetary geomorphology. Cambridge University Press, Cambridge. https://doi.org/10.1017/CBO9781139020961

Hawkes PW (2012) Lens aberrations. In: Focusing of charged particles, vol 1, pp 411–468

He Z, Xu R, Li C, Yuan L, Liu C, Lv G, Jin J, Xie J, Kong C, Li F, Chen X (2021) Mars mineralogical spectrometer (MMS) on the Tianwen-1 mission. Space Sci Rev 217:1–36. https://doi.org/10.1007/s11214-021-00804-z

Hechler M, Yáñez A (2003) Mars express orbit design. Acta Astronaut 53(4–10):497–507. https://doi.org/10.1016/S0094-5765(03)80010-3

Holsclaw GM, Deighan J, Almatroushi H, Chaffin M, Correira J, Evans JS, Fillingim M, Hoskins A, Jain SK, Lillis R, Lootah FH (2021) The Emirates Mars ultraviolet spectrometer (EMUS) for the EMM mission. Space Sci Rev 217:1–49. https://doi.org/10.1007/s11214-021-00854-3

Hou Y, Geng X, Xing S, Tang Y, Xu Q (2016) Radiometric calibration of mars HiRISE high resolution imagery based on FPGA. Int Arch Photogramm Remote Sens Spat Inf Sci XLI-B4. In: XXIII ISPRS congress, Prague, Czech Republic, 12–19 July 2016

Howell SB (2006) Handbook of CCD astronomy, 2nd edn. Cambridge University Press, New York, 218 pp

Iqbal M (1983) An introduction to solar radiation. Elsevier, Academic Press, Canada, 387 pp

Irie K, Mckinnon AE, Unsworth K, Woodhead IM (2008) A model for measurement of noise in CCD digital-video cameras. Meas Sci Technol 19:045207

Janesick JR, Elliott T, Collins S, Blouke MM, Freeman J (1987) Scientific charge-coupled devices. Opt Eng 26:692–714

Kinch KM, Madsen MB, Bell JF, Maki JN, Bailey ZJ, Hayes AG, Jensen OB, Merusi M, Bernt MH, Sørensen AN et al (2020) Radiometric calibration targets for the Mastcam-Z camera on the Mars 2020 rover mission. Space Sci Rev 216:1–51. https://doi.org/10.1007/s11214-020-00774-8

Kreslavsky MA, Head JW III (2000) Kilometer-scale roughness of Mars: results from MOLA data analysis. J Geophys Res Planets 105:26695–26711

Lillesand TM, Kiefer RW (2000) Remote sensing and image interpretation. Wiley, USA, 724 pp

Malin MC, Bell III JF, Cantor BA, Caplinger MA, Calvin WM, Clancy RT, Edgett KS, Edwards L, Haberle RM, James PB, Lee SW (2007) Context camera investigation on board the Mars reconnaissance orbiter. J Geophys Res Planets 112. https://doi.org/10.1029/2006JE002808

McCluney R (1994) Introduction to radiometry and photometry. Artech House, Boston

McEwen AS, Eliason EM, Bergstrom JW, Bridges NT, Hansen CJ, Delamere WA, Grant JA, Gulick VC, Herkenhoff KE, Keszthelyi L, Kirk RL (2007) Mars reconnaissance orbiter's high resolution imaging science experiment (HiRISE). J Geophys Res Planets 112. https://doi.org/10.1029/2005JE002605

Metzger SM, Carr JR, Johnson JR, Parker TJ, Lemmon MT (1999) Dust devil vortices seen by the Mars Pathfinder camera. Geophys Res Lett 26:2781–2784

Murchie S, Arvidson R, Bedini P, Beisser K, Bibring JP, Bishop J, Boldt J, Cavender P, Choo T, Clancy RT, Darlington EH (2007) Compact reconnaissance imaging spectrometer for Mars (CRISM) on Mars reconnaissance orbiter (MRO). J Geophys Res Planets 112. https://doi.org/10.1029/2006JE002682

Nasibov A, Kholmatov A, Nasibov H, Hacizade F (2010) The influence of CCD pixel binning option to its modulation transfer function. In: Optics, photonics, and digital technologies for multimedia applications, vol 7723. International Society for Optics and Photonics, p 77231A. https://doi.org/10.1117/12.854738

Pain CF (2005) Size does matter: relationships between image pixel size and landscape process scales. In: MODSIM 2005 international congress on modelling and simulation, Dec 2005. Modelling and Simulation Society of Australia and New Zealand, pp 1430–1436

Poletto L, Boscolo A, Pelizzo MG. Placentino L, Tondello G (1997) Quantum efficiency measurement on a CCD detector in the 0.3-to-1100 nm spectral region. In: EUV, X-ray, and gamma-ray instrumentation for astronomy VIII, Oct 1997, vol 3114. SPIE, pp 617–624

Reibel Y, Jung M, Bouhifd M, Cunin B, Draman C (2003) CCD or CMOS camera noise characterisation. Eur Phys J Appl Phys 21:75–80. https://doi.org/10.1051/epjap:2002103

Salva AR (2018) Design of refractive optical lenses for an infrared camera. Master's thesis, University of Liege, Facility of Applied Sciences, France

Schade U, Wäsch R, Moroz L (2004) Near-infrared reflectance spectroscopy of Ca-rich clinopyroxenes and prospects for remote spectral characterization of planetary surfaces. Icarus 168:80–92. https://doi.org/10.1016/j.icarus.2003.10.016

Sharaf O, Amiri S, Almheiri S, Almatroushi H, Alshamsi M, AlTeneiji E, Lootah F, McGrath M, Withnell P, Ferrington N, Reed H (2017) Emirates Mars mission (EMM) science overview. In: 48th annual lunar and planetary science conference, Mar 2017, no. 1964, p 1992

Soderblom JM, Bell JF, Johnson JR, Joseph J, Wolff MJ (2008) Mars exploration rover navigation camera in-flight calibration. J Geophys Res 113. https://doi.org/10.1029/2007JE003003

Švab A, Oštir K (2006) High-resolution image fusion: methods to preserve spectral and spatial resolution. Photogramm Eng Remote Sens 72:565–572

Wertz J, Everett DF, Puschell JJ (2011) Space mission engineering: the new SMAD. Microcosm Press, CA, 1029 pp

Chapter 4
Optical Remote Sensing

Visible-light imagers have been the most common remote sensor sent to Mars since Mariner 4 captured the first images of the planet near the start of the space age. The first successful lander, Mars 3, used a cyclotronic imager to attempt to return the first image from the surface; with every successful surface mission carrying cameras in the decades since (Chap. 1). On-orbit and surface-based visible light imaging technology remains a predominant tool for past and present surface process interpretation and change monitoring (Barlow 2008). This type of imagery is also critical for engineering functions, such as safe landing site selection (Golombek et al. 2012), as well as identifying safe navigation paths for moving vehicles to follow (e.g. Rankin et al. 2021).

4.1 Missions and Sensors

4.1.1 Orbiter Imager Types

Early imagers were rudimentary, reflecting the limited technology and bandwidth available. They relied on vacuum tube vidicon technology, where a gridded target converted pixels into a television picture that was digitally sampled, recorded on magnetic tape and returned to Earth by a high-gain antenna. The Mariner 4 camera returned 200 × 200 pixels at 64 shades of grey (Leighton 1970) while the Viking orbiter cameras, representing the most advanced vidicon technology ever sent to Mars, returned 1182 × 1056 pixels of 128 shades of grey across six colour filters (Carr et al. 1972). Each filter was cycled between snapshot exposures to build up colour and multispectral images.

The decades long hiatus to the next successful mission, Mars Global Surveyor (MGS) Mars Orbiter Camera (MOC), allowed its imagers to exploit CCD technology. The silicon-based sensors enabled lighter, more robust and sensitive imagers coupled

S. Hobbs, *Sensing Mars: Remote Sensing to Understand the Red Planet*,
https://doi.org/10.1007/978-3-032-24416-1_4

with long focal length optics that were able to return high resolution imagery orders of magnitude greater than the previous Viking imagers (1.5 m vs. 150–300 m GSD, Malin et al. 2010).

The exploration of Mars at this resolution created a step change increase in understanding the planet and allowed for discovery of small-scale features, such as glacier-like forms, gullies and sedimentary layering (Barlow 2008). The MOC consisted of linear CCDs operating in a pushbroom mode. This technology was first used in space by some terrestrial remote sensing satellites (Hartley and Gupta 1994). In contrast to the snapshot operation employed by Viking Orbiter cameras, and similar to commercial camera operation (Fig. 4.1a), the pushbroom technology exploits the orbital motion of the spacecraft to generate an image. The linear CCD sensor is “pushed” along the surface through orbital motion, where sequential exposures build a two-dimensional swath line by line (Fig. 4.1b).

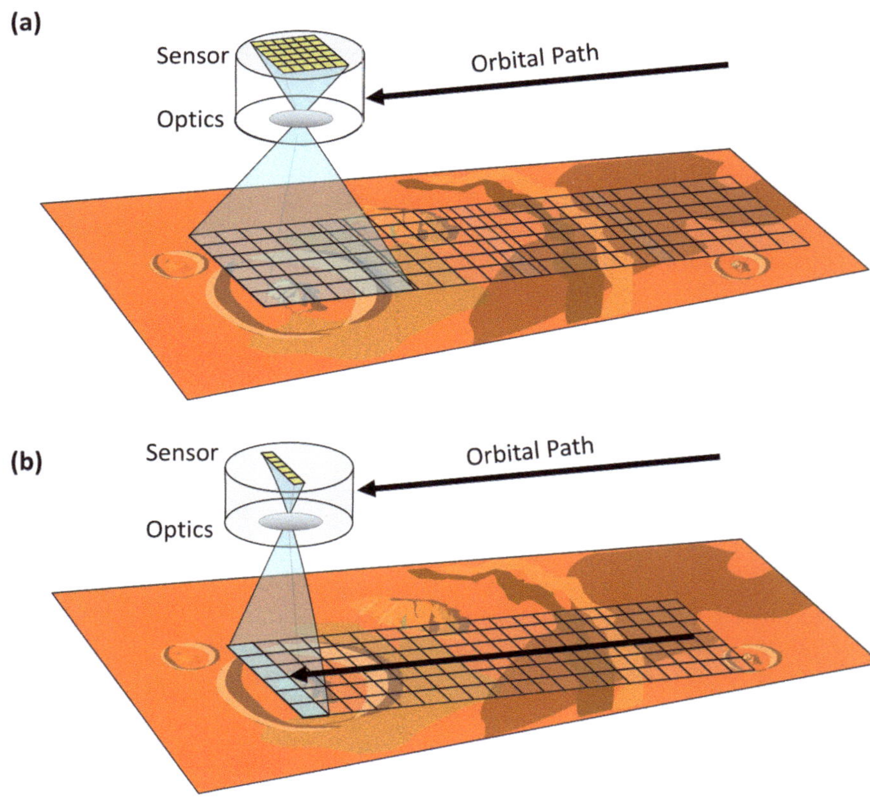

Fig. 4.1 **a** Snapshot image capture where a two-dimensional array of pixels is exposed, capturing an area of the planet’s surface. **b** Pushbroom operation where the spacecraft’s orbital motion drags a linear sensor array across the surface, building up an image swath

This mode is also used for the Mars Express, MGS CTX and HiRISE plus Chinese Mars mission imagers, where single pixel rows representing individual colour bands (or panchromatic sensor in the case of MRO CTX) use pushbroom to generate individual channels that can be combined into true-or false colour images (Meng et al. 2021).

Snapshot imaging remains a viable option for Mars remote sensing, depending on mission operational constraints where pushbroom may not be preferable. This is the case where spacecraft orbit is not circular, leading to image distortion if pushbroom were used. The Indian MOM was such an example. Its primary imager, the 2048 × 2048-pixel Indian Mars Color Camera, operated in snapshot mode (Arya et al. 2015).

4.1.2 *Popular Sensor Types*

Since 1996, CCD or more recently, complementary metal-oxide-semiconductor (CMOS) chips have been the most common types of imaging sensors for visible light remote sensing. These are sensitive to <400–1000 nm with a peak sensitivity centred around 500 nm (green wavelengths, Fig. 4.2).

Quantum efficiency (QE) is useful for determining the sensitivity of a given sensor. It measures the ability of a sensor to convert incident photons to electrons; the higher the QE the more effective this occurs. A sensor with 100% QE would theoretically convert all photons it receives into electrons. The sensitivity of CCD and CMOS sensors (Fig. 4.2) makes them ideal for detecting VIS/NIR wavelengths, however

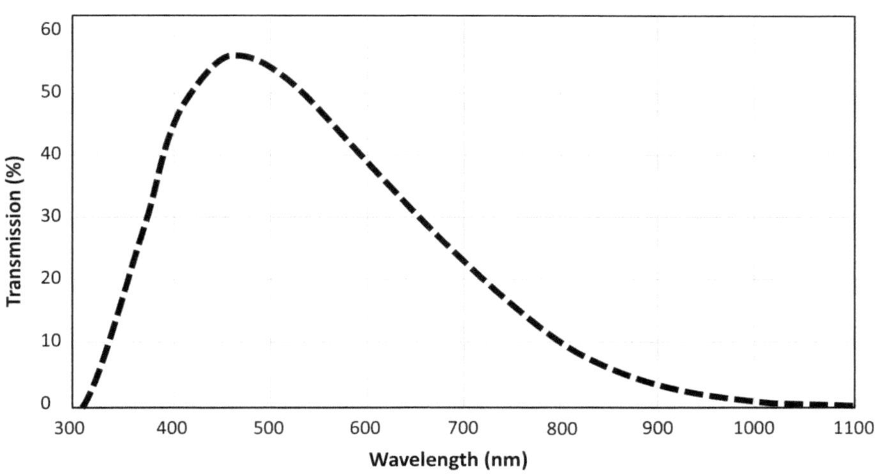

Fig. 4.2 Example CCD QE as represented by the Curiosity Rover Mastcam sensor

sensitivity for short wave infra-red (SWIR, 0.9–1.8 μm), investigated using Martian hyperspectral sensors, requires a different technology.

CCDs have a longer space heritage than CMOS sensors for space applications, though both sensor types have flown to Mars. CCDs consist of light sensitive pixels capable of collecting photons and generating electrons which can then be converted to a voltage reading. The size of these CCD pixels must be at least the same size as the wavelength of radiation to be detected. For example, green light at 550 nm would need a sensor pixel at least that size to be detected; 700 nm red light would require at least a 700 nm-sized pixel detector. In practice larger pixel sizes are often employed to increase sensitivity. Larger sizes provide more surface area to collect incident photons, but come at the expense of a larger sensor size for the same resolution. The Perseverance rover Mastcam-Z pixels are $7.4 \times 7.4\ \mu m^2$, larger than the ~ $1.2 \times 1.2\ \mu m^2$ required for detecting the VIS/NIR wavelengths the camera is sensitive to, but provides more light gathering capability (Maki et al. 2020). Earlier Mars lander examples of CCD sensors were the Spirit and Opportunity Rovers (Maki et al. 2003).

An exposure time is set through imager operation, which governs how long the pixels will collect photons for before being read and reset for the next exposure. Electrons stored within each pixel stored increase in proportion to the number of photons collected. Exposures are optimised to ensure sufficient photons are collected to provide meaningful information while not saturating the sensor. Following exposure to incident photons (Fig. 4.3a [1, 2]), pixel electron counts are moved across and down the chip (Fig. 4.3b [3]), read and converted to voltages that are digitised into a 2D array of values, forming an image (Fig. 4.3b [4]). A Bayer filter consisting of an array of red, green and blue or other colour filters may be overlaid on the sensor to produce multiband images (Fig. 4.3c [5]; Burt 1974). This technology has featured in most consumer-grade cameras. To simulate increased sensitivity to green of the human eye and produce a similar response, Bayer filters typically have double the number of green filters than red or blue.

CMOS, like CCDs, are also light sensitive imaging devices, but include an amplifying circuit for each pixel, reducing input power requirements at the expense of increased image noise. Integrated circuitry within the CMOS chip allows for faster processing and frame rates, and lowers manufacturing costs. The Perseverance Mastcam-Z is a modern example of a flown CMOS imager (Maki et al. 2020).

4.1.3 Surface Imagers

To date, Viking was the only Mars lander to successfully operate a whiskbroom camera (Fig. 4.4). Later landers used snapshot type imagers employing two-dimensional gridded arrays to return information. These imagers have employed two technologies for image capture. In the first, panchromatic sensors use filter wheels to return multispectral imagery across a number of selected bands as determined by the filters. The second technology, full colour cameras able to simultaneously return red, green and blue bands, as flown have been flown with Mars Insight, generated

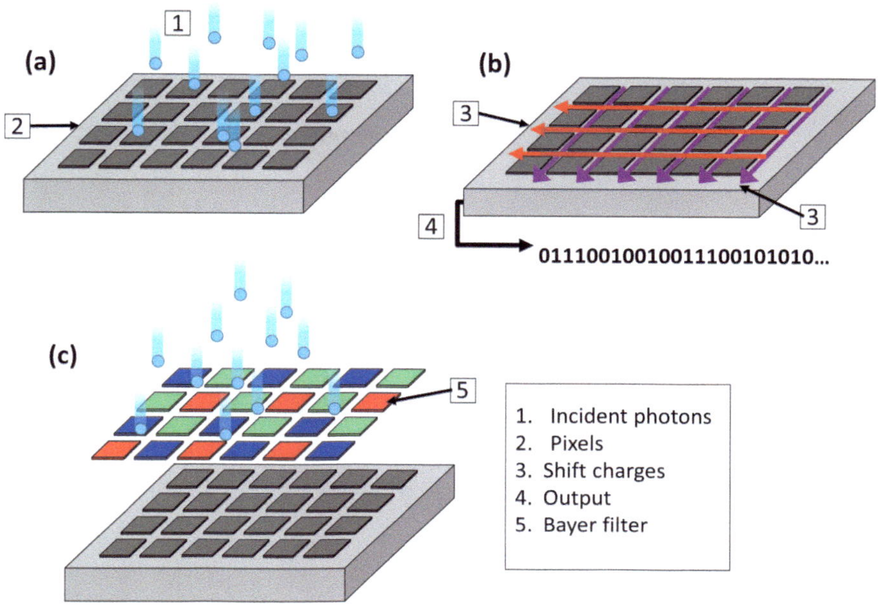

Fig. 4.3 **a** CCD overview where incident photons [1] generate voltage charges on light sensitive pixels [2]. **b** Charges are moved across and down ("shifted", 3) and sent to a processing unit [4]. **c** Bayer filter [5] of red, green and blue channels overlaid on a CCD sensor to generate an RGB colour image

colour imagery. Insight's camera was used where detailed mineralogical discrimination through multispectral filters was not a priority (Maki et al. 2018). Mastcam-Z, employing a combination of RGB camera and filter wheels represents the highest resolution surface imager sent to the planet to date. Its 1648 × 1214-pixel imager is complemented by an active 3.6:1 zoom optical system, allowing the capture of wide and narrow angle imagery from the same camera (Bell et al. 2021). The Remote Micro Imager (RMI) on board the Curiosity Rover is another versatile surface imager as part of the ChemCam payload suite. RMI captures images illuminated by the ChemCam laser beam to provide geological context. Additionally, the long (700 mm) focal length of the RMI optics allows for detailed imaging of formations several kilometres distant from the rover (Le Mouélic et al. 2015; Caravaca et al. 2021).

Fig. 4.4 Viking lander flight spare with two cameras in the Smithsonian Air and Space Museum (Public Domain)

4.2 Multi-band Imagery

4.2.1 Multispectral Imagery

Multispectral sensors are able to collect multiple spectral data using two or more discrete bands. They are often of lower resolution than panchromatic counterparts due to requirements for incorporating additional photosensitive elements in accommodating colour separation not needed for single-band sensors (e.g. Hnatushenko et al. 2015). Additional download time is also required to transmit some types of multi-band images from Mars to Earth, limiting the number of images that can be sent in each communication pass. This may include visible light ranges as collected by red, green, blue (RGB) cameras through to sampling wavelengths in near-infrared (NIR) or other spectral ranges (Barlow 2008).

The Viking lander cameras returned scientifically useful multispectral imagery from the surface of Mars and operated using a six-band sensor operating in a whiskbroom mode. A scanning mirror within each camera turret cycled vertically, providing one-dimensional information, and a step rotation mechanism mounted in the camera turret allowed a two-dimensional scene to be generated (Fig. 4.4; Wellman et al. 1976).

Fig. 4.5 The Mastcam used for the Curiosity Rover is an example of a flown multispectral camera (NASA/JPL)

CCD and CMOS technologies are popular choices for deep space multispectral applications (e.g. Mastcam, Fig. 4.5). The sensitivity of VIS/NIR wavelength ranges as a target wavelength range for multispectral cameras is of particular application not only to general surface characterisation but analysis of Mars mineralogy (Farrand et al. 2016).

Minerals commonly found on the planet include gypsum, iron oxide, olivine and hematite and possess spectral absorption and peaks that can be exploited in this range (Fig. 4.6; Bell et al. 2003).

Spectral discrimination has either been provided by using specific colour filters emplaced on the sensor or as a filter wheel positioned in front of a panchromatic sensor, where individual filters are cycled through each exposure. Mars Pathfinder and MER imagers used snapshot panchromatic cameras with colour filter wheels for generating multispectral imagery. The filter band passes were chosen to equate to spectral absorption features found on Mars, such as those caused by transition metal ions such as iron and magnesium (Fe^{2+}, Fe^{3+}, Mg^{2+}, Bell et al. 2008). Experimentation to identify the most appropriate filter selection has also been conducted during Phase A and Phase B mission planning. This involved comparing deviations between actual spectra and that collected through a filter set. Identifying unique spectral signatures of minerals of interest (e.g. Spectral curves in Fig. 4.6) could generate a trade between highly accurate returns over a specific mineral (e.g. Hematite curve, Fig. 4.6), or lower overall accuracy, but able to discriminate between more diverse mineralogy types (Cousins et al. 2011).

Table 4.1 provides exemplar filter bandpass values for Mars exploration as used by the Mastcam-Z instrument (Bell et al. 2021). The 476, 542 and 643 nm Bayer

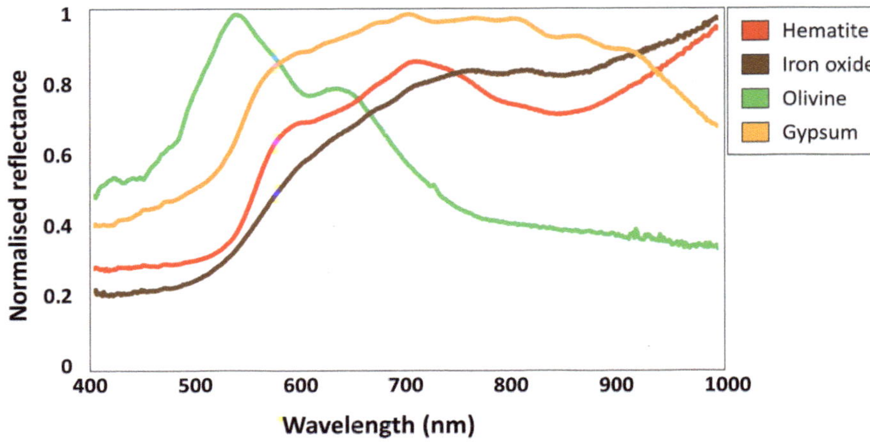

Fig. 4.6 Representative VIS/NIR spectra of minerals found on Mars. Each mineral exhibits unique spectral responses in the 400–1000 nm range that can be exploited

filters are specifically used to generate RGB colour imagery representative of what a human astronaut would see on the Martian surface.

The remaining filters within Mastcam/Z are used to characterise minerals of interest found on Mars, with spectral peaks and troughs coinciding with filter band-pass values (Fig. 4.7). An example of this is Hematite, where the 754 nm and 866 nm filters correlate with this mineral's spectral peak and trough respectively (Fig. 4.7; Table 4.1).

In filter wheel operation, individual filters are cycled over the sensor for different spectral bands to be images (Fig. 4.8a). While Mars Insight's camera used Bayer filter technology to capture simultaneous red, green and blue imagery that could be radiometrically corrected for multi-band (Maki et al. 2018), the latest Mastcam/Z multispectral imagers of Curiosity and Perseverance rovers employ Bayer filters and filter wheels (overview, Fig. 4.8b, c). Bayer filtered cameras can produce colour

Table 4.1 Mastcam-Z multispectral filters

Filter centre wavelength and bandpass	
476 ± 46 (Blue Bayer)	754 ± 10 (right camera)
542 ± 42 (Green Bayer)	801 ± 9 (right camera)
634 ± 43 (Red Bayer)	866 ± 10 (right camera)
442 ± 12(left camera)	910 ± 12 (right camera)
528 ± 11 (left camera)	939 ± 12 (right camera)
605 ± 9 (left camera)	978 ± 10 (right camera)
677 ± 11 (left camera)	1017 ± 18 (right camera)
880 ± 10 (neutral density filter, right camera)	

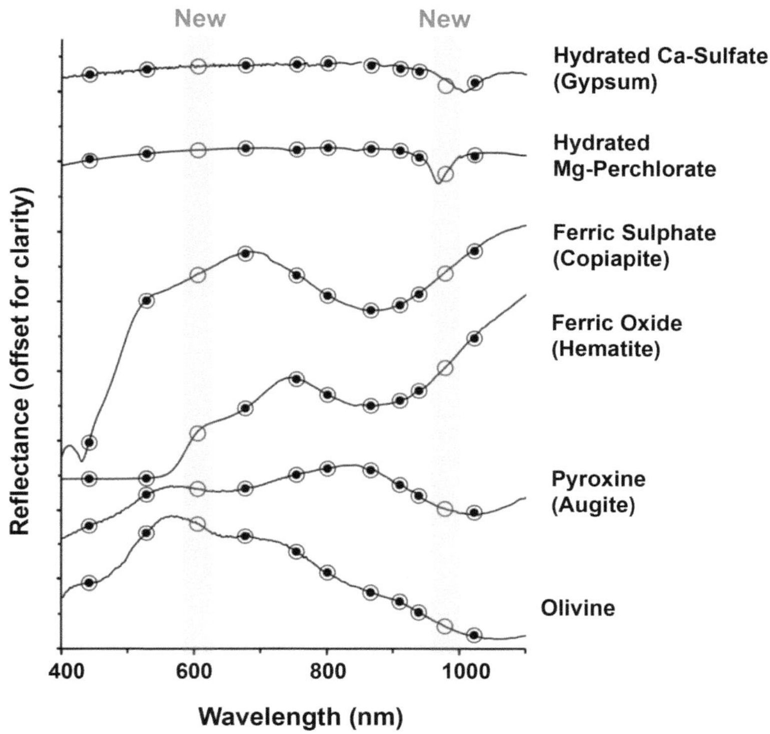

Fig. 4.7 VIS/NIR spectra of minerals found on Mars with the Mastcam centre bandpass wavelengths shown as dots. The two grey bands denote additional filters installed on the Mastcam-Z instrument versus Mastcam (NASA/JPL)

images without the use of a filter wheel (Fig. 4.8b). Along with Mastcam/Mastcam-Z, flown Bayer-filtered cameras include the MarsCO engineering cameras and Mars Insight colour camera (Maki et al. 2018; Bell et al. 2021).

Whether imagery is captured through a filter wheel or Bayer filter, individual monochrome channels, such as red, green and blue examples illustrated in Fig. 4.8c are captured. These must be subsequently combined, either as part of internal camera software processing or through an image processing pipeline, to generate a multispectral image (Fig. 4.8d).

The addition of a Bayer filter, while enabling simultaneous colour-image capture, introduces a trade-off in the form of reduced photon sensitivity. Significant absorption regions occur where the Bayer filter bands overlap (red/green, green/blue), and undesired leakage of NIR wavelengths through each channel can degrade spectral quality when compared to monochrome cameras. Additionally, the inclusion of Bayer filters reduces the overall pixel resolution of the camera by requiring three or more sensor pixels to generate a pixel of a single colour. This makes this technology less suitable for high spectral resolution applications (Hubel et al. 2004).

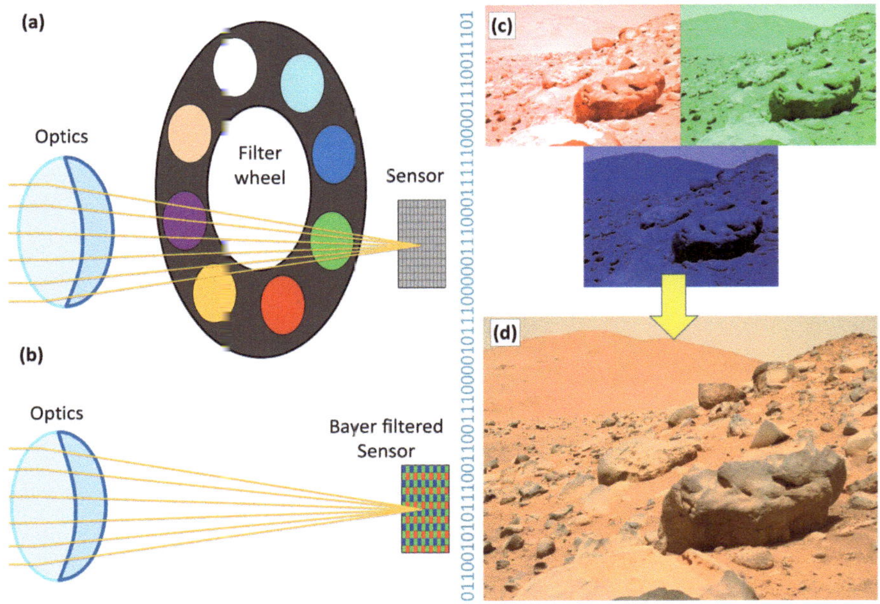

Fig. 4.8 **a** Schematic of a filter wheel-based camera, where individual frames are captured through a series of spectral filters. **b** A Bayer filtered camera uses a series of spectral filters overprinted on the sensor; for colour applications, the Bayer filter is a series of red, green and blue filters. **c** Exposed camera sensor output is converted to single channel images, red, green and blue shown as examples (NASA/JPL). **d** The single channels can then be combined to produce a multispectral (or RGB colour) image (NASA/JPL)

Despite this, these sensors have returned unprecedented imagery of the Gale and Jezero Crater landing sites. An additional filter wheel employed on both Curiosity's and Perseverance's cameras allows these imagers to produce narrower band-pass spectral investigations that include NIR spectrum sensitivity.

Custom Bayer filters can be developed to provide additional spectral bands in lieu of the standard RGB range. This can be achieved through replacing red, green and blue filters with custom wavelength ranges or increasing the number of filters (Tisserand 2019); though none have yet flown to Mars. Instead, using filter wheels remains the favoured approach to Mars VIS/NIR multispectral imagers.

4.2.2 *Multispectral Band Ratios*

Having access to more than one discrete spectral range increases the ability to discriminate surface mineralogy and better describe surficial geology. Comparing the results of one band against the other (either as a ratio or index) has quantified the spectral differences between different wavelengths. In the terrestrial context, band ratios terrestrial applications to study and characterise vegetation. A simple example is the green/red ratio, exploiting the higher reflectance vegetation has in the green band over the red band (Motohka and Nishida 2009).

Red/blue ratios are popular and exploit the higher reflectivity iron oxide-related minerals found on Mars have in the red (~ 650 nm) than the blue (~ 450 nm) bands. Increased red/blue ratios indicate a "redder" material and can be used to infer the level of dust coating on a Mars rock as well as its degree of oxidization. These ratios were used in multiple Mars surface mission to distinguish between rocks of different geological makeup (Farrand et al. 2008).

Band ratios were recently used to critically analyse the inferred presence of hematite at Gale Crater. The discovery of this mineral at Meridiani Planum was seen as evidence for conditions favouring a past aqueous environment, of interest to astrobiologists (Barlow 2008). Analysis of surficial results suggested that red crystalline hematite had not been found in Gale Crater. Instead, band ratio analysis identified that other mineral oxidisation events better fit the multispectral data, having implications for a potential past aqueous environment on this part of Mars (Jacob et al. 2020).

As most multispectral cameras sent to Mars have possessed not only visible, but NIR sensitivity, other band calculations are possible. A Martian band ratio that uses NIR has exploited the depth of iron absorptions affecting the slope of the 750–840 nm (NIR) and 535 nm (green) ranges responds to the presence of crystalline ferric oxides (Fig. 4.7; Morris et al. 1994). As described in Chap. 2, NDVI is a popular VIS/NIR band ratio used extensively for monitoring vegetation health in terrestrial applications with values between − 1 and 1. Can it also be used for Mars? While custom band ratios between specialized filters may be employed (e.g. between 801 and 754 nm band for Hematite, Table 4.1), the standard NDVI ratio may be considered to reduce design constraints by trading out the requirement for multiple narrow pass filters or even the need for a hyperspectral instrument (Peñuelas et al. 1997), as it requires only two bands for processing. Such a sensor is the single-filter low-cost camera detailed in the high-altitude balloon case study (Chap. 2).

Many minerals found on Mars exhibit spectral phenomenology in the red/NIR range, and NDVI is also able to return useful information for Mars applications. Examples derived from Mars Pathfinder results (Farrand et al. 2008) and laboratory specimens of minerals commonly found on the planet ("Terrestrial Sample") are shown in Table 4.2. For terrestrial context, a value for broad-leaf vegetation is also included. Vegetation NDVI values are very high, highlighting its extensive NIR reflectance compared with red, Gypsum and iron oxide-related values are in the low positives reflecting their higher NIR reflectance compared to red, while olivine, possessing lower NIR reflectance has a negative NDVI (Table 4.2). These examples and other returned Martian band data (Table 4.2) show that NDVI can provide a broad indicator to support distinctions between major classes of minerals.

Table 4.2 Normalised Difference Vegetation Index ratios for terrestrial and Martian materials

Terrestrial sample	NDVI	Martian sample	NDVI
Olivine	– 0.350563809	MPF "pink rock"	0.125
Hematite	0.016	MPF "gray rock"	0.125
Iron oxide	0.132	Rottnest dusty surface	0.19
Gypsum	0.114	MPF "red rock"	0.2
Carbonate rock	0.162	MPF "maroon rock"	0.333
Vegetation (broad leaf)	0.57		

4.2.3 Hyperspectral Sensors

Despite being immensely popular, multispectral sensors (multiple discrete bands) are able to collect data in a limited set of spectral bands (e.g. Jensen 2015). Although able to return information across parts of the spectrum in broad channels, multispectral cameras are limited in that subtle absorption characteristics are often missed, as this information is merged with other data within the channels to which the sensor is sensitive. This in turn limits the effectiveness of these instruments in identifying and characterizing mineralogy (Jensen 2015).

Hyperspectral sensors provide detailed spectral accuracy by detecting up to hundreds of bands of high spectral resolution data (Fig. 4.9a–c). The sensors form a data cube of information where each pixel contains detailed spectra of a region of interest over a given range. These instruments represent the most sophisticated method for fine-detail discrimination for planetary remote sensing. The complexity in designing such instruments to image swaths of planetary surfaces in such detail delayed their use at Mars until 2003 with the arrival of the Mars Express OMEGA instrument (Bibring et al. 2004).

Hyperspectral sensors may consist of more than one detector type to increase sensitivity in different parts of the electromagnetic spectrum within the one instrument. This provides the advantage of greater science return at the expense of more complicated optical and detector design and layout, and associated mass increase. While exotic hyperspectral sensors have flown to Mars, this section will focus on VIS/NIR/SWIR designs that have proved to be popular with recent Mars missions.

Hyperspectral Designs

Obtaining a hyperspectral data cube has required specialised optics and configurations that typically use objective optics to collect radiation, a slit to provide spectral resolution, a diffractive element to generate spectra and one or more detectors to record the information. In terrestrial spectrometers, diffraction elements may consist of gratings or prisms. Gratings possess rows of equally spaced elements that disperse incident radiation in a consistent pattern that is then detected and processed. Prisms are the oldest form of dispersive element with their physical shape refracting radiation, but their weight and tendency to provide inconsistent band spacing across the

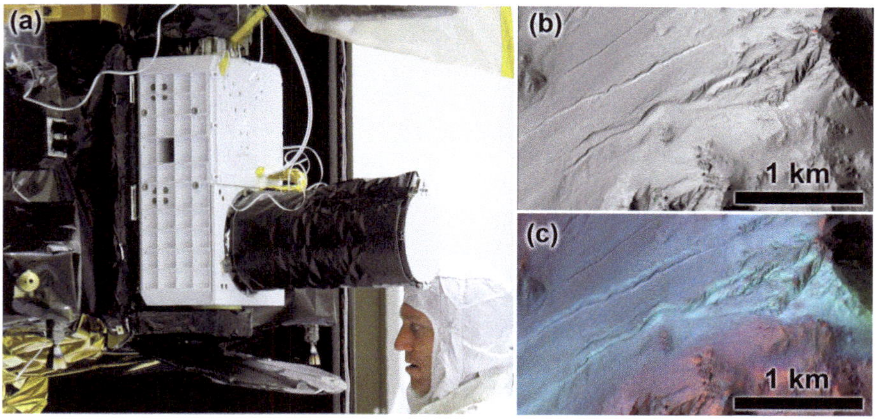

Fig. 4.9 **a** The MRO CRISM instrument, example of a hyperspectral sensor (NASA/JPL). **b** Context image of a Martian gully system (PIA20763, NASA/JPL). **c** Hyperspectral analysis of the same region from CRISM data revealing spectral differences in the landscape (PIA20763, NASA/JPL)

spectral range means that most flown Mars hyperspectral images have favoured using diffraction gratings (Sellar and Boreman 2005).

The OMEGA and CRISM instruments flown to Mars orbit have each employed objective optics, slits, diffraction gratings and detectors to generate hyperspectral data (optical diagrams, Fig. 4.10a, b). OMEGA, flown on-board Mars Express, was the first successfully flown hyperspectral imager sent to Mars. CRISM, the NASA-flown example of a VIS/NIR/SWIR Mars orbiter hyperspectral instrument, arriving in Martian orbit in 2005. Both instruments employed two distinctive spectrographs to cover the VIS/NIR and SWIR ranges (Table 4.3). CRISM has complemented the OMEGA hyperspectral instrument flown on board Mars Express.

A type of integration sphere in-flight calibration system, such as that used in CRISM (Fig. 4.10b), was used with the MMS on the Tianwen-1 Mission. MMS used two independent slit/grating dispersive VIS/NIR and SWIR spectrographs that share the same objective optics. Like CRISM, the MMS SWIR spectrograph is thermally controlled, with this instrument shunting waste heat to a cold plate on the orbiter. On-orbit pixel binning is used in this instrument, where spectral information from adjacent pixels is combined. This results in a small loss of spectral resolution but substantially reduces the amount of data returned to Earth, saving communication bandwidth (He et al. 2021).

The SWIR-sensitive element of the Mars Express OMEGA instrument uses objective optics plus a slit to pass a point spectral pixel. A concave holographic grating mirror is used to reduce the number of optical elements (such as a separate grating and lenses or mirrors) required in the instrument. A scanning mirror, rotating in a side-to-side action perpendicular to the spacecraft orbit, returns cross-track information, and the motion of the spacecraft then builds up the two-dimensional scene

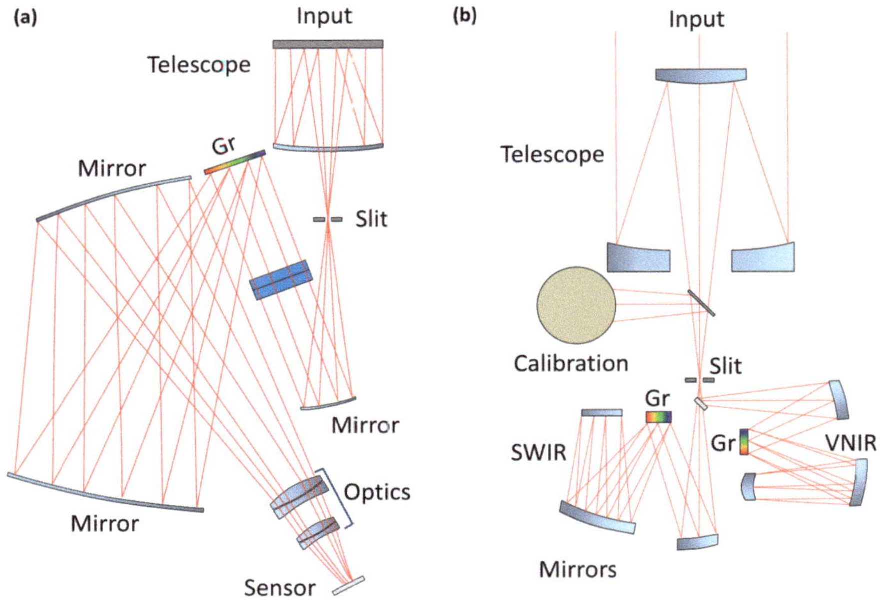

Fig. 4.10 **a** The SWIR optical pathway for OMEGA. Radiation is collected by a Cassegrain telescope, passes through a slit, dispersed (grating labelled "Gr" in the figure) and collimated through a series of lenses and mirrors to a SWIR detector. **b** Optical diagram of the CRISM hyperspectral sensor. Following input through the telescope, the incident radiation is split and routed into the VNIR and SWIR spectrographs. Each spectrograph consists of a grating (labelled as "Gr" in the figure) that disperse the incident radiation into component wavelengths. Calibration is performed by an included integration sphere

Table 4.3 Comparison of spectral and GSD attributes of flown Mars hyperspectral imagers with prime image sensors carried on board the same mission

Hyperspectral sensor, GSD/spectral resolution	Corresponding imaging sensor
MMS: 2 km/pixel/0.379–1.076 μm with 2.73 nm/band; 1.033–3.425 μm at 7.5 nm/band	HiRIC: 2.5 m/pixel (PAN), 10 m/pixel (colour)
CRISM: 18.4 m/pixel/6.55 nm/band	HiRISE: 0.025 cm/pixel
OMEGA: 300–5000 m/pixel/13.4 nm/band	HRSC: 10 m/pixel

in a whiskbroom operation (optical diagram, Fig. 4.10a). If desired, the mirror scan angle may be increased to provide a wider field of view.

Pushbroom methods (Fig. 4.1b) have also been used in Martian hyperspectral instruments, with lack of moving parts considered an engineering advantage. As an example, CRISM collected incident radiation by a specialised "Ritchey-Chretien" telescope, and passed through a slit to create a one-dimensional representation of the scene (Fig. 4.10b). The radiation is then split across dedicated VNIR and SWIR detectors, with a calibration source built into the instrument (labelled "calibration" in Fig. 4.10b). For each detector, the resulting spectra are focussed onto an imaging element, where the one dimension of spatial image is converted to two dimensions of spectral information. MRO's orbit then provides the pushbroom motion for CRISM to build up a hyperspectral data cube. In contrast to OMEGA's scanning mirror, CRISM's field of view is permanently restricted by its fixed slit height and accompanying optics.

Other hyperspectral designs, not yet flown to Mars, include Fabry–Perot interferometer (FPI) and snapshot imagers. FPIs use two opposing reflective surfaces separated by an air gap to act as a Fabry–Perot resonator. The high reflectivity of the surfaces only allows transmission of constructively interfering light at a specific wavelength (Rissanen et al. 2017). Modifying the gap between the surfaces determines the wavelength transmitted, and modern, small designs have been proposed for some future Mars missions (e.g. Ge et al. 2003; Hieta et al. 2022).

Snapshot hyperspectral imagers extend on the three channel RGB sensor and use custom, multi-band Bayer filters on the underlying sensor chip (Antonucci et al. 2019). These negate the need for additional motion or scanning to build the spectral image, allowing spectral cubes to be captured at a discrete point in time. Spectral filters deposited on the sensing chip also removes the need for diffractive optics or tuneable filters, further reducing instrument size and weight. These sensors are instead constrained by the limited number of filters available for these instruments (36 filters for a 6×6 instrument, Geelen et al. 2014) compared with hundreds of bands for flown hyperspectral sensors (e.g. 544 bands for CRISM, Murchie et al. 2007).

Detector Considerations

The choice of detector for hyperspectral sensors is derived through the Pre-Phase A-Phase B systems engineering process, with technologies chosen to accommodate the identified goal of the instrument (Chap. 2). Analysis of atmospheric properties may favour different detectors compared to characterisation and monitoring of Mars mineralogy (Burns 1993; Bell et al. 2008). Flown VIS/NIR hyperspectral detectors have included CCD and CMOS detectors, exploiting the spectral sensitivity of silicon substrates (Fig. 4.2). Silicon-based detector sensitivity degrades severely past 1000 nm, so detectors sensitive to SWIR wavelengths use a Mercury Cadmium Telluride (MCT) substrate to provide desired sensitivity. In the case of CRISM, MCT detectors provided spectral sensitivity in the 1002–3920 nm range (Murchie et al. 2007). These detectors can be less robust than silicon counterparts, incurring bad pixels due to increased sensitivity to thermal stresses such as those found in the deep space environment. Active thermal control to alleviate these stresses through maintaining constant temperatures have been used, but adding to the complexity of the instrument. Thermal cooling was used for the SWIR spectrograph within the CRISM instrument to optimise hyperspectral detection though the instrument reached end of life once the thermoelectric coolers failed (Seelos et al. 2023).

Spectral Resolution

Spectral resolution is also an important consideration for a hyperspectral sensor and measures the width of each band captured in nm. The spectral resolution may vary across the sensor's wavelength range, particularly between VIS/NIR and SWIR. High spectral resolutions are favourable for these sensors however produce trade costs of decreased incident radiation sensitivity and higher returned data volumes.

Spectral resolution for each band may be calculated using full width half maximum (FWHM) where the width of the band is measured at half the height of its maximum (Sigernes et al. 2018):

$$SR = (a \cdot \cos\alpha / k \cdot f_1) w_s \quad (4.1)$$

where a is the groove spacing of the diffraction grating, α is the incidence angle (set at 0°), k is the first spectral order (1), f_1 the focal length of the objective lens and w_s is the slit width. As an example, for the OMEGA VIS/NIR spectrometer with focal length of 57.6 mm, slit width of 50 μm, and grating groove spacing of 15.4 μm provides a spectral resolution of 13.4 nm.

GSD of hyperspectral sensors is governed by similar considerations for other orbital sensors (Chap. 3), depending on focal length, orbital height and pixel size (Eqs. 3.5 and 3.6). As with other sensor designs, desired hyperspectral sensor GSD may be subject to trade constraints due to the complexity of the optics required, high spectral resolution and resulting generation of large volumes of data returned from these instruments (2–2.5 GBits for CRISM, Seelos et al. 2023). As a result, GSD values are an order of magnitude lower than imaging sensors. This is highlighted in Table 4.3 that provides a comparison between carried hyperspectral sensors and imagers on board recently flown Mars orbiters (Oberst et al. 2004; He et al. 2021).

Calibration Methods

Hyperspectral image calibration is essential for accurate mineralogical analysis and comparison. It is similar to radiometric calibration methods is covered in Chap. 3, but entails the use of materials of known reflectance, such as an on-board calibration target, or parts of the Martian surface. The CRISM and MMS instruments possessed on-board calibration, though other methods have provided supplementary information. One such method is imaging the bottom and top of Mars' highest feature, Olympus Mons. In this method, features from the same geologic unit are imaged at the dense regions of the Martian atmosphere (bottom of Olympus Mons) and compared with images from the top, where the atmosphere is thinnest (McGuire et al. 2009). This method provides information on the CO_2 and dust absorption that can be removed to obtain reflectance (Chap. 3). Additional calibration techniques, particularly for areas distant from Olympus Mons, is to use custom-generated atmospheric transmission models specific to localised climate conditions experienced at the site being sensed (Bakker et al. 2014). UV spectrometer calibration may employ the use of bright stars, where the spectrum is well understood and can be compared with independent observations. Both the Mars Express SPICAM and UAE Hope

MMC used stellar sources to assist with instrument calibration (Bertaux et al. 2006; Holsclaw et al. 2021).

Surface Mission Spectrometers

Spectral sensing has also been conducted on landed surface missions. While not hyperspectral imagers, ChemCam and SuperCam on board the Curiosity and Perseverance Mars rovers respectively use slit and grating dispersive elements to return spectral information of the Martian surface. These instruments overcome spectral mixing caused by obscuration of dust by firing lasers into targets of interest and reading the resulting spectra from the resulting vapourised materials (Wiens et al. 2012).

In application, hyperspectral sensors have revolutionised the understanding of past and present climates of Mars. When combined with other remotely sensed datasets, hyperspectral sensors have supported or suggested alternative explanations for previously postulated presence of minerals with implications of surface process alteration by liquid water or other volatiles (e.g. Stockstill et al. 2005; Massé et al. 2014). Global maps of the planet have now been produced, relying heavily on data returned from the sensors described above.

4.2.4 Case Study, Design and Development of a Low-Cost Hyperspectral Sensor for Deep Space Applications

Recent compact optical and sensor advancements have produced increasingly sophisticated remote sensors able to return useful scientific data in Space 2.0 applications. The complexity of hyperspectral imagers has prevented their development for these applications until quite recently, with flown instruments needing to be carried on board full-scale missions (above). Recently, small reflective and refractive, dispersive hyperspectral sensors are currently being designed for deep space missions (e.g. Bosanac et al. 2018; Huq et al. 2019). This case study will investigate the utility of design and development of a low-cost hyperspectral sensors using COTS components for Mars applications (proposed flown design, Fig. 4.11).

Mineralogy analysis of Mars has been of high-interest for decades, with results providing insight into the geologic history of the planet, including potential habitability for life. In particular, many minerals, such as olivine, pyroxene, and plagioclase with absorption bands related to Fe^{2+} ions in their structure and hydrated minerals such as serpentine exhibit spectral phenomenology in the VIS/NIR range (Burns 1993; Schade et al. 2004). Could a low-cost hyperspectral sensor be constructed that would be responsive to spectral signatures of these minerals, while also fitting within a small payload space of a 6U cubesat?

A tailored design engineering approach was undertaken to identify a high-level concept and design to satisfy the overriding constraints of minimal space engineering expertise required to design, build and test the sensor, and overall cost. These were

Fig. 4.11 Hypothetical deep space cubesat mission in Mars orbit. Such a vehicle would need efficient, long-range communication and power generation systems

based on making it possible for low-budget organisations to design and fly a hyperspectral mission with limited space heritage experience, while still returning useful scientific data. A similar approach was undertaken for MarsCube One (Schoolcraft et al. 2017) and the ability for the sensor to fit within a 6U cubesat size was identified as a prime requirement. This would bound costs and complexity to within the limited goals of a Space 2.0 mission.

Based on the constraints and science goal listed above, the following requirements were developed for the sensor to:

- Provide a spectral resolution of at least 10 nm within the 550–850 nm range to discriminate mineral phenomenology;
- Possess sufficient light gathering ability to detect and characterise minerals in Martian orbit or on the surface;
- Fit within a 2U payload space of a 6U cubesat;

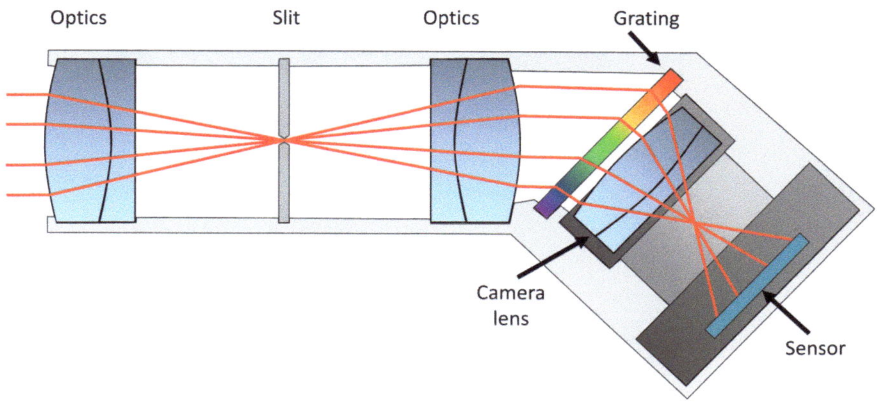

Fig. 4.12 Design overview of the custom hyperspectral pushbroom instrument showing the principal optics, slit, diffraction grating and camera elements

- Have weight and operating power constraints of less than 1 kg and 10 W;
- Be of simple construction to be within the engineering capabilities of limited space heritage organisations; and
- Use low-cost, COTS components.

Based on these requirements, a dispersive slit-based pushbroom sensor design was chosen that used refractive optics. This design required no moving parts, such as a scanning mirror or electronics, besides that required to operate the imager. Four 25 mm, 50 mm focal length achromat lens pairs (focal length of 25 mm) for the objective and collimator lens were chosen as a compromise between diffraction quality, light-gathering ability, and space constraints. A 100 μm width, 3 mm diameter mounted slit plus 600 lines/mm transmission diffraction grating completed the optical train to the camera (cutaway, Fig. 4.12).

To help refine the design and constrain the choice of sensor, mathematical estimations of signal to noise ratio and photon flux were obtained. Signal to noise ratio is an important factor for hyperspectral sensors to ensure sufficient discrimination between spectra. This ratio can be calculated as (Prentice et al. 2021):

$$\mathrm{SNR} = \mathrm{PF} \times \mathrm{Q_e} \times \mathrm{E_p} / \left((\mathrm{PF} \times \mathrm{E_p} \times \mathrm{Q_e}) + (\mathrm{I_{dark}} \times \mathrm{E_p}) + (\mathrm{E_p} \times \mathrm{E_r^2}) \right)^{1/2} \quad (4.2)$$

where PF is the photon flux, Q_e is the sensor quantum efficiency, E_p is the exposure time, I_{dark} is the dark current, E_r is the readout noise (Chap. 3).

As an example, assume the following values for Formula 4.2 at an incident light wavelength of 550 nm:

- PF = 2,000,000 photons/s
- $Q_e = 0.7$ (assuming a 70% QE)
- $E_p = 0.02$ s

Table 4.4 Estimated properties of hyperspectral sensor

Wavelength (nm)	Q_e	SNR
550	0.7	167.3
650	0.55	148.3
750	0.35	118.3
850	0.25	100.0

- $I_{dark} = 50$ electrons/s
- $E_r = 4$ electrons.

The signal is given by:

- $PF \times Q_e \times E_p = 2{,}000{,}000 \times 0.7 \times 0.02$
- $= 28{,}000$ electrons.

Noise is given by:

- $((PF \times E_p \times Q_e) + (I_{dark} \times E_p) + (E_p \times E_r{}^2))^{1/2}$
- $= ((2{,}000{,}000 \times 0.02 \times 0.7) + (50 \times 0.02) + (0.02 \times 16))^{1/2}$
- $= 167.34$.

$$SNR = 28{,}000/167.3 = 167.3.$$

Table 4.4 shows the calculated SNR ratios at 550, 650, 750 and 850 nm bandwidths for a common, silicon-based imaging sensor based on Formula 4.2, the same values in the worked examples above, with the exception of Q_e values varying in response to wavelength. SNR values decline as the sensor and grating efficiencies lower with respect to increasing wavelength; from peak QE of 550 nm (Fig. 4.2).

Informed by the inferred SNR values in Table 4.4, a COTS Arducam OV9281 camera with 1280 × 900-pixel resolution was chosen as the primary imaging sensor. It's simple, single-lane interface, sensitivity to the desired spectral range and ease of use with open access software made it a suitable candidate for the design. In addition, versions of this camera have flown in stratospheric balloons and operated successfully in the near-space environment (Chap. 2). Command and data handling (C&DH) used to operate the instrument was provided by a Field Programmable Gate Array (FPGA) small-factor board, providing protection against radiation upsets that may occur in the deep space environment, plus some mitigation against hardware obsolescence through the ability to reconfigure the hardware in flight (e.g. Tsigkanos et al. 2019).

Baseline Prototype and Test

A prototype engineering version of the pushbroom sensor was used to test optical layout, fitment and performance to refine flight versions. Ground-based calibration and performance analysis was also conducted to baseline performance of the push-broom prototype. This analysis would also assist in generating values of exposure, data compression and other attributes associated with in-flight operations. Spectral and spatial calibration of the pushbroom sensor was conducted using a compact fluorescent lamp (CFL) and halogen light with Lambertian reflector, respectively. Once warmed up and stable, the CFL emitted discrete bands of known wavelengths that were used to assign a wavelength range to the sensor output (Dominec 2010).

Additional calibration and performance metrics were obtained through imaging a series of seven colour swatches, mars analogue minerals and vegetation using a halogen light source to provide a spectrum of visible and NIR wavelengths in a similar manner as that undertaken for past Mars missions (Bell et al. 2003; Kinch et al. 2020).

As described in Chap. 3, reflectance values can only be obtained once influences from the light source, atmospheric absorption and imager sensitivities are accounted for. For this purpose, a white Lambertian reflector was also imaged to provide a calibration light source reference. Dark frames were also captured to reflectance, however, as image noise and dark current at the desired exposures proved negligible, reflectance was approximated to (Bougot-Robin et al. 2016):

$$R_f = R_s/R_r \tag{4.3}$$

where R_f is the calculated reflectance R_s is the reflected energy from the target and R_r is the reflected energy from the Lambertian reflector. This formula was used to obtain spectra for the samples using the average brightness of 50-pixel values collected at 5 nm step intervals between 550 and 850 nm. The samples were also sampled using a Spectral Evolution SR-3500 spectro-radiometer. Intensity mean absolute difference between this value and the equivalent spectrum sampled by the SR-3500 was used as a metric of comparison between the two spectra. Two vegetation samples were also included in the analysis to provide additional opportunities to characterize the sensor performance. FWHM measurements of the pushbroom were conducted using three lasers (red, green, violet) illuminating the Teflon target and measuring the width of the spectral peak. Performance was characterized over the visible part of the spectrum.

Testing Results

Colour swatch comparison results for the pushbroom and SR-3500 were in close agreement with the SR-3500 results with average differences ranging from less than 0.01 to 0.11 (Fig. 4.13a–g). Average standard deviations for each of the curves was 0.04. The FWHM results with the red, green and violet lasers are shown in Fig. 4.13h, showing a stable spectral resolution of 5.3–5.5 nm for the custom sensor.

The Mars analogue materials and representative vegetation showed similar agreement between the pushbroom and SR-3500 spectrometer (Fig. 4.14a–i). Average fit values ranged from 0.01 to 0.06.

The results show 1–4% accuracy between 550 and 800 nm in characterizing spectra from materials exhibiting VIS/NIR phenomenology in a controlled environment (Figs. 4.13 and 4.14). The binned spectral resolution (maximum 5.5 nm) returned from this instrument provided 55 discrete bands useful for identifying spectral peaks in minerals such as iron oxide families and olivine (Fig. 4.14). This result is less than the spectral range offered by flown instruments (Table 4.3) but were

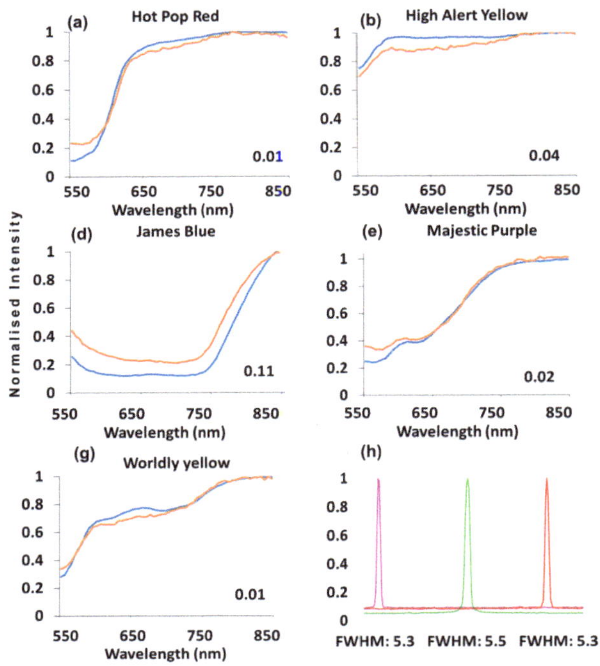

Fig. 4.13 **a–g** Comparison of pushbroom hyperspectral outputs (orange curves) with SR-3500 values (blue curves) for the colour swatch materials. **h** Measured, normalised FWHM values for purple, green and red lasers

of sufficient quality for it to be useful as a sensor on a low-cost Mars mission by detecting mineral phenomenology from orbit or the Martian surface. The ease of construction and readily-available COTS components would allow development of a version suitable for a 6U cubesat payload in a Space 2.0 application.

Optical remote sensing is the most popular technique for Mars applications since the first images returned by Mariner 4 (Chap. 1). It is readily used for interpretation of surface processes as well as engineering tasks such as landing site selection and rover navigation. Advances from early low-resolution vidicon cameras to modern CCD and CMOS sensors enabled dramatic improvements in image resolution and sensitivity, driving major discoveries of small-scale geological features.

Bayer filters or filter wheels allow geology discrimination using multispectral imaging, while the latest generation of optical sensors have returned hyperspectral across hundreds of bands. Mineralogical analysis is greatly enhanced through these latter sensors, and miniaturisation is making hyperspectral remote sensing a popular choice for Mars exploration.

Martian remote sensing has used other portions of the EMS and these will now be discussed.

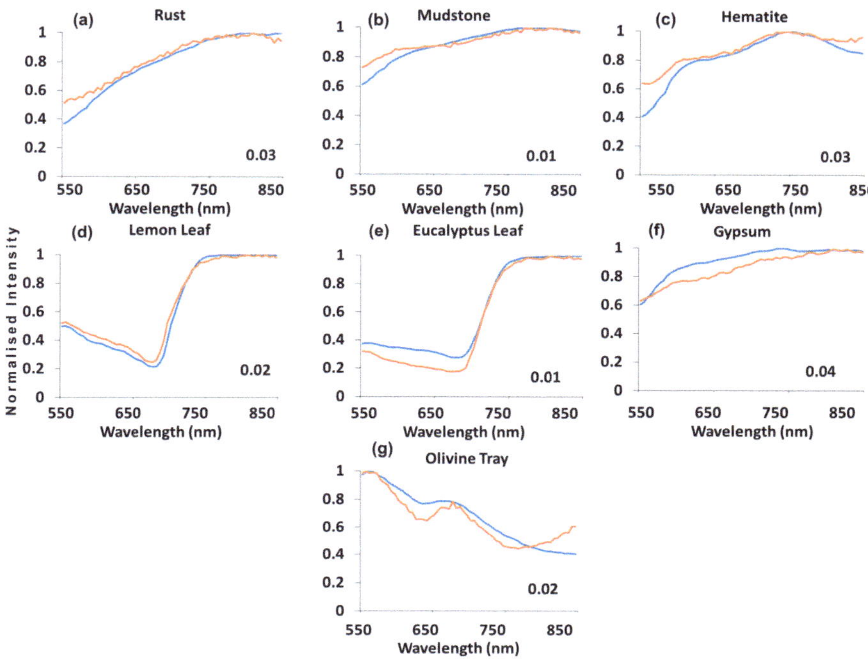

Fig. 4.14 **a–g** Comparison of pushbroom hyperspectral outputs (orange curves) with SR-3500 values (blue curves) for the Mars analogue and vegetation materials

References

Antonucci GA, Vary S, Humphreys D, Lamb RA, Piper J, Tanner J (2019) Multispectral snapshot demosaicing via non-convex matrix completion. arXiv preprint arXiv:1902.11032. https://doi.org/10.1109/DSW.2019.8755561

Arya R, Rudravaram PR, Bhatnagar SR, Sur K, Chauhan P, Sarkar S, Subbiah M, Srinivas A, Patel V, Roy S, Misra I, Gambhir R, Patel K, Drm S, Seelin K (2015) Mars color camera onboard Mars orbiter mission: initial observation. In: 46th lunar and planetary science conference, Houston, Texas, abstract 2123

Bakker WH, van Ruitenbeek FJA, van der Werff HMA, Zegers TE, Oosthoek JHP, Marsh SH, van der Meer FD (2014) Processing OMEGA/Mars Express hyperspectral imagery from radiance-at-sensor to surface reflectance. Planet Space Sci 90:1–9. https://doi.org/10.1016/j.pss.2013.11.007

Barlow N (2008) Mars: an introduction to its interior, surface and atmosphere. Cambridge University Press, New York, 264 pp

Bell III JF, Squyres SW, Herkenhoff KE, Maki JN, Arneson HM, Brown D, Collins SA, Dingizian A, Elliot ST, Hagerott EC, Hayes AG (2003) Mars exploration rover Athena panoramic camera (Pancam) investigation. J Geophys Res Planets 108. https://doi.org/10.1029/2003JE002070

Bell III JF, Glotch TD, Hamilton VE, McConnochie T, McCord T, McEwan A, Christensen PR, Arvidson RE (2008) Visible to near-IR multispectral orbital observations of Mars. In: Bell III JF (ed) The Martian surface: composition, mineralogy and physical properties. Cambridge University Press, New York

Bell JF, Maki JN, Mehall GL, Ravine MA, Caplinger MA, Bailey ZJ, Brylow S, Schaffner JA, Kinch KM, Madsen MB, Winhold A (2021) The Mars 2020 perseverance rover mast camera zoom (Mastcam-Z) multispectral, stereoscopic imaging investigation. Space Sci Rev 217:1–40. https://doi.org/10.1007/s11214-020-00755-x

Bertaux JL, Korablev O, Perrier S, Quemerais E, Montmessin F, Leblanc F, Lebonnois S, Rannou P, Lefèvre F, Forget F, Fedorova A (2006) SPICAM on Mars Express: observing modes and overview of UV spectrometer data and scientific results. J Geophys Res Planets 111. https://doi.org/10.1029/2006JE002690

Bibring JP, Soufflot A, Berthé M, Langevin Y, Gondet B, Drossart P, Bouyé M, Combes M, Puget P, Semery A (2004) Mars Express: the scientific payload. ESA Spec Publ 1240:37–58

Bosanac N, Cox AD, Howell KC, Folta DC (2018) Trajectory design for a cislunar CubeSat leveraging dynamical systems techniques: the lunar IceCube mission. Acta Astronaut 144:283–296. https://doi.org/10.1016/j.actaastro.2017.12.025

Bougot-Robin K, Paget J, Atkins SC, Edel JB (2016) Optimization and design of an absorbance spectrometer controlled using a Raspberry Pi to improve analytical skills. J Chem Educ 93:1232–1240. https://doi.org/10.1021/acs.jchemed.5b01006

Burns RG (1993) Mineralogical applications of crystal field theory. Cambridge University Press (CUP), New York, USA, pp 460–470

Burt DJ (1974) Basic operation of the charge coupled device. In: International conference on technology and applications of charge coupled devices, Sept 1974. University of Edinburgh, Centre for Industrial Consultancy and Liaison, Edinburgh

Caravaca G, Le Mouélic S, Rapin W, Dromart G, Gasnault O, Fau A, Newsom HE, Mangold N, Le Deit L, Maurice S et al (2021) Long-distance 3D reconstructions using photogrammetry with curiosity's ChemCam remote micro-imager in Gale Crater (Mars). Remote Sens 13:4068. https://doi.org/10.3390/rs13204068

Carr MH, Baum WA, Briggs GA, Masursky H, Wise DW, Montgomery DR (1972) Imaging experiment: the Viking Mars orbiter. Icarus 16:17–33

Cousins CR, Griffiths AD, Crawford A, Prosser BJ, Coates AJ (2011) Selection of the geological filters of the ExoMars Pancam instrument. In: 42nd lunar and planetary science conference, abstract 1826

Dominec F (2010) Design and construction of a digital CCD spectrometer. Research thesis, Czech Technical University in Prague, Czech Republic

Farrand WH, Bell III JF, Johnson JR, Bishop JL, Morris RV (2008) Multispectral imaging from Mars Pathfinder. In: Bell III JF (ed) The Martian surface: composition, mineralogy and physical properties. Cambridge University Press, New York, NY, USA, pp 265–280. ISBN: 978-0-521-86698-9

Farrand WH, Johnson JR, Rice MS, Wang A, Bell JF III (2016) VNIR multispectral observations of aqueous alteration materials by the Pancams on the spirit and opportunity Mars exploration rovers. Am Miner 101:2005–2019. https://doi.org/10.2138/am-2016-5627

Ge J, Ren D, Lunine JI, Brown RH, Yelle RV, Soderblom LA (2003) Compact high-resolution 3D imaging spectrometer for discovering oases on Mars. In: Instruments, methods, and missions for astrobiology V, vol 4859, Feb 2003. SPIE, pp 45–56

Geelen B, Tack N, Lambrechts A (2014) A compact snapshot multispectral imager with a monolithically integrated per-pixel filter mosaic. In: Proceedings of SPIE 8974. Advanced fabrication technologies for micro/nano optics and photonics VII, p 89740L. https://doi.org/10.1117/12.2037607

Golombek M, Grant J, Kipp D, Vasavada A, Kirk R, Fergason R, Bellutta P, Calef F, Larsen K, Katayama Y, Huertas A (2012) Selection of the Mars Science Laboratory landing site. Space Sci Rev 170:641–737. https://doi.org/10.1007/s11214-012-9916-y

Hartley RI, Gupta R (1994) Linear pushbroom cameras. In: Computer vision—ECCV'94: third European conference on computer vision, Stockholm, Sweden, 2–6 May 1994. Proceedings, vol I 3. Springer Berlin Heidelberg, pp 555–566

He Z, Xu R, Li C et al (2021) Mars mineralogical spectrometer (MMS) on the Tianwen-1 mission. Space Sci Rev 217:27. https://doi.org/10.1007/s11214-021-00804-z

Hieta M, Genzer M, Haukka H, Kestilä A, Rodríguez IA, Palacio VA, Aranda MR, Melchor AG, Oter JM, Bartolomé MGG, Ortega C (2022) State-of-the-art miniaturised science instruments of the MiniPINS landers (No. EPSC2022-820). Copernicus meetings

Hnatushenko VV, Hnatushenko V, Kavats OO, Shevchenko VY (2015) Pansharpening technology of high resolution multispectral and panchromatic satellite images. Sci Bull Natl Min Univ

Holsclaw GM, Deighan J, Almatroushi H, Chaffin M, Correira J, Evans JS, Fillingim M, Hoskins A, Jain SK, Lillis R, Lootah FH (2021) The emirates Mars ultraviolet spectrometer (EMUS) for the EMM mission. Space Sci Rev 217:1–49. https://doi.org/10.1007/s11214-021-00854-3

Hubel PM, Liu J, Guttosch RJ (2004) Spatial frequency response of color image sensors: Bayer color filters and Foveon X3. In: Electronic imaging. International Society for Optics and Photonics, Santa Clara, CA. https://doi.org/10.1117/12.561568

Huq R, Islam M, Das P, Siddique S (2019) System design concept of Mars intelligent imaging & atmospheric research CubeSat constellation using distributed deep learning (MIIAR). In: Proceedings of iCubesat 2019, the 8th interplanetary cubesat workshop, Milan

Jacob SR et al (2020) Spectral, compositional, and physical properties of the upper Murray formation and Vera Rubin ridge, Gale Crater, Mars. J Geophys Res Planets 125. https://doi.org/10.1029/2019JE006290

Jensen JR (2015) Introductory digital image processing: a remote sensing perspective, 4th edn. Prentice-Hall Inc., Upper Saddle River, NJ, USA. ISBN: 0132058405

Kinch KM, Madsen MB, Bell JF, Maki JN, Bailey ZJ, Hayes AG, Jensen OB, Merusi M, Bernt MH, Sørensen AN, Hilverda M (2020) Radiometric calibration targets for the Mastcam-Z camera on the Mars 2020 rover mission. Space Sci Rev 216:1–51. https://doi.org/10.1007/s11214-020-00774-8

Le Mouélic S, Gasnault O, Herkenhoff KE, Bridges NT, Langevin Y, Mangold N, Maurice S, Wiens RC, Pinet P, Newsom HE, Deen RG (2015) The ChemCam remote micro-imager at Gale crater: review of the first year of operations on Mars. Icarus 249:93–107. https://doi.org/10.1016/j.icarus.2014.05.030

Leighton RB (1970) The surface of Mars. Sci Am 222:26–41

Maki JN, Bell III JF, Herkenhoff KE, Squyres SW, Kiely A, Klimesh M, Schwochert M, Litwin T, Willson R, Johnson A, Maimone M (2003) Mars exploration rover engineering cameras. J Geophys Res Planets 108. https://doi.org/10.1029/2003JE002077

Maki JN, Golombek M, Deen R, Abarca H, Sorice C, Goodsall T, Schwochert M, Lemmon M, Trebi-Ollennu A, Banerdt WB (2018) The color cameras on the InSight lander. Space Sci Rev 214:1–34. https://doi.org/10.1007/s11214-018-0536-z

Maki JN, Gruel D, McKinney C, Ravine MA, Morales M, Lee D, Willson R, Copley-Woods D, Valvo M, Goodsall T, McGuire J (2020) The Mars 2020 engineering cameras and microphone on the perseverance rover: a next-generation imaging system for Mars exploration. Space Sci Rev 216:1–48. https://doi.org/10.1007/s11214-020-00765-9

Malin MC, Edgett KS, Cantor BA, Caplinger MA, Danielson GE, Jensen EH et al (2010) An overview of the 1985–2006 Mars Orbiter Camera science investigation. Int J Mars Sci Explor 4:1–60. https://doi.org/10.1555/mars.2010.0001

Massé M, Beck P, Schmitt B, Pommerol A, McEwen A, Chevrier V, Brissaud O, Séjourné A (2014) Spectroscopy and detectability of liquid brines on Mars. Planet Space Sci 92:136–149. https://doi.org/10.1016/j.pss.2014.01.018

McGuire PC, Bishop JL, Brown AJ, Fraeman AA, Marzo GA, Morgan MF, Murchie SL, Mustard JF, Parente M, Pelkey SM, Roush TL (2009) An improvement to the volcano-scan algorithm for atmospheric correction of CRISM and OMEGA spectral data. Planet Space Sci 57:809–815

Meng Q, Wang D, Wang X, Li W, Yang X, Yan D, Li Y, Cao Z, Ji Q, Sun T, Yan W (2021) High resolution imaging camera (HiRIC) on China's first Mars exploration Tianwen-1 mission. Space Sci Rev 217:1–29. https://doi.org/10.1007/s11214-021-00823-w

Morris HR, Hoyt CC, Treado PJ (1994) Imaging spectrometers for fluorescence and Raman microscopy: acousto-optic and liquid crystal tuneable filters. Appl Spectrosc 48:857–866. https://doi.org/10.1366/0003702944029820

Motohka T, Nishida K (2009) A simple and robust method for remote sensing of phenology using green and red reflectance: GRVI-method. AGU Fall Meet Abs 2009:B43C-0382

Murchie S, Arvidson R, Bedini P, Beisser K, Bibring JP, Bishop J, Boldt J, Cavender P, Choo T, Clancy RT, Darlington EH (2007) Compact reconnaissance imaging spectrometer for Mars (CRISM) on Mars reconnaissance orbiter (MRO). J Geophys Res Planets 112. https://doi.org/10.1029/2006JE002682

Oberst J, Roatsch T, Giese B, Wählisch M, Scholten F, Gwinner K, Matz K-D, Hauber E, Jaumann R, Albertz J, Gehrke S, Heipke C, Schmidt R, Ebner H, Spiegel M, Gasselt S, Neukum G, HRSC Team (2004) The mapping performance of the HRSC/SRC in Mars orbit. ISPRS, Istanbul, Turkey, 12–23 July 2004

Peñuelas J, Pinol J, Ogaya R, Filella I (1997) Estimation of plant water concentration by the reflectance water index WI (R900/R970). Int J Remote Sens 18:2869–2875. https://doi.org/10.1080/014311697217396

Prentice EF, Grøtte ME, Sigernes F, Johansen TA (2021) Design of a hyperspectral imager using COTS optics for small satellite applications. In: Proceedings of SPIE 11852, international conference on space optics—ICSO 2020, vol 1185258, 11 June 2021. https://doi.org/10.1117/12.2599937

Rankin A, Maimone M, Biesiadecki J, Patel N, Levine D, Toupet O (2021) Mars curiosity rover mobility trends during the first 7 years. J Field Robot 38:759–800. https://doi.org/10.1002/rob.22011

Rissanen A, Bin Guo B, Saari H, Näsilä A, Mannila R, Akujärvi A, Ojanen H (2017) VTT's Fabry–Perot interferometer technologies for hyperspectral imaging and mobile sensing applications. In: Proceedings volume 10116, MOEMS and miniaturized systems XVI; 101160I. SPIE OPTO, 2017, San Francisco, California, United States. https://doi.org/10.1117/12.2255950

Schade U, Wäsch R, Moroz L (2004) Near-infrared reflectance spectroscopy of Ca-rich clinopyroxenes and prospects for remote spectral characterization of planetary surfaces. Icarus 168:80–92. https://doi.org/10.1016/j.icarus.2003.10.016

Schoolcraft J, Klesh A, Werne T (2017) MarCO: interplanetary mission development on a CubeSat scale. In: Space operations: contributions from the global community, pp 221–231. https://doi.org/10.2514/6.2016-2491

Seelos FP, Seelos KD, Murchie SL, Novak MAM, Hash CD, Morgan MF, Arvidson RE, Aiello J, Bibring JP, Bishop JL, Boldt JD (2023) The CRISM investigation in Mars orbit: overview, history, and delivered data products. Icarus 115612. https://doi.org/10.1016/j.icarus.2023.115612

Sellar RG, Boreman GD (2005) Classification of imaging spectrometers for remote sensing applications. Opt Eng 44:2861–2868. ISSN: 0091-3286

Sigernes F, Syrjasuo M, Storvold R, Fortuna J, Grotte ME, Johansen TA (2018) Do it yourself hyperspectral imager for handheld to airborne operations. Opt Express 26. https://doi.org/10.1364/OE.26.006021

Stockstill KR, Moersch JE, Ruff SW, Baldridge A, Farmer J (2005) Thermal emission spectrometer hyperspectral analyses of proposed paleolake basins on Mars: no evidence for in-place carbonates. J Geophys Res Planets 110. https://doi.org/10.1029/2004JE002353

Tisserand S (2019) Custom Bayer filter multispectral imaging: emerging integrated technology. In: 2019 10th workshop on hyperspectral imaging and signal processing: evolution in remote sensing (WHISPERS), Sept 2019. IEEE, pp 1–4. https://doi.org/10.1109/WHISPERS.2019.8921288

Tsigkanos A, Kranitis N, Theodorou GA, Paschalis A (2019) A 3.3 GBs CCSDS 123.0-B-1 multispectral and hyperspectral image compression hardware accelerator on a space-grade SRAM FPGA. IEEE Trans Emerg Top Comput. https://doi.org/10.1109/TETC.2018.2854412

Wellman JB, Landauer FP, Norris DD, Thorpe TE (1976) The Viking orbiter visual imaging subsystem. J Spacecr Rocket 13:660–666. https://doi.org/10.2514/3.57128

Wiens RC, Maurice S, Barraclough B et al (2012) The ChemCam instrument suite on the Mars Science Laboratory (MSL) rover: body unit and combined system tests. Space Sci Rev 170:167–227. https://doi.org/10.1007/s11214-012-9902-4

Chapter 5
Other Martian Remote Sensors

5.1 Sensor Types

5.1.1 Radar Remote Sensing

Radar sensors differ from passive sensors in that they direct energy to the surface and measure returns. Radar uses much longer wavelengths (cm to metres) than optical or thermal sensors and require more power to operate. These place constraints on spacecraft design where more energy must be generated from solar or nuclear sources for the radar to transmit its directed energy to the surface, along with larger antennas needed to receive the returned signal. Terrestrial radar missions typically use synthetic aperture radar (SAR), where orbital spacecraft motion enables multiple radar returns from a smaller antenna to be combined into a higher resolution product (Li 2021). This gives the appearance of being collected with a much larger antenna, hence providing more detail. Such a sensor was flown to Venus as part of the Magellan mission and mapped 98% of the planet's surface (Fig. 5.1, Saunders et al. 1990).

Early Earth-based radar mapping of Mars in the 1960's was more difficult than for Venus, with the planet being smaller and further away from terrestrial systems. Initial observations inferred a smoother planet than Venus, with areas of high reflective return (Goldstein 1965).

Additional Earth-based radar identified strong returns from the ice of the south pole and virtual absence of reflections from the Tharsis volcanic region (Muhleman et al. 1991).

As a sensor for deep space consideration, the Magellan mission to Venus highlighted the utility of radar to penetrate through a planet's dense atmosphere. Its longer wavelengths can pass through dust, clouds and weather on Earth. Additionally, being an active sensor, radar remote sensors can operate regardless of solar illumination. In the case of Mars, where lack of oceans and thin atmosphere provide far better

S. Hobbs, *Sensing Mars: Remote Sensing to Understand the Red Planet*,
https://doi.org/10.1007/978-3-032-24416-1_5

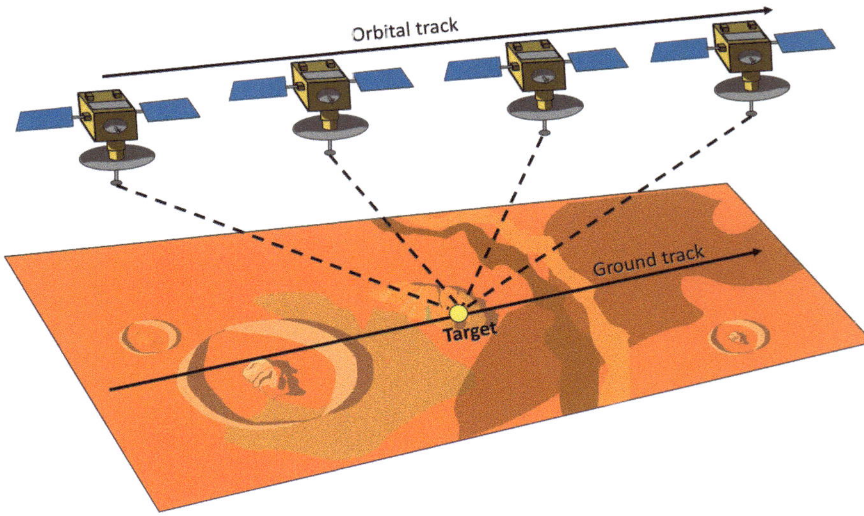

Fig. 5.1 Hypothetical SAR orbiter operation. Multiple radar collections over the same target along the orbital path increase the resolution of the resulting return

optical observation opportunities than Venus or Earth, subsurface exploration is the primary reason for radar sensors being included in mission design (Barlow 2008).

Ground penetrating radar, so far operated from orbit (e.g. Fig. 5.2) and on the surface, work on the principle that energy is susceptible to dielectric permittivity. This attribute characterises how much charges within a material will orient and polarize under the influence of an external field, provided by the transmitted radar energy. Conductive materials such as metals and water-laden materials provide a strong reflective return to the radar detector, with other permittivity properties, refracting or scattering the signal to reveal subsurface properties. Terrestrial Synthetic Aperture Radar (SAR) frequencies can range between 3 MHz and 300 GHz depending on application—higher frequencies typically generate higher resolution returns but can be susceptible to extreme weather (Wertz et al. 2011). Dedicated terrestrial meteorological radars reflect from storm events and are used to track weather such as rain or hail. Ground penetrating radar use lower frequencies to penetrate beneath the surface at the expense of less detailed discrimination (Li 2021).

5.1.2 Factors Affecting Martian Radar Remote Sensors

Martian regolith is typically much drier than that of Earth, meaning that radar returns are more dependent on the density of material than the dielectric properties of water. More dense materials, such as rock, have higher dielectric constant values and show slower wave propagation than loosely consolidated regolith. Differences in dielectric

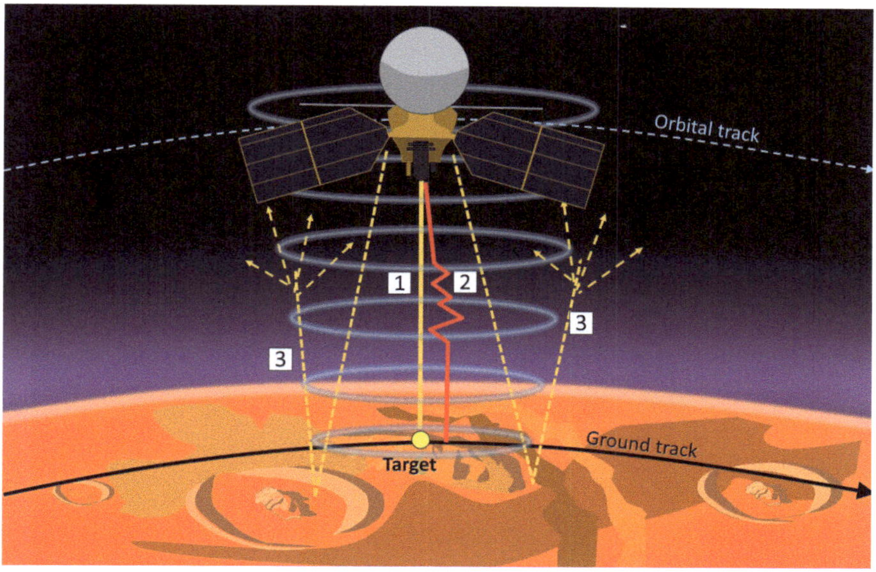

Fig. 5.2 Overview of environmental influences to orbital ground penetrating radar (represented by MRO SHARAD) showing (1) direct return from the nadir area of interest. (2) interference from the Martian ionosphere and (3) clutter returns from off-nadir topography

values produce subsurface reflections that are then detected by the radar system (Hamran et al. 2022).

Dry material dielectric values typically range between 2 and 5 units whereas brines raise this value to approximately 80 units, providing a useful indicator for potential past habitable environments (Grant et al. 2003). Recently, the presence of subsurface ice and potential liquid water was detected by MRO SHallow RADar (SHARAD) due to exploitation of the low dielectric values compared to dry materials, enforcing this as a useful property to investigate via radar remote sensing (Orosei et al. 2018).

Radar in Mars orbit is also susceptible to interference from the Martian ionosphere that extends as a single layer from 100 km to several hundred km from the surface (in comparison, Earth's ionosphere has four distinct layers). The increased distance of Mars from the Sun results in plasma densities an order of magnitude lower than terrestrial values though this is still sufficient to degrade orbital radar sensor performance (nadir return represented by Fig. 5.2 [1] while ionospheric interference represented by Fig. 5.2 [2]). Both Mars Express and MRO have also used their respective radar instruments to exploit this property and study the Martian ionosphere, returning useful science from a phenomenon that otherwise hinders radar operations.

Ideally, returned signals should originate directly from the target region located in the nadir of the instrument. This is typical for smooth regions, though areas with high surface roughness such as regions in the southern highlands may suffer off-nadir "clutter" (represented in Fig. 5.2 [3]). This unwanted situation occurs when radar energy is reflected from nearby topography in addition to the target and arrives at the same time to the detector, obscuring the desired return. Like MARSIS, SHARAD is sensitive to ground clutter, and the integration with other remote sensing data, particularly elevation data, is useful in assisting to interpret returned radar data (Desage et al. 2023).

Recent work has also exploited the traditionally unwanted clutter returns (Fig. 5.2 [3]) by correcting for ionosphere and altitude and comparing against "true" nadir echoes. The result has enhanced the usefulness of returned radar signals through extracting information from the clutter returns. This used power law trends through attenuation by the Martian ionosphere measured over varying periods of the planet's year. Enough data was collected to enable surface roughness to be estimated at a 75 m scale, making useful reference for future research (Campbell et al. 2018).

Ground penetrating radars measure the depth D of a reflective object through (Orosei et al. 2018; Letamendia et al. 2023):

$$D = vt/2 \tag{5.1}$$

where v is the wave velocity and t is the two-way travel time.

As an example for radar propagating through water ice, $v = 1.7 \times 10^8$ m/s (Picardi et al. 2005; Orosei et al. 2018).

- MARSIS two-way travel time $t = 40 \times 10^{-6}$ s, providing
- $D = (1.7 \times 10^8 \times 40 \times 10^{-6})/2$
- $D = 3400$ m

Resolution D_r is dependent on the transmit frequency bandwidth B, given by (Noon et al. 1998):

$$Dr = v/2B \tag{5.2}$$

Using MARSIS bandwidth B of 1.0 MHz (Jordan et al. 2009) and the water ice propagation velocity of 1.7×10^8 m/s listed above:

- $Dr = 1.7 \times 10^8/(2 \times 1{,}000{,}000)$
- $Dr = 85$ m.

Example Radar Sensors

MARSIS flown on Mars Express uses the lowest radar frequencies practicable (1.8, 3, 4 and 5 MHz) to maximise subsurface penetration but at the expense of providing a lower vertical resolution of 75 m. MARSIS was used to detect and characterise the base of the North Pole layered deposits by penetrating deep enough to discover the base of the ice deposits (Seu et al. 2007).

The SHARAD radar sounder on the NASA Mars Reconnaissance Orbiter complements MARSIS by providing higher resolution but at lower ground penetration. The 20 MHz centre frequency plus 10 MHz bandwidth provides a vertical resolution of 15 m and 10 W transmission power provides an operational penetrating depth

Fig. 5.3 Illustration of perseverance RIMFAX in operation (PIA24047, NASA/JPL-Caltech/FFI)

of hundreds of metres. SHARAD investigations through provide higher resolution radar remote sensing in the first several hundred metres of Martian regolith.

Surface radar investigations were made possible with the Radar Imager for Mars' Subsurface Experiment, (RIMFAX), a radar system small enough to be deployed on the Perseverance Mars Rover with a ground penetration of 10 m (Fig. 5.3). In contrast to orbital radars, RIMFAX exploits high frequencies of 150 MHz–12 GHZ to produce a much higher theoretical vertical resolution (12.5 cm). The shallower depth is far more appropriate for the smaller exploration footprint of a surface mission. The higher resolution also produces detailed subsurface data to complement and assist in interpreting measurements taken by other instruments of the surface.

Recent findings provide an example of this process where analysis of orbital and ground imagery indicated that Perseverance's landing site was the bottom of an ancient lake. RIMFAX's discovery of tilted sediment layers, not discoverable by other methods, suggest more complex past geologic history for the site (Fig. 5.3, Hamran et al. 2022). Strictly fluvial deposits deposited at the bottom of a standing lake were expected to be more horizontally layered (Patruno and Helland-Hansen 2018). The RIMFAX data of slanting reflectors suggests that the observed layering could be aqueous, or of volcanic origin or a complex process of volcanic and multiple fluvial events.

The Rover Penetrating Radar (RoPeR) on the first Chinese Mars mission is a second example of rover-borne ground penetrating radar. Using high (450–2150 MHZ) and low (15–95 MHz) frequencies, the instrument provided high vertical resolution to depths up to 10 m, 100 m in lower resolution modes. RoPeR failed to detect appreciable subsurface water or ice, but analysis of returned data suggested late Hesperian to Amazonian infilling of the Utopia by likely fluvial-based sediments (Li et al. 2022).

Results from the rover-borne ground penetrating radars show how these systems are able to provide fine scale "ground truth" of underlying geology. In absence of these instruments, researchers are constrained by above-ground observations of features that may be too small to be detected from orbit, such as the tilted layers in Jezero Crater (Fig. 5.3).

5.1.3 Thermal Infra-Red

Thermal infrared (TIR) sensors use wavelengths of the electromagnetic spectrum shorter than for radar (3–35 μm). Instruments designed to operate in this range rely on materials, rather than the Sun, emitting this radiation for the remote sensor to collect. Emissivity is the measure of how efficiently an object will radiate TIR, materials with high emissivity values provide a more accurate indication of temperature than low emissivity objects. Reflected solar energy at TIR is much less intense than that emitted by highly emissive objects in daylight. Highly polished metal, such a mirror, is an example of a low emissivity object, where measured temperature is more heavily influenced by its surroundings. As such, mirrors are used for TIR sensors due to their ability to reflect captured energy with little unwanted introduction of thermal noise (e.g. Christensen et al. 2004).

Terrestrial thermal remote sensing has many applications, including monitoring the effects of climate change through oceanic temperature fluctuations or influence on vegetation Terrestrial environments are largely dictated by soil moisture, vegetation and urban development, all of which significantly influence thermal properties (Farella et al. 2022). In the case of planetary geology, thermal heating and cooling of the natural environment is of interest, consisting of geology and in the case of Mars, volatiles such as water or CO_2 frost and ice. Diurnal solar heating and cooling on Mars is able to penetrate the upper tens of centimetres of the planet's regolith; below this depth, heat flow is governed through geothermal energy released from the Martian interior into the crust (Parro et al. 2017).

Unlike radar, which is influenced somewhat through Mars' ionosphere, thermal remote sensing is subject to greater atmospheric absorption. On Earth water, ozone and CO_2 act to constrain wave bands to 3–5 μm (higher temperature volcanism), 8.1–9.3 μm, 10.1–12.2 μm (Fig. 3.3a). The much drier and oxygen poor Martian atmosphere provides an advantage of wider TIR wavelength range over terrestrial counterparts, with the main absorption region being 12.5–16.5 μm for CO_2 (Fig. 3. 3b).

Mars thermal remote sensing is made difficult with a colder environment compared to Earth, with less energy available to heat materials during the planet's sol. This increases the likelihood of solar reflection during daylight mixing with emission readings, and sensitivities to surface roughness (Ramsey et al. 2016).

As with terrestrial sensors, flown Martian thermal sensors may be cooled as the sensor needs to be at a lower temperature than the sample under study for it to be able to detect it. Cooling also reduces random voltages within the sensor defined as noise (Chap. 3), increasing accuracy of readings, and can be accomplished using cryogenic refrigeration. Detection of longer IR wavelengths at lower temperatures (long wave temperatures: 20–40 K, medium wave temperatures: 77 K, shorter wave temperatures: 150 K) typically require colder temperatures than higher (to short wave infrared, SWIR, and above) wavelengths (Schueler et al. 1998). Not all flown sensors have used active cooling systems, with mission operation parameters instead adjusted to maintain sensor temperature. MGS Thermal Emission Spectrometer (TES) and Mars Odyssey THEMIS are examples of uncooled thermal detectors, where the absence of cryogenic cooling was chosen to simplify mission design (Christensen et al. 2004).

Applications of Thermal Remote Sensing to Mars

Thermal remote sensors have been used to infer properties of Martian surface features and underlying material where optical sensors are unable to penetrate. Mineralogy discrimination is possible for Mars, as on Earth, based on the principle that specific elemental compositions and crystal structure will vibrate at fundamental frequencies at thermal wavelengths (in addition to other wavelengths of the electromagnetic spectrum). Detection instruments sensitive to these bands would be able to discriminate mineral composition based on this phenomenology. These properties were used to generate maps that classified mineralogy of the planet (detailed examples in Chap. 7).

Along with ground penetrating radar, thermal remote sensing provides useful information to infer underlying geology through thermal inertia studies (Global Mars map, Fig. 5.4). In this method, thermal inertia measures a material's resistance to temperature change during a planet's day/night heating cycle. Surface features consisting of loosely consolidated, fine-grained material, such as dust or sand dunes, will heat and cool quickly, and possess low thermal inertia values. Larger-grained materials, such as pebbles or rocky outcrops, retain heat received from the sun longer and would have high thermal inertia values. This is also the case for compacted materials that would possess a higher thermal inertia value for the same material in a loosely consolidated form. In using this method, it is important to note that thermal inertia is a useful data to infer underlying geology to 30 cm depth and can be affected by local topographic effects, mixed particle sizes and atmospheric deviations from chosen thermal models (Fergason et al. 2006).

Thermal inertia itself is calculated by (Putzig and Mellon 2007):

$$\mathrm{I} = (k\rho c)^{1/2}\left[\mathrm{tiu} = \mathrm{Jm^{-2}K^{-1}s^{-1/2}}\right] \tag{5.3}$$

where:

- k = thermal conductivity [$\mathrm{W\ m^{-1}\ K^{-1}}$]
- ρ = density [$\mathrm{kg\ m^{-3}}$]
- c = heat capacity [$\mathrm{J\ kg^{-1}\ K^{-1}}$]

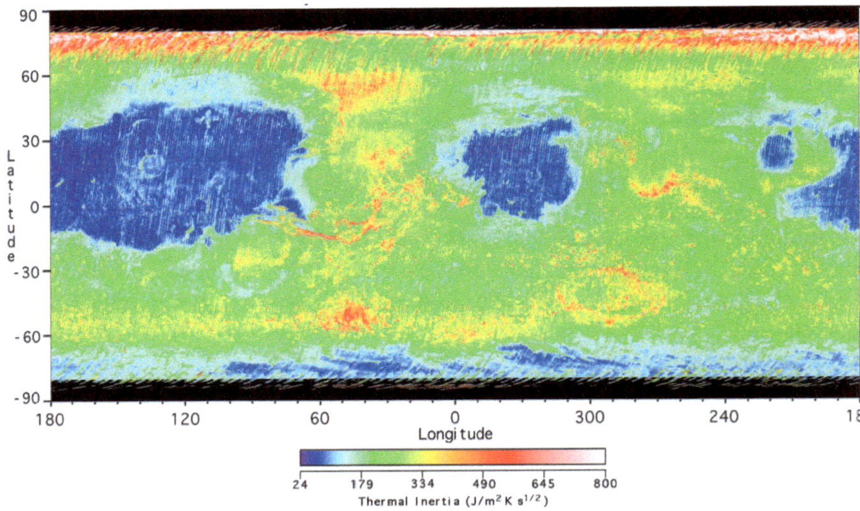

Fig. 5.4 Global thermal inertia-derived map of Mars from MGS TES (Putzig and Mellon 2007)

For dense Martian basaltic rock, the following values are appropriate (Baratoux et al. 2014; Vu et al. 2019):

- $k = 2\ W\ m^{-1}\ K^{-1}$
- $\rho = 3100\ kg\ m^{-3}$
- $c = 800\ Jkg^{-1}\ K^{-1}$

Using formula 5.3 to calculate thermal inertia provides:

- $I = (2 \times 3100 \times 800)^{½}$
- $I = 2227.11\ J\ m^{-2}\ K^{-1}\ s^{-½}$

Formula 5.3 for loosely consolidated geological material, such as sand, would be (Presley and Christensen 1997; Putzig and Mellon 2007):

- $k = 0.08\ Wm^{-1}\ K^{-1}$ (typically 0.05–0.1 $Wm^{-1}\ K^{-1}$)
- $\rho = 1500\ kg\ m^{-3}$ (typically 1200–1600 $Wm^{-1}\ K^{-1}$)
- $c = 700\ Jkg^{-1}\ K^{-1}$ (typically 600–800 $J\ kg^{-1}\ K^{-1}$)

Using formula 5.3 to calculate thermal inertia provides:

- $I = (0.08 \times 1500 \times 700)^{½}$
- $I = 289.83\ J\ m^{-2}\ K^{-1}\ s^{-½}$.

This value aligns with orbital thermal inertia measurements for Martian sandy regions (Presley and Christensen 1997; Putzig and Mellon 2007).

Flown Thermal Sensors

To directly measure thermal inertia would require physical placement of a bolometer over a feature of interest and monitor temperature changes throughout a Martian sol; to date this has only been attempted at Jezero Crater so thermal inertia modelling is a popular method for obtaining this data (Martínez et al. 2023).

Indirect thermal inertia measurements have exploited flyby or orbital thermal remote sensors flown to Mars, with Mariners 6, 7, 9 and Viking orbiter carrying early instruments. MGS TES captured global data at 3 km GSD (and whose data created the map at Fig. 5.4), collecting over 206 million IR spectra over its operational life. It used a Michelson Interferometer to cover 6.25–50 μm spectral range, solar reflectance and broad band radiance covering 0.2–3.9 μm and 0.3–100 μm respectively, providing a visible light sensitivity range in addition to thermal infrared.

Michelson interferometry operates using a beam splitter to divide incident radiation to a fixed and movable mirror (Fig. 5.5). Amplitude combination of the two radiation beams generates an interference pattern at the detector that was recorded and exploited generate spectra (Christensen et al. 2001).

Detections of the above wavelength bands were sufficient for TES to enable development of dust, daytime and nighttime thermal maps that were the highest resolution to that time. In addition, TES data was used to identify the hematite-rich region in Meridiani Planum (Christensen et al. 2000).

Themis

The THEMIS imager on board the Mars Odyssey orbiter represents the highest resolution thermal orbital sensor sent to Mars and is representative of orbital

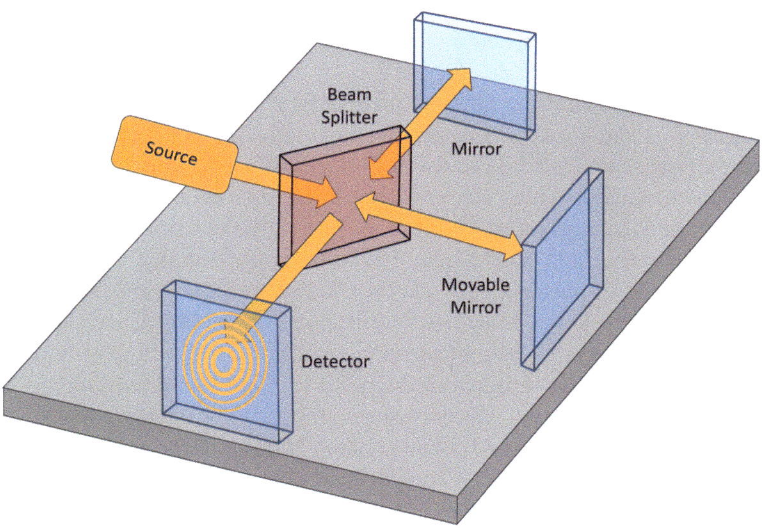

Fig. 5.5 Overview diagram of the Michelson interferometry principle

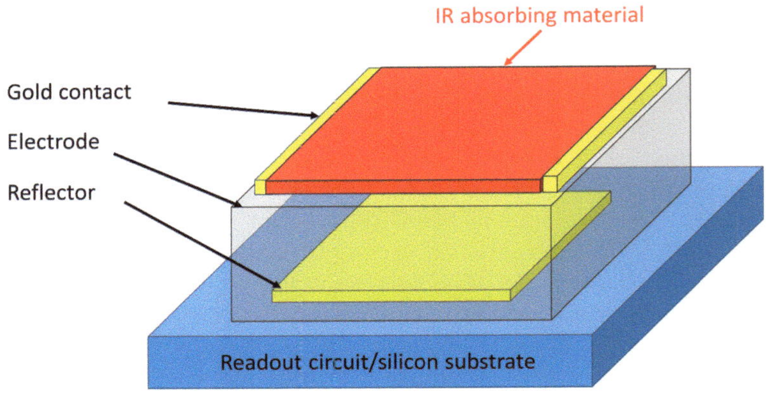

Fig. 5.6 Overview diagram of a microbolometer design using IR absorbing material to generate a detectable voltage

thermal remote sensing for the planet. THEMIS IR has returned a global dataset of 100 m/pixel resolution. The instrument comprises a three mirror astigmat telescope (diagrammatic example shown in Fig. 3.5c). Thermal detectors share the optical path with a visible sensor, and a beam splitter directed thermal energy to a 320 × 240 pixel uncooled but temperature stabilized microbolometer.

Each microbolometer pixel operates by allowing an IR absorbing material to heat up, changing its resistive properties and resulting voltage from a readout circuit (Fig. 5.6). Reflective material placed beneath the IR absorbent material maximises thermal heating and resulting read out voltage. To add a wavelength band discrimination capability to the microbolometers, spectral filters are included in the design and provide 10 bands of 1 μm spectral resolution between 6.8 and 14.9 μm (Fergason et al. 2006).

To convert microbolometer detections to surface temperatures, THEMIS shares a similar internal blackbody source referencing method for calibration as terrestrial airborne thermal scanners (Lillesand and Kiefer 2000). A calibration flag within the THEMIS instrument provides a continuous and known thermal source to provide flat fielding, and calibration of the sensor. These two activities remove artefacts induced by the sensor and optical path that would otherwise degrade the quality of thermal detections. Calibration with a known source ties detector values as obtained through bolometer voltages to actual temperatures of the Martian surface. During operation, THEMIS periodically images the flag every 3–10 min to provide consistency in thermal mapping throughout the mission. This also providing a reference source for thermal calibration (Christensen et al. 2004). An example thermal inertia result from THEMIS data analysis conducted on Noachis Terra gullies (visible light image: Fig. 5.7a) is shown in Fig. 5.7b. THEMIS-derived thermal inertia was helpful in identifying geological differences between equator and pole-facing walls of the gullies' host crater.

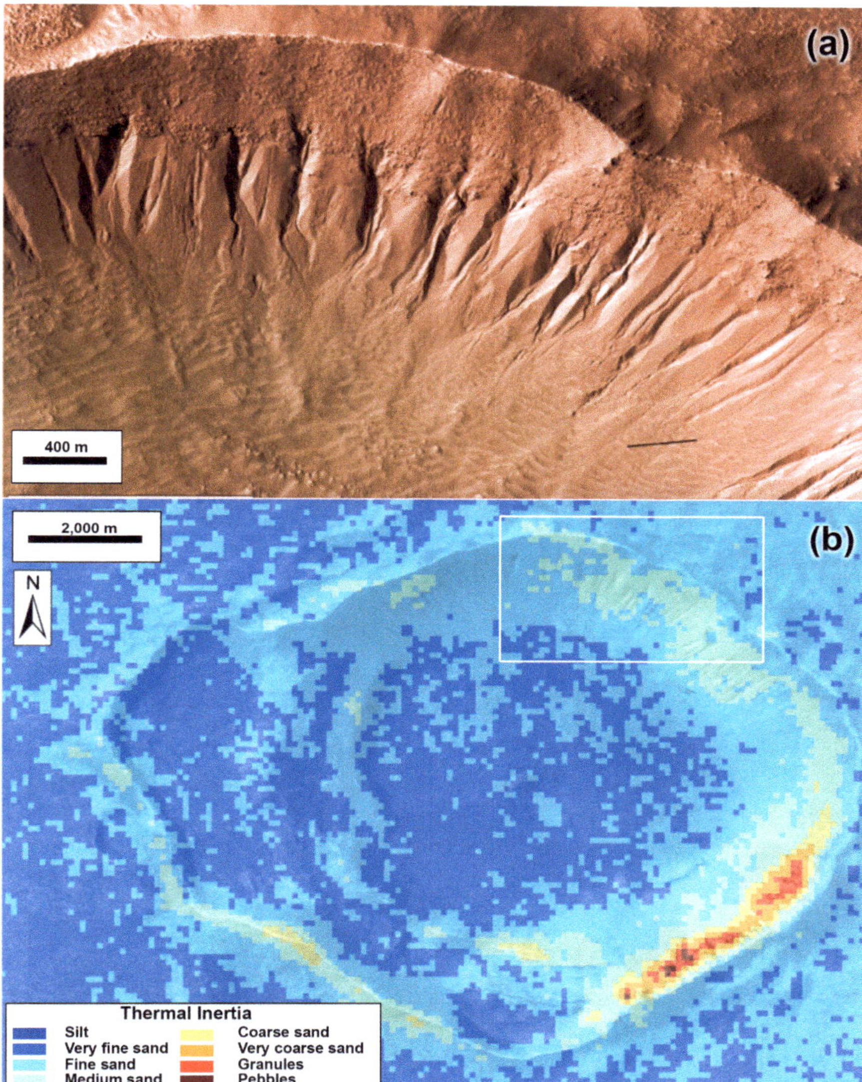

Fig. 5.7 **a** Colourised Martian gullies west of Hellas Basin in Noachis Terra (ESP-011817-1395, NASA/JPL). **b** THEMIS thermal inertia analysis of the crater in which the gullies (white inset box) reside. The gullies located within the northern crater rim exhibit low thermal inertia values, indicating finer-grained regolith. In contrast, the south-east portion of the crater wall possesses much higher thermal inertia values and coarser regolith (author processed, NASA/JPL)

5.2 High Energy Radiation—UV, X-Ray and Gamma Rays

Martian remote sensing of high-energy radiation (below 400 nm wavelength) has revealed additional geophysical and chemical properties of the planet. This radiation ionises and in the case of X-rays to gamma rays has enough energy to break apart molecules. UV remote sensing measures wavelengths between 10 and 400 nm (extreme UV, 10–100 nm, far UV, 100–200 nm, middle UV, 200–300 nm and near UV, 300–400 nm). Terrestrial applications provide information regarding ozone, sulphur dioxide and trace gasses in Earth's upper atmosphere (Krueger 2014). Interactions of oxygen, hydrogen and CO with far UV radiation are detectable in the upper portions of the Martian atmosphere (100–200 + km).

These are important for understanding how the planet's atmosphere, unprotected by an ozone layer field, has suffered degradation through heating and UV-induced disassociation. Additionally, while the CO_2 atmosphere absorbs UV radiation below 190 nm, Mars receives significantly more UV flux between 200 and 390 nm at its surface than Earth. This splits water molecules into base atoms, altering regolith chemistry to induce its present-day oxidising nature, and rapidly degrades organic material (Cockell et al. 2000).

5.2.1 *Flown UV Sensors*

Orbital UV sensors have so far flown on board Mariners 6, 7 and 9 (Hord 1972), Mars Express (Bertaux et al. 2006), MAVEN (Jain et al. 2015, Fig. 5.8a–c), and the Emirates Mars Mission (EMM, Holsclaw et al. 2021). The Scanning Habitable Environments with Raman and Luminescence for Organics and Chemicals (SHERLOC) instrument represents the current UV-sensitive surface instrument on board the Perseverance Rover (Fries et al. 2022).

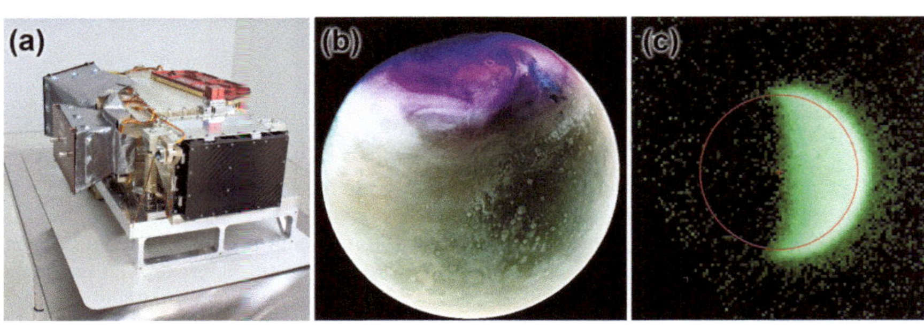

Fig. 5.8 **a** MAVEN UV instrument (NASA/JPL). **b** False-colour MAVEN output showing UV interactions with the upper Martian atmosphere (NASA/JPL). **c** Atomic oxygen as detected by UV remote sensing (PIA18613, NASA/JPL)

The orbital UV sensors have relied on detecting interactions between solar UV radiation and the Martian atmosphere. Similar to hyperspectral instruments described in Chap. 4, a dispersive optical grating and slit design was used for the MAVEN and EMM spectrometers. The Mars Express Spectroscopy for the Investigation of the Characteristics of the Atmosphere of Mars (SPICAM) combined UV and IR into one sensor. This provided wavelength sensitivities between 118–320 nm (far to middle UV) with spectral resolution of 1.5 nm, and 1.0–1.5 μm IR sensitivity with a spectral resolution of 0.5–1.0 nm (Bertaux et al. 2006).

SHERLOC employs active UV illumination of the planet's surface using a 248.6 nm laser to obtain fluorescence signatures of organic material and mineralogy (Fig. 5.9a–c). Illuminated organic material absorbs the UV radiation and reemits at a higher wavelength allowing the difference between the two used to identify the material through its Raman scattering pattern (Fig. 5.9a). More complex molecules produce more electron transitions, which are then detected though the Raman spectrometer (Perseverance flight instrument, Fig. 5.9b). Flown calibration targets assist in classifying SHERLOC Raman returns (Fig. 5.9c). To date, SHERLOC has identified mineralogy consistent with complex alteration, suggesting that the Jezero Crater floor once hosted an environment capable of supporting microbial life (Corpolongo et al. 2023).

5.2.2 High Energy Detectors

Higher energy photon detectors have flown to Mars, with the APXS exploiting radioactive decay-produced alpha particles and 14–18 keV X-rays at a target. Curium-244 or plutonium-240 may be employed as radioactive materials from which emitted energy can be detected and characterised (Fig. 5.10a). Derived from X-Ray fluorescence spectrometers carried on board Viking, APXS employs two active energy sources to detect lighter elements such as carbon and oxygen (alpha backscatter) and elements heavier than sodium (X-Ray fluorescence). The Mars Sojourner Rover was the first to use an APXS on the Martian surface, with subsequent instrument versions included in the MER and Curiosity. The instruments may be operated day or night and often require extensive periods (up to three hours for the Curiosity instrument) to accumulate sufficient detections for mineral identification (Berger et al. 2020).

The Mössbauer spectrometer is another example of an instrument using high-energy radiation, in this case, gamma radiation, to characterise the iron-rich materials on the Martian surface (Fig. 5.10b, c). Flown instruments exploit the Mössbauer effect, where atoms struck by gamma photons recoil, emitting measurable energy. Energy loss also occurs where atoms are fixed within a crystal lattice, as occurs with types of Martian minerals (Fig. 5.10d). As the Mössbauer spectrometer is targeted specifically for iron-rich materials, composition and abundance measurements can be determined to a high level of accuracy. This comes at a cost of time, where MER measurements took up to 12 h duration to complete (Morris et al. 2008).

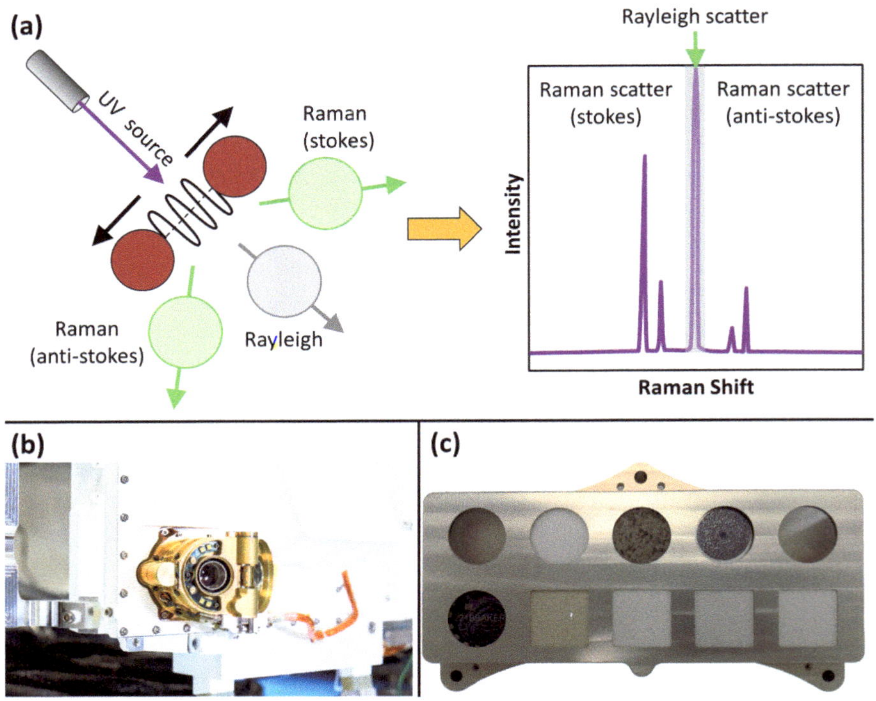

Fig. 5.9 **a** Overview of SHERLOC Raman spectroscopy operation. Stimulation by high-energy UV laser causes atomic vibration, producing stokes and anti-stokes Raman as well as Rayleigh (unwanted) scattering. These results may be graphically plotted and assist in mineral identification. **b** Flight view of SHERLOC instrument (NASA/JPL). **c** SHERLOC calibration target flown to Mars on the Perseverance Rover (NASA/JPL)

An orbital high-energy remote sensor was flown on Mars Odyssey, the Gamma Ray Spectrometer (GRS). This instrument consists of three detectors: a gamma sensor, neutron spectrometer and high-energy neutron detector. As opposed to Earth, where application for this technology is limited, the thin Martian atmosphere allows cosmic rays to reach the surface. Gamma photon impacts with the atoms of the surface geology release neutrons, colliding with adjacent atoms that release gamma radiation of discrete and classifiable energy levels (Fig. 5.11). The type of element can then be inferred from the emitted gamma ray and neutrons via the respective detectors in the GRS instrument (Boynton et al. 2004).

Hydrogen, being of similar mass to released neutrons, is able to moderate or retard their escape to space and, if present in the top metre of Martian regolith, is readily detectable by the GRS. Mars Odyssey, the spacecraft hosting GRS, is in a sun-synchronous orbit around Mars, hence providing the ability to map the entire planet. Returning neutron and gamma spectrum every 20 s, the GRS data was binned to increase statistical accuracy, and used to map hydrogen abundance across the planet (Fig. 5.12). Maps identifying the increasing abundance of water content closer to

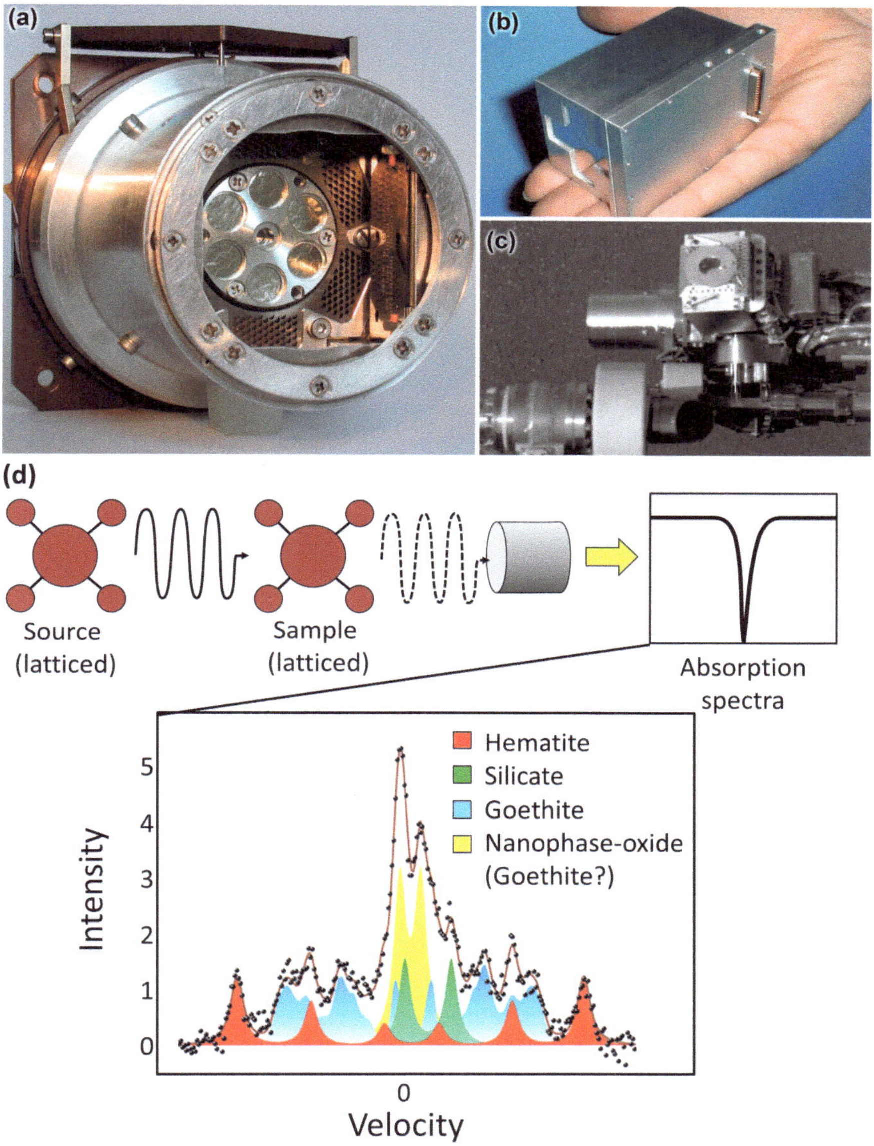

Fig. 5.10 **a** APXS sensor (NASA/JPL). **b** Flight model of Mossbauer spectrometer (NASA/JPL). **c** The MER Mossbauer spectrometer on the surface of Mars (NASA/JPL). **d** Overview of Mossbauer principle where identical source and sample nuclei contained within a lattice structure will not recoil but instead generate absorption spectra at the detector. The enlarged example shows Mossbauer spectra from Columbia Hills and identifies mineral classes at the Spirit Rover site

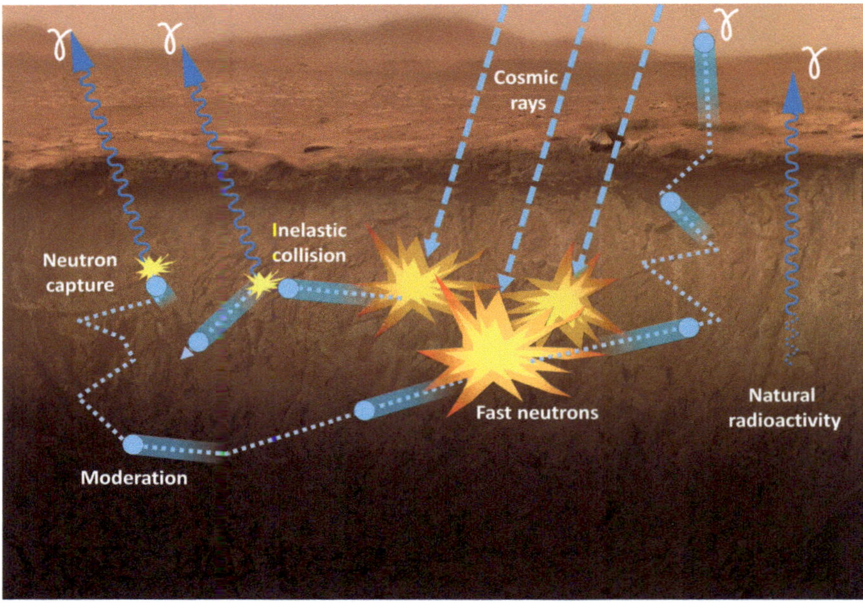

Fig. 5.11 Nuclear radiation interaction overview on Mars. Cosmic radiation is less impeded by the planet's thin atmosphere. Reaching the subsurface, neutron collisions or capture release detectable gamma radiation. Natural radioactive decay of heavy elements within the Martian crust also emits gamma radiation to space

the poles were generated from this data, providing important context for interpreting past fluvial action on the planet (Mitrofanov et al. 2002).

Exploitation of multiple parts of the electromagnetic spectrum beyond the narrow bands visible to the human eye have produced insights not otherwise possible. Radar remote sensing has allowed remote penetration deep beneath the Martian surface, an ability not possible with the current generation of landers. Thermal sensing and high energy radiation have demonstrated utility in discriminating between Martian surface geology types, while high energy radiation detectors have the additional benefit of studying solar interactions with the planet's atmosphere. Future missions will almost certainly continue to capitalise on terrestrial remote sensing technology advances, with an integration and analysis of data collected from multiple parts of the electromagnetic spectrum becoming standard practice.

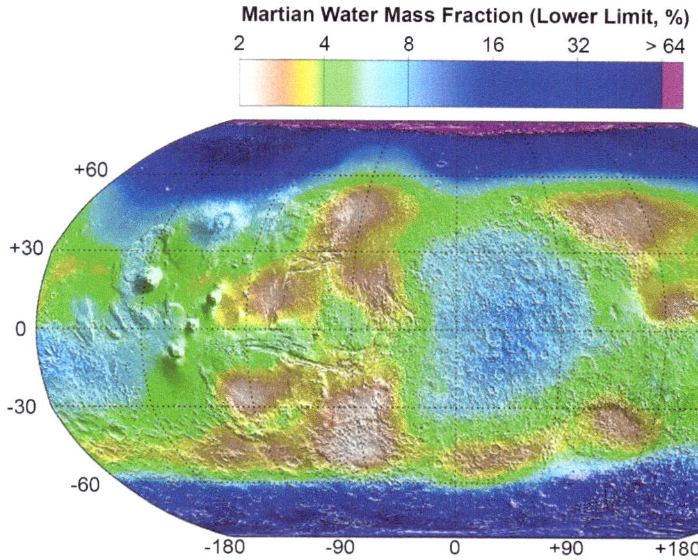

Fig. 5.12 Output from Mars Odyssey GRS using neutron abundance to infer the presence of water (NASA/JPL)

References

Baratoux D, Samuel H, Michaut C, Toplis MJ, Monnereau M, Wieczorek M, Garcia R, Kurita K (2014) Petrological constraints on the density of the Martian crust. J Geophys Res Planets 119:1707–1727. https://doi.org/10.1002/2014JE004642

Barlow N (2008) Mars: an introduction to its interior, surface and atmosphere. Cambridge Planetary Science, New York, 264 pp

Berger JA, Schmidt ME, Campbell JL, Flannigan EL, Gellert R, Ming DW, Morris RV (2020) Particle induced X-ray emission spectrometry (PIXE) of Hawaiian volcanics: an analogue study to evaluate the APXS field analysis of geologic materials on Mars. Icarus 345:113708. https://doi.org/10.1016/j.icarus.2020.113708

Bertaux JL, Korablev O, Perrier S, Quemerais E, Montmessin F, Leblanc F, Lebonnois S, Rannou P, Lefèvre F, Forget F, Fedorova A (2006) SPICAM on mars express: observing modes and overview of UV spectrometer data and scientific results. J Geophys Res Planets. https://doi.org/10.1029/2006JE002690

Boynton WV, Feldman WC, Mitrofanov IG, Evans LG, Reedy RC, Squyres SW, Starr R, Trombka JI, d'Uston C, Arnold JR, Englert PAJ (2004) The Mars Odyssey gamma-ray spectrometer instrument suite. Space Sci Rev 110:37–83. https://doi.org/10.1023/B:SPAC.0000021007.76126.15

Campbell BA, Schroeder DM, Jennifer L, Whitten JL (2018) Mars radar clutter and surface roughness characteristics from MARSIS data. Icarus 299:22–30. https://doi.org/10.1016/j.icarus.2017.07.011

Christensen PR, Bandfield JL, Clark RN, Edgett KS, Hamilton VE, Hoefen T, Kieffer HH, Kuzmin RO, Lane MD, Malin MC, Morris RV (2000) Detection of crystalline hematite mineralization on Mars by the Thermal Emission Spectrometer: evidence for near-surface water. J Geophys Res Planets 105:9623–9642

Christensen PR, Bandfield JL, Hamilton VE, Ruff SW, Kieffer HH, Titus TN, Malin MC, Morris RV, Lane MD, Clark RL, Jakosky BM (2001) Mars Global Surveyor Thermal Emission Spectrometer experiment: investigation description and surface science results. J Geophys Res Planets 106:23823–23871

Christensen PR, Jakosky BM, Kieffer HH, Malin MC, McSween HY, Nealson K, Mehall GL, Silverman SH, Ferry S, Caplinger M, Ravine M (2004) The thermal emission imaging system (THEMIS) for the Mars 2001 Odyssey Mission. Space Sci Rev 110:85–130

Cockell CS, Catling DC, Davis WL, Snook K, Kepner RL, Lee P, McKay CP (2000) The ultraviolet environment of Mars: biological implications past, present, and future. Icarus 146:343–359. https://doi.org/10.1006/icar.2000.6393

Corpolongo A, Jakubek RS, Burton AS, Brown AJ, Yanchilina A, Czaja AD, Steele A, Wogsland BV, Lee C, Flannery D, Baker D (2023) SHERLOC Raman mineral class detections of the Mars 2020 Crater Floor Campaign. J Geophys Res Planets. https://doi.org/10.1029/2022JE007455

Desage L, Herique A, Douté S, Zine S, Kofman W (2023) Resolving ambiguities in SHARAD data analysis using high-resolution digital terrain models. Remote Sens 15:764. https://doi.org/10.3390/rs15030764

Farella MM, Fisher JB, Jiao W, Key KB, Barnes ML (2022) Thermal remote sensing for plant ecology from leaf to globe. J Ecol 110:1996–2014. https://doi.org/10.1111/1365-2745.13957

Fergason RL, Christensen PR, Kieffer HH (2006) High-resolution thermal inertia derived from the Thermal Emission Imaging System (THEMIS): thermal model and applications. J Geophys Res Planets. https://doi.org/10.1029/2006JE002735

Fries MD, Lee C, Bhartia R, Razzell Hollis J, Beegle LW, Uckert K, Graff TG, Abbey W, Bailey Z, Berger EL, Burton AS (2022) The SHERLOC calibration target on the Mars 2020 Perseverance rover: design, operations, outreach, and future human exploration functions. Space Sci Rev 218:46. https://doi.org/10.1007/s11214-022-00907-1

Goldstein RM (1965) Mars: radar observations. Science 150:1715–1717. https://doi.org/10.1126/science.150.3704.1715

Grant JA, Schutz AE, Campbell BA (2003) Ground-penetrating radar as a tool for probing the shallow subsurface of Mars. J Geophys Res. https://doi.org/10.1029/2002JE001856

Hamran SE, Paige DA, Allwood A, Amundsen HE, Berger T, Brovoll S, Carter L, Casademont TM, Damsgård L, Dypvik H, Eide S (2022) Ground penetrating radar observations of subsurface structures in the floor of Jezero crater, Mars. Sci Adv. https://doi.org/10.1126/sciadv.abp8564

Holsclaw GM, Deighan J, Almatroushi H, Chaffin M, Correira J, Evans JS, Fillingim M, Hoskins A, Jain SK, Lillis R, Lootah FH (2021) The emirates Mars ultraviolet spectrometer (EMUS) for the EMM mission. Space Sci Rev 217:79. https://doi.org/10.1007/s11214-021-00854-3

Hord CW (1972) Mariner 6 and 7 ultraviolet spectrometer experiment: photometry and topography of Mars. Icarus 16:253–280

Jain SK, Stewart AIF, Schneider NM, Deighan J, Stiepen A, Evans JS, Stevens MH, Chaffin MS, Crismani M, Mcclintock WE, Clarke JT (2015) The structure and variability of Mars upper atmosphere as seen in MAVEN/IUVS dayglow observations. Geophys Res Lett 42:9023–9030. https://doi.org/10.1002/2015GL065419

Jordan R, Picardi G, Plaut J, Wheeler K, Kirchner D, Safaeinili A, Johnson W, Seu R, Calabrese D, Zampolini E, Cicchetti A (2009) The Mars express MARSIS sounder instrument. Planet Space Sci 57:1975–1986

Krueger A (2014) Ultraviolet remote sensing. In: Njoku EG (eds) Encyclopedia of remote sensing. Encyclopedia of earth sciences series. Springer, New York, NY. https://doi.org/10.1007/978-0-387-36699-9_185

Letamendia U, Navarro F, Benjumea B (2023) Ground-penetrating radar as a tool for determining the interface between temperate and cold ice, and snow depth: a case study for Hurd-Johnsons glaciers, Livingston Island, Antartica. Ann Glaciol 64:343–351. https://doi.org/10.1017/aog.2023.73

Li C et al (2022) Layered subsurface in Utopia Basin of Mars revealed by Zhurong rover radar. Nature. https://doi.org/10.1038/s41586-022-05147-5

Li J (2021) Satellite remote sensing technologies. Space Science and Technologies, Beijing. https://doi.org/10.1007/978-981-15-4871-0

Lillesand TM, Kiefer RW (2000) Remote sensing and image interpretation. Wiley, New York, NY, p 724

Martínez GM, Sebastián E, Vicente-Retortillo A, Smith MD, Johnson JR, Fischer E, Savijärvi H, Toledo D, Hueso R, Mora-Sotomayor L, Gillespie H (2023) Surface energy Budget, Albedo, and Thermal Inertia at Jezero Crater, Mars, as observed from the Mars 2020 MEDA Instrument. J Geophys Res Planets. https://doi.org/10.1029/2022JE007537

Mitrofanov I, Anfimov D, Kozyrev A, Litvak M, Sanin A, Tret'Yakov V, Krylov A, Shvetsov V, Boynton WV, Shinohara C, Hamara D (2002) Maps of subsurface hydrogen from the high energy neutron detector, Mars Odyssey. Science 297:78–81. https://doi.org/10.1126/science.1073616

Morris RV, Klingelhoefer G, Schröder C, Fleischer I, Ming DW, Yen AS, Gellert R, Arvidson RE, Rodionov DS, Crumpler LS, Clark BC (2008) Iron mineralogy and aqueous alteration from Husband Hill through Home Plate at Gusev crater, Mars: results from the Mössbauer instrument on the Spirit Mars Exploration Rover. J Geophys Res Planets. https://doi.org/10.1029/2008JE003201

Muhleman DO, Butler BJ, Grossman AW, Slade MA (1991) Radar images of Mars. Science 253:1508–1513

Noon DA, Stickley GF, Longstaff D (1998) A frequency-independent characterisation of GPR penetration and resolution performance. J Appl Geophys 40:127–137. https://doi.org/10.1016/S0926-9851(98)00008-1

Orosei R et al (2018) Radar evidence of subglacial liquid water on Mars. Science. https://doi.org/10.1126/science.aar7268

Parro LM, Jiménez-Díaz A, Mansilla F, Ruiz J (2017) Present-day heat flow model of Mars. Sci Rep 7:45629. https://doi.org/10.1038/srep45629

Patruno S, Helland-Hansen W (2018) Clinoforms and clinoform systems: review and dynamic classification scheme for shorelines, subaqueous deltas, shelf edges and continental margins. Earth Sci Rev 185:202–233. https://doi.org/10.1016/j.earscirev.2018.05.016

Picardi G, Plaut JJ, Biccari D, Bombaci O, Calabrese D, Cartacci M, Cicchetti A, Clifford SM, Edenhofer P, Farrell WM, Federico C (2005) Radar soundings of the subsurface of Mars. Science 310:1925–1928. https://doi.org/10.1126/science.1122165

Presley MA, Christensen PR (1997) Thermal conductivity measurements of particulate materials. 2. Results. J Geophys Res Planets 102:6551–6566

Putzig NE, Mellon MT (2007) Apparent thermal inertia and the surface heterogeneity of Mars. Icarus 191:68–94. https://doi.org/10.1016/j.icarus.2007.05.013

Ramsey MS, Harris AJL, Crown DA (2016) What can thermal infrared remote sensing of terrestrial volcanoes tell us about processes past and present on Mars? J Volcanol Geoth Res 311:198–216

Saunders RS, Pettengill GH, Arvidson RE, Sjogren WL, Johnson WTK, Pieri L (1990) The Magellan Venus radar mapping mission. J Geophys Res Solid Earth 95:8339–8355

Schueler CF, Blasius KR, Silverman SH (1998) Advanced multispectral TIR remote sensing. In: 1998 IEEE Aerospace Conference Proceedings (Cat. No. 98TH8339), vol 5, pp 131–142. IEEE

Seu R, Phillips J, Biccari RD, Orosei R, Masdea A, Picardi G, Safaeinili A, Campbell BA, Plaut JJ, Marinangeli L, Smrekar SE, Nunes DC (2007) SHARAD sounding radar on the Mars Reconnaissance Orbiter. J Geophys Res. https://doi.org/10.1029/2006JE002745

Vu TH, Piqueux S, Choukroun M, Edwards CS, Christensen PR, Glotch TD (2019) Low-temperature specific heat capacity measurements and application to Mars thermal modeling. Icarus 321:824–840. https://doi.org/10.1016/j.icarus.2018.10.004

Wertz J, Everett DF, Puschell JJ (2011) Space mission engineering: the new SMAD. Microcosm Press, CA, p 1029

Chapter 6
Mapping, Cartography and Data Processing for Mars Missions

6.1 Image Processing and Mapping

The ability to have a standardised method for identifying a point on a planet's surface is fundamental to navigation and mapping. A popular terrestrial example is the coordinate system by which a planet is divided into lines of latitude, north and south of the equator, plus longitude, that transect vertically while converging at both poles. The zero line of latitude is designated at the equator with $+ -90°$ latitude being at the north and south poles respectively. Zero longitude for Earth, the Prime Meridian, is centred on Greenwich, England.

The coordinate system developed for Mars uses similar lines of latitude and longitude. The Martian equator and hence zero latitude was readily identified via the planet's spin axis. Historic definitions of longitude were based on using detailed features that could be readily identified through telescopic observation. This was refined as observing equipment became more sophisticated. W. Beer and J. H. Mädler defined the first prime meridian in 1830–32, used later by Schiaparelli in 1877. A crater within this region, Airy, was defined as the Prime Meridian during the space age, however increased Mariner 9 and Viking image resolutions forced more precise location for Zero longitude (Fig. 6.1a, b). Airy-0, a smaller crater nested within larger Airy (Fig. 6.1c), was subsequently defined as the zero point (Duxbury et al. 2014).

6.1.1 *Map Projections*

Projecting remotely sensed data onto the correct part of Mars is essential for understanding the planet at a fundamental level. Projection entails a trade in accuracy, as three-dimensional images need to be emplaced on a two-dimensional surface (Hargitai et al. 2017).

S. Hobbs, *Sensing Mars: Remote Sensing to Understand the Red Planet*,
https://doi.org/10.1007/978-3-032-24416-1_6

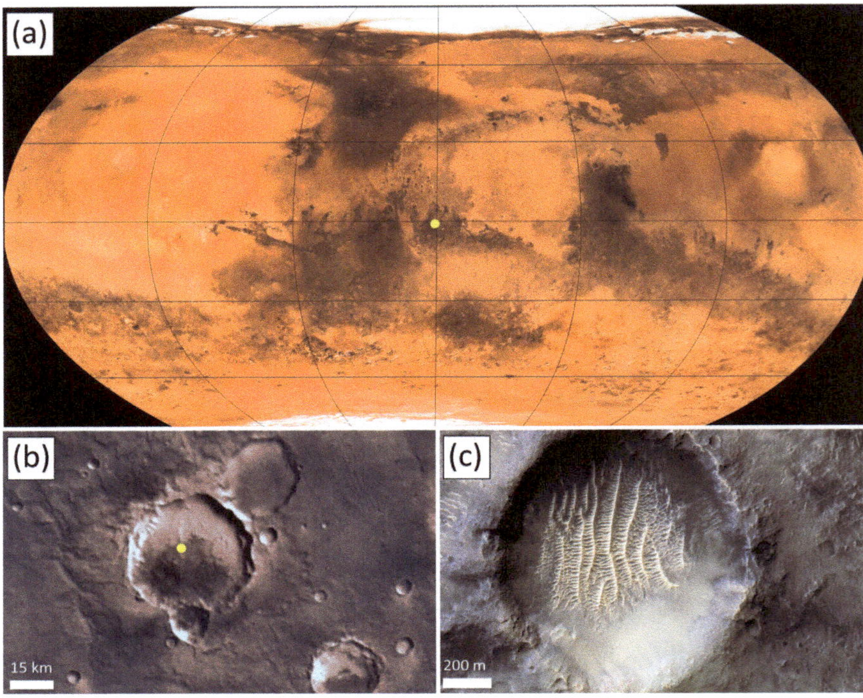

Fig. 6.1 **a** Position of Airy Crater as central Meridian on Mars highlighted by yellow dot (NASA/JPL). **b** Viking mosaic image of Airy Crater, with yellow dot highlighting the position of Airy-0 (NASA/JPL). **c** HiRISE image of Airy-0 Crater (PIA25090, NASA/JPL)

Map projections work to preserve a particular attribute, such as area, shape, direction, angle or distance, at the expense of distorting others. A popular and one of the oldest terrestrial projections is the Mercator. This represents a planetary sphere as a two-dimensional plane, with distortion increasing towards the poles. The effect of this distortion makes high-latitude countries, such as Greenland, appear much larger than their actual size. A similar effect occurs with this projection on Mars, with craters from the mid-latitudes appearing more elliptical than their physical round shape (Fig. 6.2a). A compromise has been to restrict the use of Mercator to latitudes of 70° N and S, or to use the Robinson projection that reduces (but not eliminate) distortion at high latitudes (Robinson projection shown in Fig. 6.2b).

Related projections are simple cylindrical and sinusoidal. A sinusoidal map provides equal area, where measurements of surface area are constant throughout the projection at the expense of shape distortion. Representation of the poles as points reduces shape distortion near the poles when compared with Mercator or simple cylindrical projections.

A polar stereographic map projection is conformal with area, distance and scale distortions at a minimum at the pole and increasing with distance from the standard

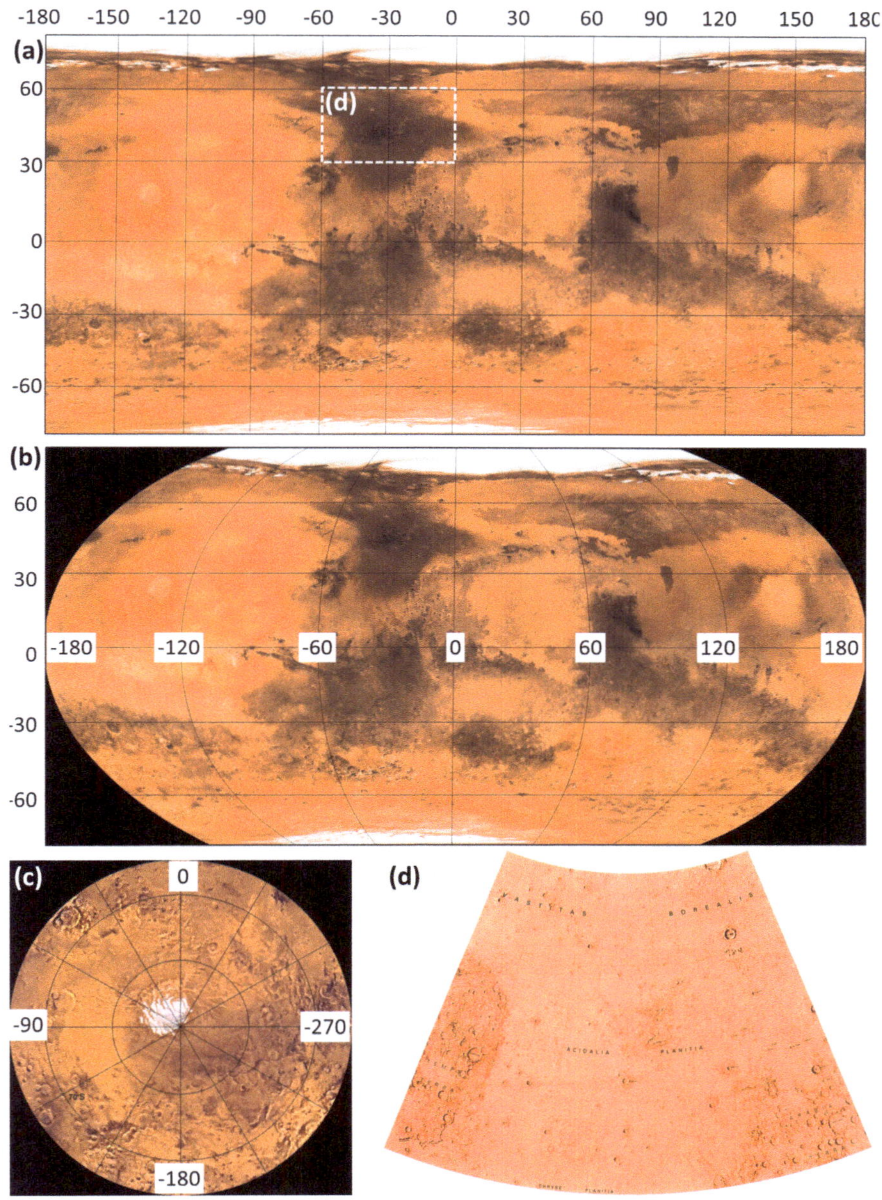

Fig. 6.2 **a** Mercator projection of Mars, showing high spatial distortion at the poles (NASA/JPL). **b** Robinson projection with reduced polar distortion (NASA/JPL). **c** Polar projection of the planet's south pole. **d** Conformal conic topographic map of the Mare Acidalium quadrangle, the location of which is shown in part (**a**) (NASA/JPL)

parallel. This projection is popular for polar visualisations, although the opposite pole cannot be seen (e.g. Figure 6.2c). Conformal projections maintain angles and shapes at the expense of directions (Fig. 6.2d). A Lambert Conformal Conic is an example, where latitude lines intersecting the Martian globe represent consistent regions of scale; regions outside these intersections suffer changes in scale (Greeley and Batson 1990).

Planetographic and planetocentric coordinate systems are two types of planet fixing methods used and approved by the International Astronomical Union in 2000. The Viking Mars mission used planetographic coordinates with longitude increasing westwards and the angle between the equator and a point on the surface at the planet's centre. The planetocentric system maps coordinates on the planet's surface and uses longitude increasing from the east, with Mars possessing 0°–360° longitude. By comparison, Earth has − 180° to − 180° longitude. Most modern Mars missions have adopted the planetocentric system, with the notable exception being Mars MAVEN. This mission used the planetographic coordinate method, equating to differences of up to 0.34° and 2 km spatial distance. These differences are sufficient to introduce errors in scientific measurements if not accounted for (Withers and Jakosky 2016).

6.1.2 Mars Ellipsoid

The process of transforming imagery and datasets onto a projection surface requires a common reference model. Off-Earth cartography has exploited spherical or ellipsoid models of celestial bodies (Greeley and Batson 1990). Modern mapping of Mars is made possible through transforming the development of an ellipsoid model of the planet. This considers the equatorial bulge caused by centrifugal forces increasing the planet's diameter in this dimension as compared to the measurement from pole to pole.

The ellipsoid thus improves on a spherical model that would have otherwise created inaccuracies between the equator and poles. The Martian ellipsoid has steadily improved through increasing accuracy of surface elevation measurements. MGS MOLA data has provided the most accurate global elevation measurements, with the International Astronomical Union (IAU) 2000 standard being the version in use today. This ellipsoid has a polar radius of 3376.2 km and an equatorial radius of 3396.19 km (Archinal et al. 2011).

In practice Mars, like Earth, is not a perfect spheroid, and exhibits significant topographic variation between the northern and southern hemisphere. Terrestrial ellipsoids have suffered similar issues with fitting, and numerous models exist to provide localised fitting such as Airy 1830, Everest 1830, Clarke 1866, WGS 1984, with WGS 1984 being a popular contemporary model. For Mars, the IAU 2000 remains the standard ellipsoid model of choice.

The ellipsoid model allows for latitude and longitude to be computed and features to be georeferenced. To exist in three dimensions, a third parameter in the form of an equipotential surface from which elevation can be calculated. The terrestrial equipotential surface or geoid is defined as the mean sea level, using our planet's oceans as a handy reference. This surface represents constant gravity values throughout, with gravity force acting perpendicular to the surface bodies (Greeley and Batson 1990). As Mars, like Mercury and Venus, possesses no oceans, zero elevation was initially defined at the level at which atmospheric pressure was 6.1 mbar. From the advent of MOLA data, the mean height from the lowest depressions to the top of Olympus Mons is used to calculate zero elevation, the equivalent of sea level on Earth. All values below this point are considered negative and elevations above are positive (Smith and Zuber 1998).

6.1.3 Mars Data Processing

As the various sensors described in Chaps. 4 and 5 have all returned, and many continue to do so, vast amounts of information is available about Mars from diverse measurements of physical phenomena. Data preparation and processing methods post-collection depends on the purpose for which this information will be used. Caution must be applied to avoid over processing of data and loss of essential information. To assist with appropriate dataset processing, multiple levels may be used including (Earthdata 2023):

- Level 0: represents reconstructed but unprocessed data with communication artifacts removed;
- Level 1: reconstructed data is at full-resolution and radiometrically corrected but not georeferenced (positioned on the Martian surface). Sub-processing stages at this level includes Level 1A—Level 0 data with radiometric and geometric calibration parameters added and an optional Level 1B that processes data to units of measurement used by the sensor.
- Level 2: data is converted to represent physical attributes of the environment under study. In this step for example, THEMIS sensor collected DNs would be converted to temperature values.
- Level 3: data is corrected geometrically. Internal sensor distortions are corrected and each pixel is projected onto an array corresponding with latitude and longitude coordinates. This process requires accurate pointing methods described in Chap. 2 (e.g. Figure 2.5).
- Level 4: data may consist of multiple inputs to generate a meaningful overlay. Thermal inertia data are an example of level 4 data, where day and night-time temperature collections are processed through levels 1–3 and combined to generate a final dataset.

Processing and Filters

In addition to calibration used to remove undesired environment and sensor-induced artefacts (Chap. 3), more than one data processing method may be used to generate Level 1 or Level 2 remote sensing datasets. This is typically dependant on the intended purpose as designed during Phase A mission concept design (Chap. 2). Performing histogram stretches as part of a processing pipeline is a traditionally simple and popular technique. In this method, DN levels are shifted in distribution changing the brightness and overall contrast of the resulting image (e.g. Fig. 6.3a, b). Portions of the histogram may be weighted in more complicated stretches to emphasise specific DN value ranges.

Image restoration in the form of pixel averaging may be used to repair pixel dropouts caused by detector signal loss or mis-calibration. This process samples, then averages the pixel DNs either side of the loss and fills the dark pixel with this value. It should be noted that this process is an approximation only and does not represent actual detected values.

Image enhancement filters may also use multiple pixels within an image to produce a desired output, with processing chains using multiple filters as needed. Low-pass

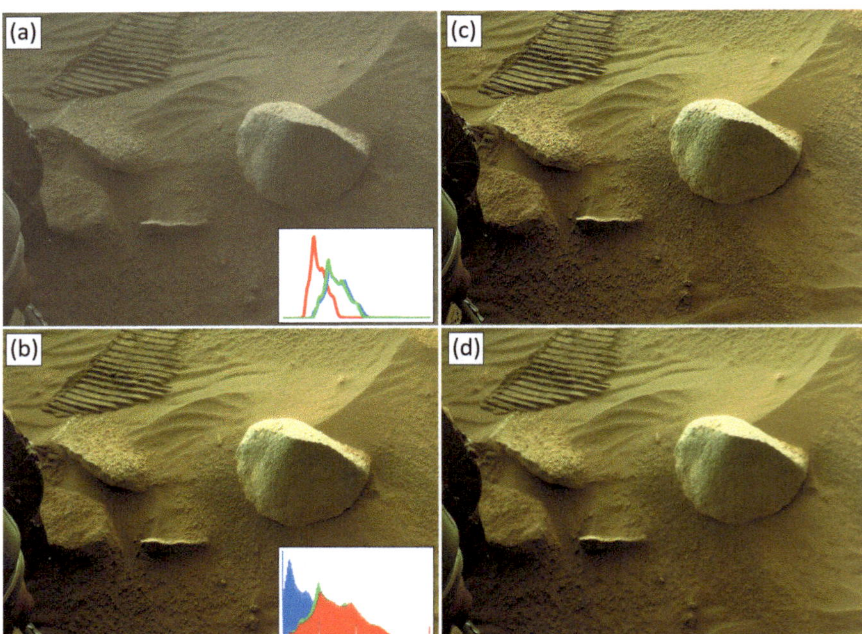

Fig. 6.3 **a** Raw image from Perseverance Navigation camera on 18 January 2024 with representative inset histogram (all images: NRF_1035_0758841684_862ECM_N0490370NCAM00710_03_095, NASA/JPL). **b** Stretched image showing enhanced contrast and corresponding change to the histogram. **c** The same image with an edge sharpening filter, enhancing high-frequency detail. **d** The same image with a smoothing filter applied, reducing fine-scale detail

and high-pass filters are examples of "boxcar" filters, where pixel values are weighted in accordance with the values of neighbouring pixels and the "boxcar" is slid over the image. The central pixel of a low pass filter is an average of the surrounding pixels, smoothing out high frequency changes. A high pass filter subtracts DNs generated by a low-pass filter from the image, enhancing high-frequency changes for a sharper image.

Edge-detection filters enhance high-contrast changes, such as at the boundaries of craters, while de-emphasising pixels of similar DN values (Fig. 6.3c). These filters are useful for automated feature extraction in AI/ML applications. The opposite effect is achieved using a low-pass filter, where high frequency variations are discarded for a smoother, overall effect (Fig. 6.3d). This filter may be useful in removing high frequency noise from elevation data. In addition, the above processing techniques are routinely performed on commercial or even open-source image management software.

Georeferencing Process

Spacecraft orienting information captured in the sensor metadata is also used to "georeference" the image to the correct region of the Martian surface. This georeferencing process has become increasingly essential in co-registering different datasets. Without careful application of this information, datasets may not accurately share position information and thus mis-register. An example occurred during early processing of the ESA High Resolution Stereo Camera (HRSC) and MRO CTX datasets that suffered mis-registration by up to 100 m, making analysis of fine-scale features, such as gullies or dunes, difficult (Tao et al. 2016).

Supplementing more automated data processing pipelines, georeferencing may also be achieved through identification of tie points using common features, such as small craters, or high-contrast areas, of a new, unregistered, dataset to an underlying layer (Fig. 6.4a–c). This was originally a manual process, where operators selected common features that were used as tie points (green and yellow crosses, Fig. 6.4a, b) to place the image on the correct portion of the Martian surface (Fig. 6.4c). Recently, image matching algorithms reduce this manual task through automated tie point selection to enable bulk georeferencing and processing of multiple datasets (Muller et al. 2016; Sidiropoulos et al. 2018).

Mosaicking is made possible through georeferencing multiple images of a feature that are too large for coverage by individual swathes. The tie-point method described above allow pixels to be reprojected for placement of each swath on the Martian surface. Image processing filters such as using pixel averaging is then used between image edges to remove these artefacts and develop seamless mosaics. Figure 6.5 provides an example of mosaicking Mariner 7 images of the Martian pole, where common features between images (highlighted by coloured boxes, Fig. 6.5a) and spacecraft data allow reprojection onto the correct portion of the Martian surface (Fig. 6.5b). Edge averaging may then be used to smooth transitions between the mosaiced images to produce a more eye-pleasing result (Fig. 6.5c).

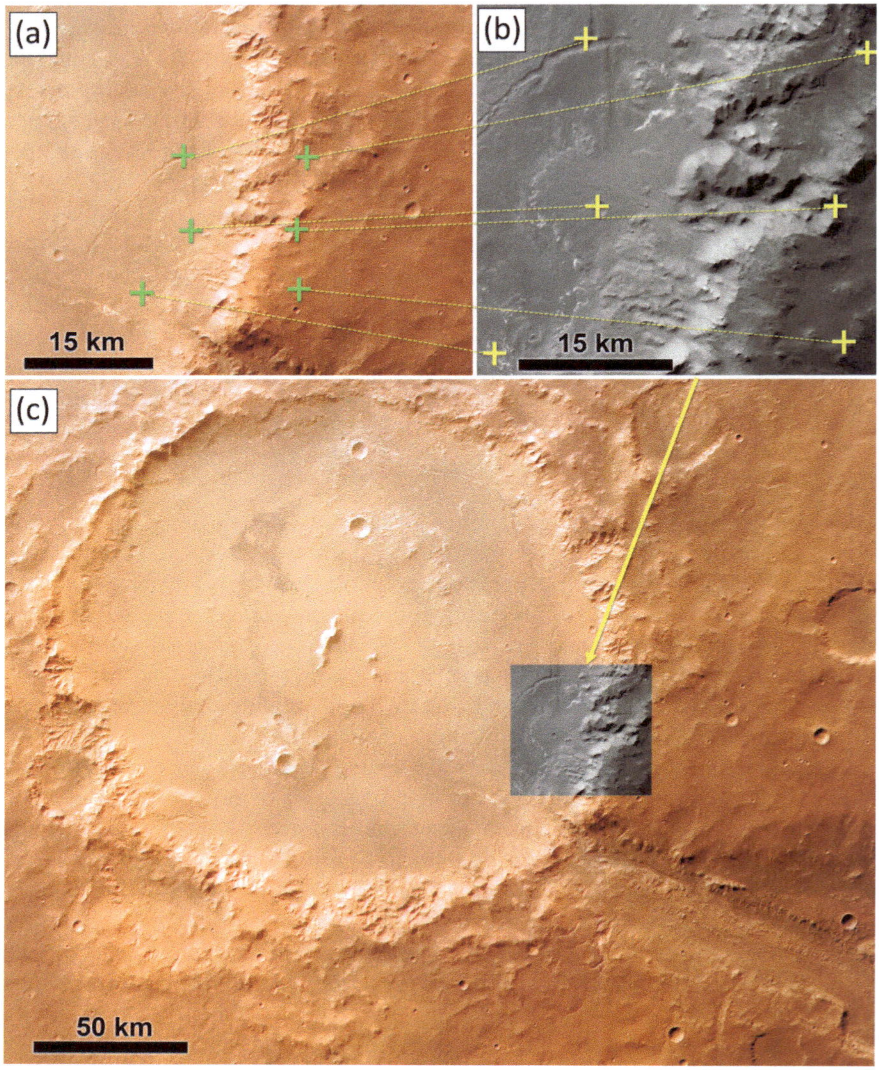

Fig. 6.4 **a** Holden Crater HRSC colour image (ESA/DLR/FU Berlin, CC BY-SA 3.0 IGO). **b** High resolution monochrome image for georeferencing. Common features used as tie points between both images are highlighted by green crosses in the base image (part **a**) and yellow crosses in the image to be georeferenced (part **b**), (NASA/JPL). **c** Final result showing high resolution monochrome image georeferenced and overlaid onto the lower resolution base image (ESA/DLR/FU Berlin, CC BY-SA 3.0 IGO, NASA/JPL)

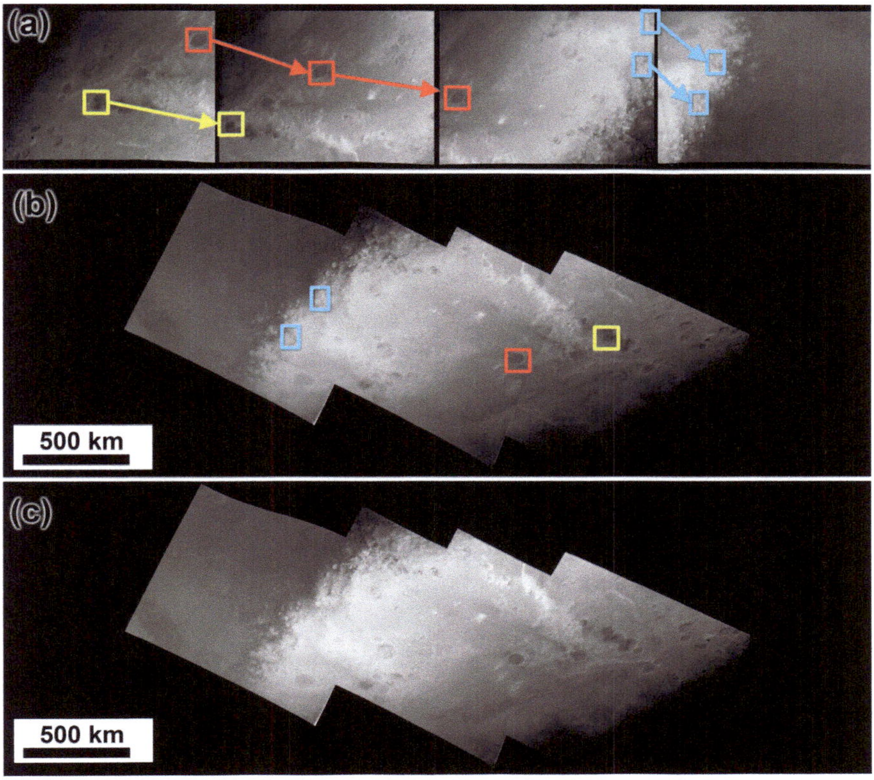

Fig. 6.5 **a** Mariner 7 camera images of the Martian pole. Common features between images are highlighted by coloured boxes (NASA/JPL). **b** Images are then overlapped and reprojected ono the Martian surface (NASA/JPL). **c** Pixel averaging removes edge artefacts to produce a seamless mosaic (NASA/JPL)

Orthorectification

On its own, georeferencing may introduce registration errors by failing to account for local topography. Features closer to the imaging sensor, such as crater walls or mountains, appear larger in the image than more distant regions, adding distortions that georectification cannot remove. An additional processing step, called "orthorectification", is required (Hu et al. 2019). This method, originated from terrestrial GIS and aims to represent every pixel of the captured image as if it were directly below the sensor and at a uniform scale. Orthorectification (Fig. 6.6a–c) uses elevation data and stored rational polynomial coefficients that provide a sensor model of the positional relationship between the image and the surface. Like georeferencing, orthorectification is automated through an open-source NASA Aims Stereo Pipeline that is capable of processing multi-resolution datasets (Broxton and Edwards 2008).

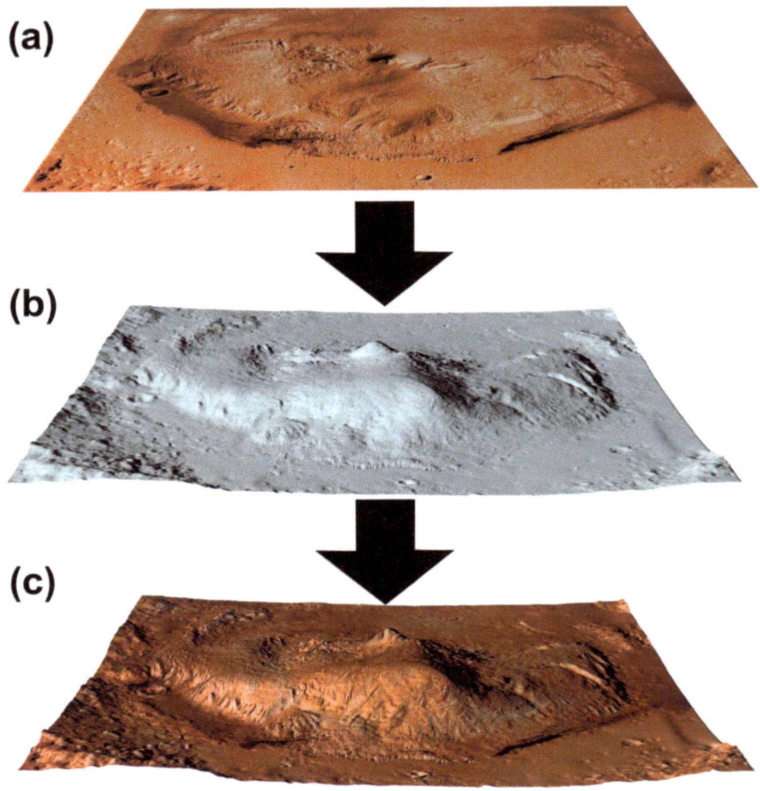

Fig. 6.6 **a** Mars orbiter image of Gale Crater (NASA/JPL). **b** Elevation model to assist with orthorectification. **c** Final orthorectified image processed using the elevation dataset in part (**b**)

6.2 Elevation Data

Mapping and characterising terrain topography are essential terrestrial remote sensing practices that are centuries old. Similarly, understanding the topography of Mars is key to interpreting the evolution of landforms on the planet. Prior to Mariner 9, scientific consensus was that Mars lacked significant topography, and was generally flat from horizon to horizon (Fig. 6.7, Chap. 1). This view was incorrectly inferred through Mars orbiting further from the Sun than Earth, adversely affecting telescopic observations. This meant that favourable viewing at opposition also meant direct illumination and minimal shadows from which to identify surface topography. While imagery from Mariner 9 provided qualitative information on the diverse topography of Mars (Fig. 1.5d–e), more direct elevation data was returned by the UV spectrometer. This instrument was able to measure atmospheric density

Fig. 6.7 For centuries, most astronomers believed that Mars was flat, with minimal topography. Early flyby missions reenforced this view until the arrival of Mariner 9 provided conclusive evidence to the contrary

and identified lower regions with higher pressure compared to elevated areas (Hord et al. 1972).

A step change in Martian topographic measurements occurred through the Mars Global Surveyor Mars Orbiter Laser Altimeter experiment that used LiDAR to map the planet. From 1997 to instrument failure in 2001 MOLA completed over 330,000,000 laser measurements with a vertical accuracy of 0.4 m over smooth terrain (Neumann et al. 2001). Returned data processed to an elevation grid spacing of 450 m/pixel provided the most accurate elevation dataset of any world in the Solar System, including the Earth, and was subsequently used to define a new Mars geoid and coordinate system (Fig. 6.8, Hirt et al. 2012).

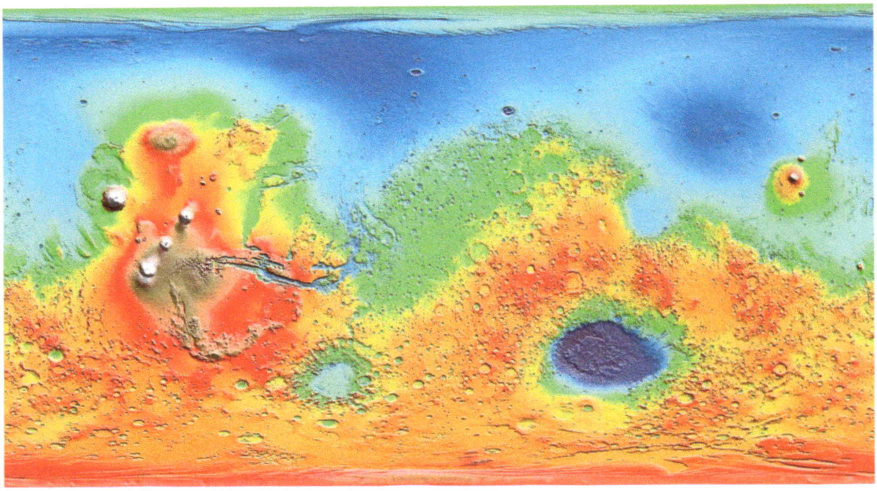

Fig. 6.8 MOLA-produced map of Mars (NASA/JPL)

6.2.1 Stereoscopic Imagery

Collection and Processing

Additional elevation data can also be produced from photogrammetric processing of stereo imagery. Images captured at an offset angle of the same scene introduce parallax, which can produce three-dimensional information. Early applications of this application were used for aerial reconnaissance during World War 1. Sequential images that overlapped were printed and viewed through a stereoscope to extract terrain features, such as topographic lines for mapping (Baltsavias 1996).

Orbital missions accomplish this through capturing the same area in a different sensor position, ensuring a high degree of overlap between image swathes (Fig. 6.9). HSRC and HiRISE imagery captured in this method have returned unprecedented high-resolution elevation information (Gwinner et al. 2009; Sutton et al. 2022). Local shadowing and atmospheric fog and dust to degrade or obscure topographic feature extraction through this process. Stereo mapping missions may thus be pre-planned or have post-processing conducted on returned data to minimise seasonal shadowing, fog or dust impacts (Kirk et al. 2021). In these instances, active sensors such as LiDAR or radar would be advantageous, but MOLA has so far been the only sensor of this type flown to Mars (Kreslavsky and Head 2000).

The HRSC on board Mars Express exploits orbital track motion to produce sequential images that are processed to produce digital terrain models (DTM). On receipt of imagery from Mars, an automated process generates DTMS plus output imagery available for exploration (Fig. 6.10, Scholten et al. 2005). Raw data are decompressed and radiometrically corrected from known sensor performance values to remove solar, atmospheric and sensor artifacts and obtain Level 2 product of

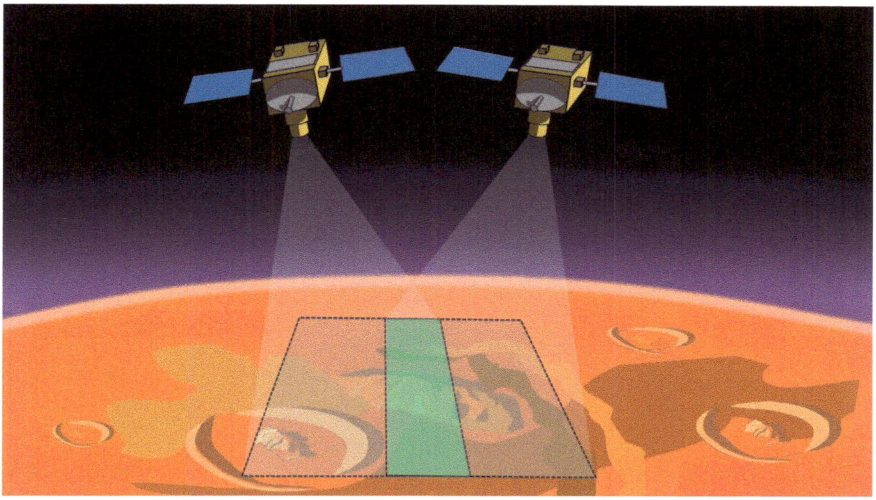

Fig. 6.9 Stereo image capture for Mars orbiters overview. Swaths captured by orbiters over subsequent paths may be made to overlap (green region, may be up to 60% overlap). This area contains parallax information from which stereo imagery and elevation models may be generated

reflectance values. Additional processing of image orientation and georeferencing, using obtained spacecraft orbital position and pointing data, plus MOLA elevation models, position the image on the correct portion of the Martian surface generates Level 3 products ("Map projected data", Fig. 6.10).

Additional processing ("Photogrammetric Product Processing", Fig. 6.10) fits returned images to a common geometry, required to reduce errors for an image matching process, where common picture elements are identified and used to generate elevation information. This is achieved through pre-rectification processing, image matching and processing to match image tie points to an underlying DTM grid. Following this process, an orthorectified image and DTM are generated as end products ("Level 4, Fig. 6.10). The resulting DTM is saved as 16-bit files and, in conjunction with the orthorectified imagery, make up Level 4 products for HRSC.

While standard RGB colour images popular with the commercial camera market, are saved as 8 bit (256 levels of grey across three channels), this is not sufficient to capture fine-scale changes in elevation. The increased bit depth (16 bit) enhances what is captured in the DTM, providing a vertical resolution of 6–8 m and making a useful tool for image orthorectification or geomorphic exploitation (Spiegel 2007).

A related process is also used to generate HiRISE DTMs. Commercial photogrammetry software or a Linux-based processing pipeline ingests overlapping HiRISE imagery, performs georeferencing, orientation and point matching to produce the highest-resolution orbital DTMs available (1 m post spacing, Beyer et al. 2018). These DTMs are used for fine-scale geomorphic analysis, being able to identify slopes of sub-1 km sized features. An example is shown for the Eberswalde Crater region of Mars (Fig. 6.11a) where orthorectified HiRISE imagery (Fig. 6.11b) is combined

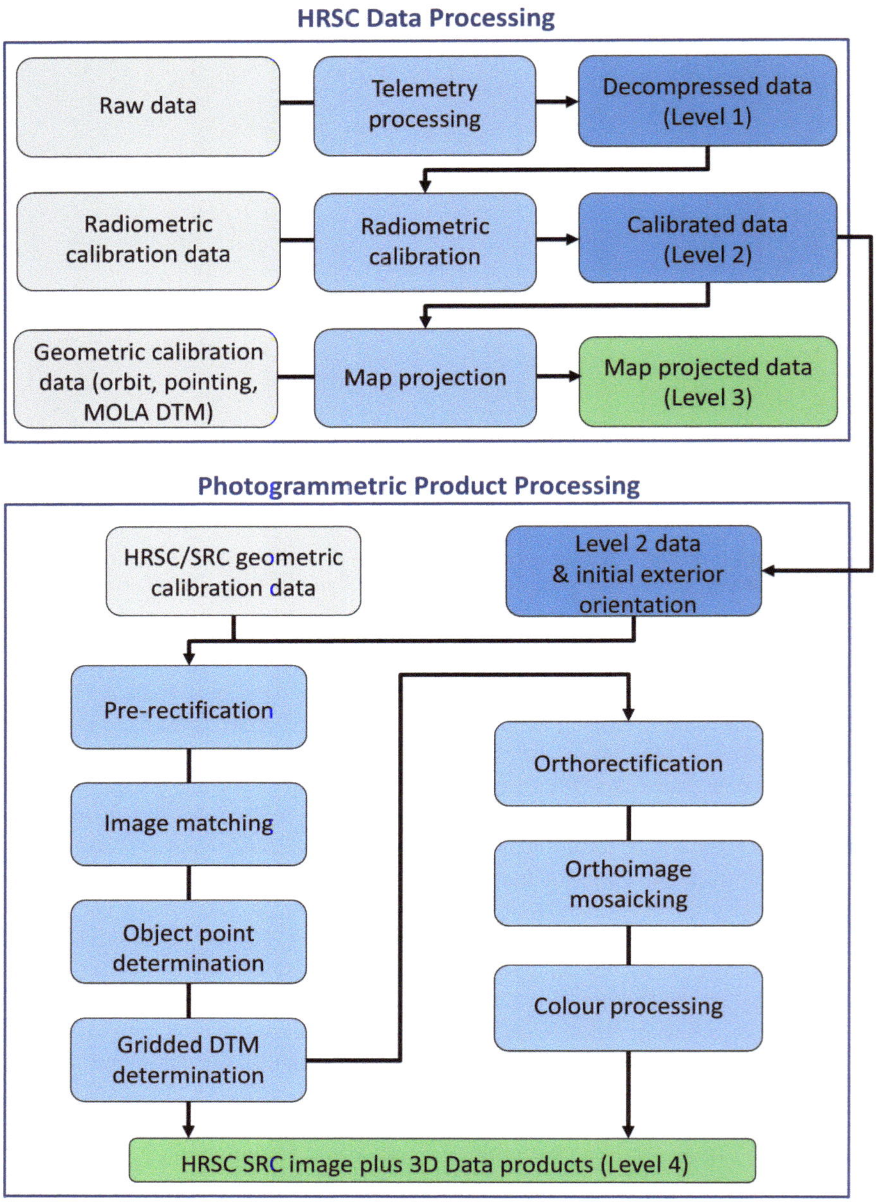

Fig. 6.10 HRSC product processing and production workflow. In the first step, raw data is radiometrically and geometrically corrected to produce Levels 1, 2 and 3 data. Level 4 data that includes elevation data products requires additional processing from Level 2 data

with a HiRISE DTM of the same extents (Fig. 6.11c) to generate a perspective view of the region (Fig. 6.11d).

For surface missions, the inclusion of additional cameras on most missions provides imagery redundancy plus the ability to generate stereoscopic images through spatial offset of the instruments. Surface-generated stereoscopic imagery is also used for geologic interpretation, albeit in a much finer and local scale than can be achieved from orbit (Li et al. 2007). The early Viking lander cameras produced wide-parallax stereo imagery that was manually exploited via hard-copy photographs. Recent developments have seen this type of imagery being captured through dedicated navigation cameras to generate route maps for rovers. Navigation and primary imaging cameras since MER have been capable of producing stereo imagery from which point matching and surface-level DTMs can be obtained (Li et al. 2007; Bogue 2012; Liang et al. 2021). These can produce millimetre-scale elevation data but are restricted by what the cameras can see. Obstructions such as rocks or topography in addition to local shadowing occlude DTM processing. In addition, DTM accuracy is dependent on feature distance from the imagers, with objects further away possessing lower resolution than closer terrain (Di et al. 2013; Paar et al. 2023).

Exploitation

Exploitation of DTMs generated in processes described above have provided critical insight into understanding Mars. Straightforward processing can generate slope maps that are not only useful for landing site selection but useful for understanding Martian surface processes. Many erosive and depositional processes are dependent on slope, and measured results from DTMS, such as along Martian gully channel profiles, can assist in inferring the type of erosion. This uses a geophysical trait of the angle of repose, the maximum slope at which material will remain in place prior to failure and downward collapse. Fluidised lubrication, such as through frost sublimation or liquid water, serve to lower this angle (Sutton et al. 2022).

DTMS can also be used to extract volume data from geological features. As an example, measuring the volume of sand dunes can provide information on Martian wind-related processes (HiRISE DTM of barchan dune example, Fig. 6.12). In this method, dunes are extracted from the surrounding plains (dashed line around dune highlighted with arrow, Fig. 6.12), and the volume calculated from the DTM elevation pixels (Boazman et al. 2021).

6.3 GIS Creation and Management

Accurate spacecraft positioning (Cassidy 1990; McEwan et al. 2007) and data processing not only enable detailed measurements of topography, and meaningful representation of key information in maps, but also facilitate overlaying of different datasets in a geographic information system (GIS). Despite the proliferation of

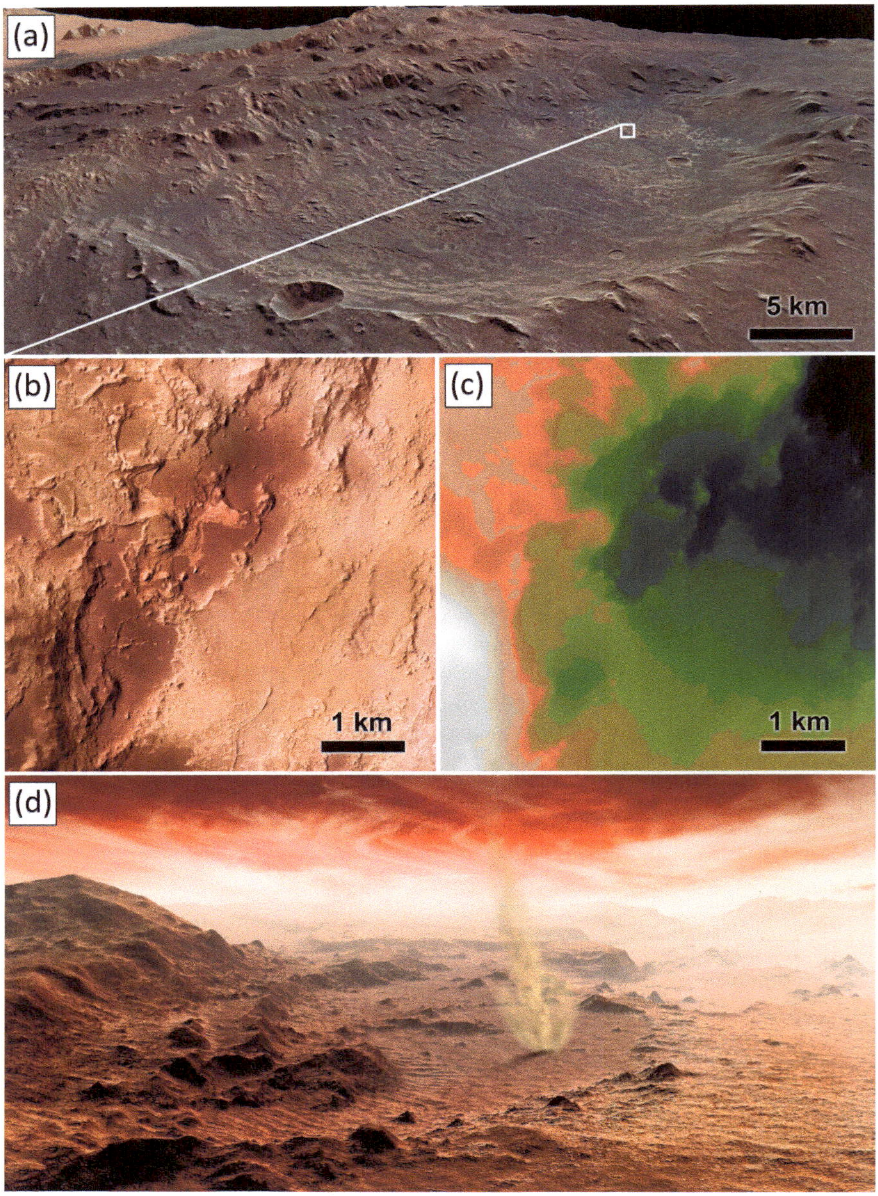

Fig. 6.11 **a** HRSC Perspective image of Eberswalde Crater region. White square identifies extent of HiRISE DTM (ESA/DLR/FU Berlin, CC BY-SA 3.0 IGO). **b** Colourised HiRISE orthorectified image of the Eberswalde delta (ESP_019190_1560, NASA/JPL). **c** HiRISE DTM produced from overlapping imagery (NASA/JPL). **d** Perspective illustration generated from the HiRISE DTM and orthorectified image

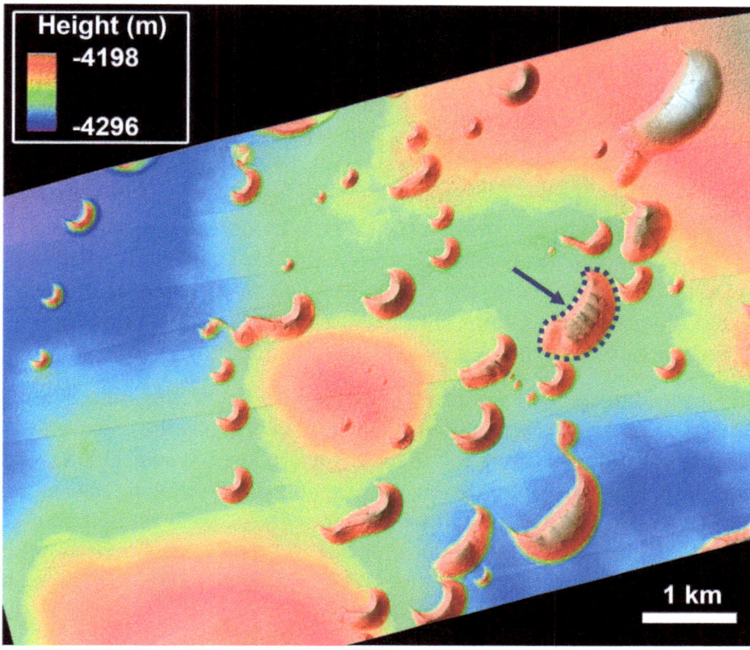

Fig. 6.12 HiRISE DTM of barchan sand dunes on the Martian northern plains (ESP_071630_2560, NASA/JPL). Post spacings of 1 m allow dune size and volume estimations in unprecedented detail. This provides insight into the strength of aeolian processes and abundance of erodible material on the planet

disparate datasets, all measurements have common attributes of location in physical space and time. These are essential for positioning and overlaying datasets that are key to developing a GIS.

Terrestrial maps such as topographic lines, used transparent overlays to combine information and create more meaning for the user (Deuchlera et al. 2004). Digital GIS enables overlaying of multiple layers and desired combinations as well as the ability to perform analysis. Freely available data repositories, such as the PDS, facilitate user access and research. Mars GIS lagged behind its terrestrial counterpart; the latter being driven by the need for natural and built environmental applications, as well as global Mars datasets being limited to Mariner 9 and Viking until 1997 (Batson and Ellianson 1995). The return of high-resolution orbital and elevation data coincided with the increased availability of network and computer power that facilitated download and exploitation. From 1997, the increasing volume of datasets made Mars GIS an attractive option for geospatial data management and research (Marco et al. 2018).

Applications enabled by GIS are numerous and include:

- Change detection. The ability to overlay spatially identical but temporally separated image data reveals changes in the Martian surface over time, such as contemporary erosion, shifting dunes, and discovery of new impact craters.
- Landing site selection. Safely landing a mission requires accurate local information about the atmosphere, elevation, slope and dust hazards while ensuring the site is scientifically interesting. Combining topographic with thermal inertia, image and other data has been standard practice for landing site selection.
- Detailed landscape evolution and geological analysis. Combining imagery, elevation, and other datasets allows more robust analysis based on multiple lines of evidence, strengthening inferences made about Mars.

Georeferenced images are stored, along with other underlying data, such as planetary base maps, in databases, enabling them to be queried by the end user. In NASA and ESA's case, these data are stored in the Planetary Data System (PDS), which are designed to provide ready search and access capabilities. Following data storage and retrieval, Mars researchers can conduct further analysis, such as digitising geology boundaries and make custom maps to infer historic processes. Spatially referenced data allows for an operator, or in some cases computer-vision segmentation algorithms, to classify terrain into human-understandable geologic features (Greeley 2013).

Recent advances in Mars GIS align to Open Geospatial Consortium (OGC) standards, simplifying data management and user interactions. This allows terrestrial-GIS experienced personnel to more easily modify pre-built ingestion, storage and processing to use Mars datasets. In addition, cloud-based data storage, as opposed to physical hardware combined to an organisation, can present information "on the fly" to the end user. Web Map Service (WMS) and Web Feature Service (WFS) allow automated requests, triggered by an operator, to a map cache consisting of rapidly-downloadable sub-elements of a much larger dataset, such as MOLA. Cached tiles called via WMS or WFS present a seamless geospatial overlay to the user.

In addition, many GIS applications are able to provide the ability to query multiple datasets at once, providing multiple lines of evidence to support or refute a hypothesis. Web-based GIS applications, such as Java Mission-planning and Analysis for Remote Sensing (JMARS, Christensen et al. 2009) and Orbital Data Explorer (ODE), the latter application downloading data directly from the PDS. Currently available applications such as MarsSI and THEMPROC also provide automated pipelines for processing imagery as described above (Quantin-Nataf et al. 2018).

These GIS applications and techniques provide continuity of geologic and geomorphic mapping essential for contemporary development of understanding of past and present Mars. The spatial referencing and near-simultaneous exploitation of disparate datasets described above are now standard practices for unravelling complex research problems. A recent exemplar is the study of alluvial fans where researchers combined imagery and elevation data was exploited by GIS processing to semi-autonomously detecting these features and analysing probable drainage and watersheds. Comparison with similar analysis of terrestrial alluvial fans provided

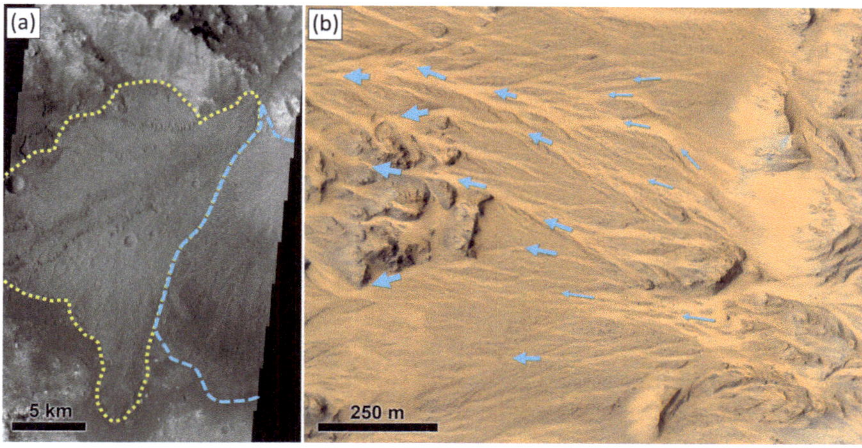

Fig. 6.13 **a** Alluvial fans (blue and yellow dashed lines, PIA14179, NASA/JPL). **b** Fluvial features at Mojave Crater. Blue arrows denote downslope movement (PIA09399, NASA/JPL)

metrics on the behaviour of past liquid water flows on Mars (Martian examples shown in Fig. 6.13a, b).

6.3.1 GIS Support in Landing Site Selection

Exploitation of multiple remote sensing datasets using a GIS is essential to support surface missions to Mars. Placing a science package on the surface of Mars in a way that it survives remains one of the most challenging engineering endeavours to date. Approximately half of all surface attempts have failed and so far, only China and the USA have successfully soft-landed surface missions returning useful scientific data. Choosing an appropriate region of Mars to land and conduct surface operations requires a combination of scientific and engineering disciplines. This entails obtaining maximum return for mission designer and researcher investment, while also minimising landing site hazards, such as excessive slopes, rocks and dust (e.g. Golombek et al. 2003a, b, 2012).

Remote sensing and space mission engineering are key to this process. Using a systems engineering process, mission designers identify scientific gaps requiring surface investigation and develop a mission concept and series of experiments (often through international collaboration) and generate a proposal for a mission (Chap. 2). Examples include Mars Insight's goal to investigate Martian geothermal temperature, or Mars Viking lander's search for life experiments (Masursky 1981; Golombek et al. 2017).

This process generates a series of engineering constraints, such as minimum surface temperature, dustiness, atmospheric pressure, that determine where a vehicle

Fig. 6.14 Illustration of the Curiosity Rover landing by parachute into Gale Crater. Such an activity is the culmination of extensive analysis of suitable landing sites based on engineering constraints and scientific value

can, and cannot land safely. Lower elevations have been favoured in historic missions, since increased atmospheric density allows more time to slow a spacecraft down via heat shield and parachute from interplanetary speeds to those manageable by descent rockets or airbags (Curiosity landing illustration, Fig. 6.14).

Following the final endorsement of engineering landing constraints, remote sensing data are used to search the surface of Mars for scientifically interesting landing sites that satisfy the identified constraints. Imagery is a prime source, where potential hazards and features of interest are identified through imagery analyst techniques. Image resolution is critical to this process, as regions appearing smooth at course resolutions may possess mission ending obstacles only visible at higher-resolution. This was the case of the Viking mission, where initial planning was performed on 1 km resolution Mariner 9 orbital imagery; the only dataset available at the time. The preferred landing site was changed following analysis of higher resolution imagery returned by the then just arrived Viking orbiters (Masursky and Crabill 1976).

Recent Mars missions have had access to a wealth of data returned from the remote sensors described in Chaps. 3–5. Slope and roughness products derived from topographic mapping describe safe or hazardous landing zones, with thermal sensor data used to infer whether a site is too dusty to do meaningful science (see thermal inertia, Chap. 5). Geomorphic and mineralogical mapping, made possible with GIS-based fusion of multiple datasets, are used for landing site proposals to suggest scientifically interesting regions. The NASA method of identification of suitable and safe sites have often used a series of workshops, where multiple proposed sites are voted on until a single, best possible area is identified and passed to mission planners (Golombek et al. 2012).

Surface Mission Location

Once a mission has landed successfully, finding its location on the ground depends on orbit and returned lander data, either from static landers such as Viking, Phoenix and Insight, or Mars Rovers. In this instance, using a GIS containing relevant georeferenced orbital imagery is of benefit, however there is the added complexity of localisation using ground-level observations. Early orbital mosaics from Mariner or Viking were resolution-constrained to hundreds of metres per pixel, making early localisation efforts difficult. Mission planning, such as the landing error ellipse, and returned telemetry during entry, descent and landing provided regional information about the spacecraft location for these earlier landers. Returned ground-level imagery from the lander was then used to identify resolvable features, such as crater rims or hills that could be identified from orbital imagery. Co-location of these features with equivalents from the orbital mosaic allowed triangulation to be performed, accurately locating the lander (Morris et al. 1978; Golombek et al. 1997).

Since the advent of MRO HiRISE imagery, lander localisation is made much easier; this sensor is capable of directly imaging landed hardware. Operators of the latest generation of Mars rovers such as Curiosity and Perseverance routinely use orbital and surface imagery in a complementary fashion for identifying scientific targets of interest from orbital imagery. The imagery plus produced elevation models are sufficient to plan safe routes for the rover to travel, before returning "in-situ" ground-level measurements (Golombek et al. 2020).

6.3.2 Case Study: Using GIS for Solar and Nuclear Rover Landing Site Selection

The solar-powered MER and nuclear-powered Curiosity and Perseverance Mars Rovers provide examples of multi-criteria decision making (MCDM) required for landing site selection. The two generations of rover technology have also highlighted differing constraints on where to land on Mars and how GIS can be used to assist in safe landing site selection.

Methods

The Spirit and Opportunity rovers (MER) were solar-powered, golf cart sized vehicles designed to search for past evidence of fluvial environments (illustration shown in Fig. 1.8c). They flew directly from interplanetary space to accomplish landings using combined heatshield, parachute, rockets and airbags. Sites were considered where remotely sensed data suggested past fluvial activity, such as the presence of hematite, a mineral associated with liquid water, or possible crater lakes such as Gusev Crater (Catling and Moore 2003; Golombek et al. 2003a, b). The requirements for abundance of MER solar power and mobility, plus landing and airbag systems placed critical constraints on the landing site, representatives of which are summarised in Table 6.1 (Golombek et al. 2003a, b).

A custom GIS that exploited the MOLA, MGS TES and visible imagery available at the time of MER site selection demonstrates the determination of suitable landing sites. The latitude constraint restricted possible landing sites to near the Martian equator (Fig. 6.15a–d). The GIS was also used to generate indicative elevation (Fig. 6.15a), slope (Fig. 6.15b), and thermal inertia (Fig. 6.15c) maps by overlaying each dataset and performing raster calculations and reclassification (Curkendall et al. 2006). A conditional statement of slopes satisfying the value shown in Table 6.1 selected pixels where the elevation and the scale slopes both had to be satisfied for a raster pixel to obtain a value of acceptable landing. All other regions were discarded. Thermal inertia constraints representing rock and dust abundance were then overlaid on the calculated dataset using a similar raster calculation, the result of which further reduced the safe landing site options (Fig. 6.15d).

Despite the extensive limitations, approximately 30 scientifically interesting sites were identified for MER in 2001. These included likely crater lakes and fluvial outflow systems observed on MOC imagery, plus a hematite-rich site in Meridiani Planum. Subsequent workshops and collection of additional orbital data down selected the initial list to Gusev Crater, thought to possess crater lake sediments, and the hematite-rich site in Meridiani Planum (Golombek et al. 2003a, b). Both rovers subsequently landed and conducted successful operations well past their planned lifetimes (Callas 2015; Callas et al. 2019).

The Curiosity and Perseverance rovers represented a step change in power, mobility and landing system technology. The sky-crane descent, plus improved environmental tolerances led to more generous engineering constraints when compared with the solar powered MER. Table 6.2 provides a representative of critical constraints proposed landing sites needed to satisfy, showing less constrained attributes compared to MER (Golombek et al. 2012; Chen et al. 2014).

The remotely sensed datasets available for Curiosity and Perseverance mission planning were unprecedented, with HiRISE imagery from which 1 m-scale DTMS were generated, plus Mars odyssey THEMIS able to map thermal inertia to 100 m/pixel. These high-resolution datasets, compared to the 1.2 m/pixel MOC data, 450 m/

Table 6.1 Representative MER constraints

Mission engineering parameter	MER constraint
Altitude	≤ -1.3 km w.r.t. MOLA defined geoid
Latitude (MER-A and MER-B)	5° N to 15° S and 10° N to 10° S
1 km/100 m/10 m length-scale slopes	< 2°, 5°, 15°
Minimum temperature at site	> − 97 °C
Rock abundance (from thermal inertia)	< 20%;
Local dust environment	> 200 thermal inertia units
Landing area size	> 2945 km^2

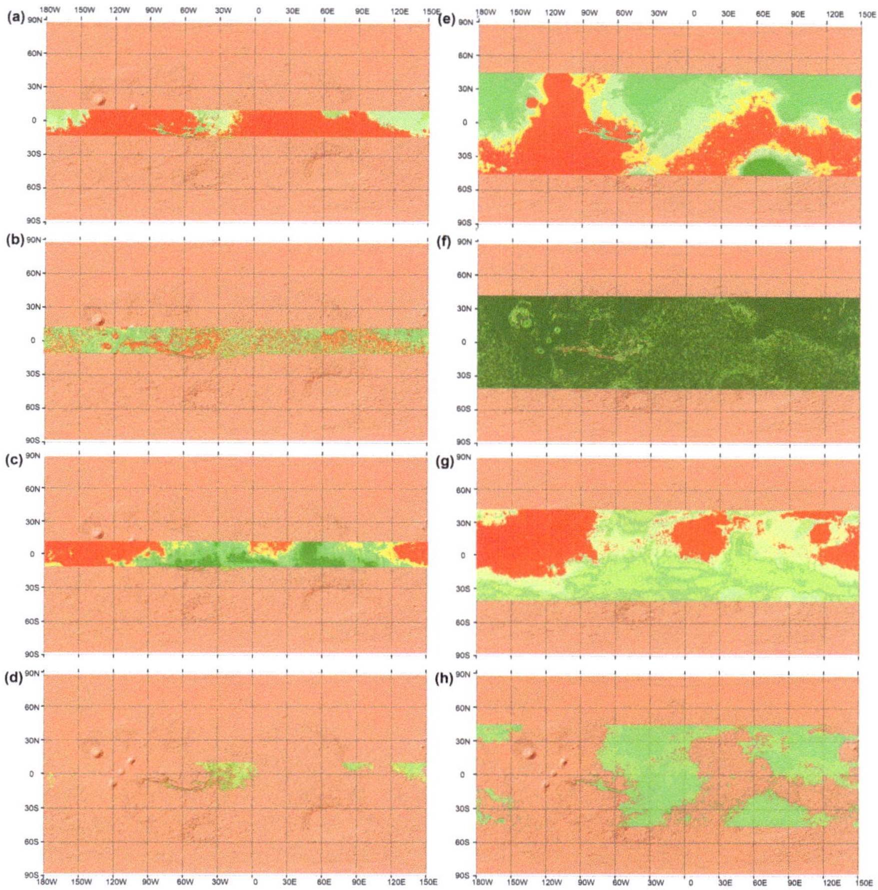

Fig. 6.15 Custom GIS analysis outputs for the solar (**a–d**) and nuclear (**e–h**) rover. Input layers included elevation (**a**, **e**), slope (**b**, **f**) and thermal inertia (**c**, **g**). All layers were cropped for maximum northern and southern latitude constraints, with output result for the solar rover (**d**) showing much less exploration potential than for the nuclear rover (**h**)

Table 6.2 Representative nuclear rover constraints

Mission engineering parameter	Nuclear rover constraint
Altitude	≤ − 0.5 km w.r.t. MOLA defined geoid (Perseverance allowed + 0.5 km above geoid
Latitude	30 °N to 30 °S
2–5 m length-scale slopes	< 30°
Minimum temperature at site	> − 120 °C
Rock abundance (from thermal inertia)	< 8%;
Local dust environment	> 100 thermal inertia units
Landing area size	> 286 km^2

pixel MOLA and 3 km/pixel TES allowed elevation, slopes, dust and rock abundance to be estimated to an accuracy not possible with MER.

Results

Figure 6.15e–h show the results of the GIS processing steps using elevation (Fig. 6.15e), slope (Fig. 6.15f) and thermal inertia-derived rock abundance (Fig. 6.15g) layers for the nuclear (Curiosity) rover landing constraints. Allowances for much larger surface slopes and higher latitude ranges provided a significantly increased area of Mars for exploration (Fig. 6.15h). Over the period 2006–2011, over 50 candidate sites were down-selected to Gale Crater for the Mars Curiosity Rover (Golombek et al. 2012), and Jezero Crater was chosen for Perseverance from 2020 (Way et al. 2020).

The above examples demonstrate the constraints and analysis that is performed for landing site selection and highlights the dependency of mission success to availability of remote sensing data. A version of the techniques above were used by the Zhurong Rover team to identify high slopes and other hazards that would have impacted the landing at Utopia Planitia (Wu et al. 2022).

6.4 Artificial Intelligence (AI) for Martian Applications

Continuous space-based observations of Mars have returned datasets of such a volume that they far exceed the ability of the small number of planetary researchers able to interpret them. Tasks such as crater counting, crucial for inferring the age of Martian landforms (described in more detail in Chap. 8), traditionally required manual identification, capture and classification of these features using GIS. Recent advancements in AI and Machine Learning (ML) have the ability to automate much of this manual work while still maintaining human quality control and oversight, vastly increasing the productivity of scientific research. AI/ML is a research area of rapidly growth and development, expected to significantly increase the exploration potential of surface and orbital missions.

6.4.1 Image Segmentation

A popular method for exploiting the terabytes of returned orbital and surface data is segmentation. The intent of this computer vision process is to use threshold values to separate pixels into meaningful classes. Figure 6.16 shows MOLA-derived slope classification of the region south of Le Verrier Crater in Noachis Terra. Drawing from this example, a threshold identification of topographic slopes greater than 5° can be conducted. The impact of high slopes on landing missions was shown in the above

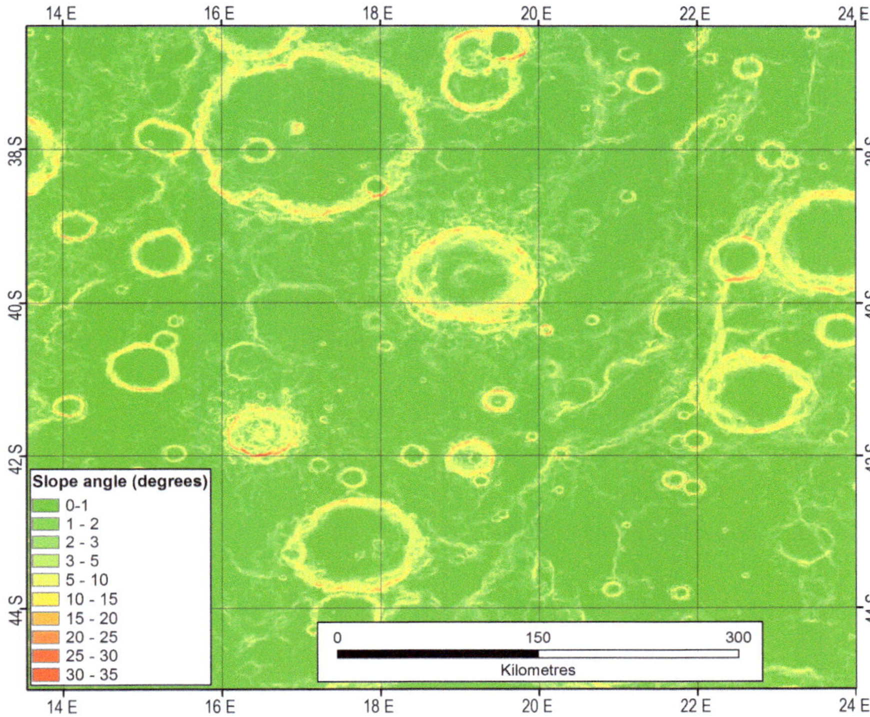

Fig. 6.16 Example MOLA-derived map of a cratered region in Noachis Terra derived from segmenting slope values. The crater walls and some interiors possess the steepest slopes

case study. The segmented MOLA result (Fig. 6.16) identifies hazardous areas for where a surface mission can only land on flat or gentle ($< 5°$) slopes (Walpert 2023).

Classification by spectral returns is another example of this method, where multispectral or hyperspectral image pixels are compared against known spectra and correlated values as classed as a particular mineral. Versions of this method have been used to generate mineral maps of the planet (see Chap. 7 for detailed examples).

Regression techniques are a simple method for undertaking classification and identifying features of interest. Linear regression is very popular, where a line of best fit is applied to plotted data points to minimise the distance between it and the data. The line of best fit is expressed with the formula:

$$Y = aX + b \tag{6.1}$$

where X is an explanatory variable and b is a dependant variable. The slope of the resulting line is given by a; b will provide the intercept when $x = 0$.

Table 6.3 Synthetic thermal inertia and albedo data

Albedo (X)	Thermal Inertia (Y, $Jm^{-2} K^{-1} s^{-½}$)
0.12	420
0.15	370
0.2	310
0.25	260
0.3	220

The following linear regression example uses five synthetic data values (Table 6.3) consistent with observed Martian trends (Putzig and Mellon 2007) to compare thermal inertia with albedo.

To fit Y = aX + B, slope (a) and intercept (b) are given by:

- $a = (N\sum XY - \sum X \sum Y)/(N\sum X^2 - (\sum X)^2)$
- $b = (\sum Y - a\sum X)/N$

Calculating individual terms provides:

- $N = 5$
- $\sum X = (0.12 + 0.15 + 0.2 + 0.25 + 0.3) = 1.02$
- $\sum Y = (420 + 370 + 310 + 260 + 220) = 1580$
- $\sum X^2 = 0.229$
- $(\sum X)^2 = 1.04$
- $\sum XY = 298.9$

Calculate "a" (rounded):

- $a = (5(298.9) - (1.02)(1580))/(5(0.229) - 1.02^2)$
- $a = (1499.5 - 1611.6)/(1.13 - 1.04) = -1098$

Calculate "b" (rounded):

- $b = (1580 - (-1098 \times 1.02))/5$
- $b = 540$

Using the above coefficients:

- $Y = -1098X + 540\ Jm^{-2} K^{-1} s^{-½}$

For every + 0.1 increase in albedo, thermal inertia increases by approximately 125. Dustier areas with high albedo have lower thermal inertia.

The goodness of fit to the line can be provided by R^2, which compares the actual fit with residual fit. R^2 values can reach a maximum value of 1, indicating a perfect fit between variables; values can range between 0 and 1, with a threshold value programmed beforehand to indicate the correlation between two variables. Using the worked example above demonstrates a significant fit between albedo and thermal inertia (Fig. 6.17).

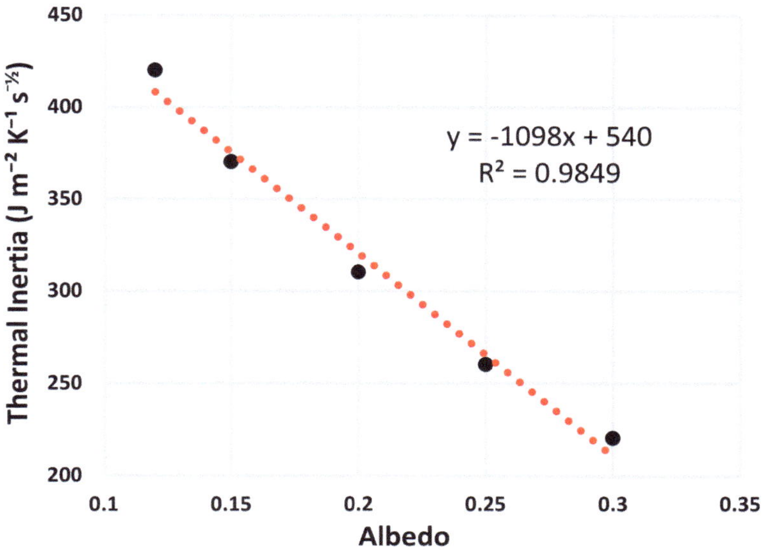

Fig. 6.17 Plotted linear regression between two spectral instruments, with the formula showing the slope and intercept, plus goodness of fit (R^2)

More complex classification methods have used convolutional neural networks (CNN). CNN are a class of neural networks that are ML constructs built to mimic the function of a human brain. The input dataset is processed by layers of artificial neurons that each receive input data, undertake a processing function using probability thresholds (if the calculated output exceeds a given threshold, then assign it a value of "X") to return an output that is fed to the next layer of neurons (Li et al. 2021). The sequential "hidden" layers of neurons accept input from the previous layer and conduct additional processing to produce an end result or to another "layer" of neurons until the desired end-state is achieved (Albawi et al. 2017).

In this method, a series of filters seek to extract information from pixel data, reducing the total information into discrete outputs that are fed into successive filters until the desired result is achieved. The first neuron layer conducts processing or "convolution" to extract features of interest, for example using filters such as identifying rapid contrast changes to find the edges of objects such as a Martian crater. To assist data processing, pooling layers reduce the amount of data required for processing by aggregating or summarising adjacent pixels (Fig. 6.18).

A classification neuron layer, fully connected with the preceding neurons takes the input from pooling layers and undertakes final analysis to identify a specific feature class, such as sand dunes, craters or other items of interest (Fig. 6.19).

For these methods to provide meaningful output, such as confident identification of geologic features of interest such as craters and dunes, the CNN algorithms must be trained, typically using over 1000 images for providing useful accuracy. This has traditionally entailed manual classifying instances of occurrences by human

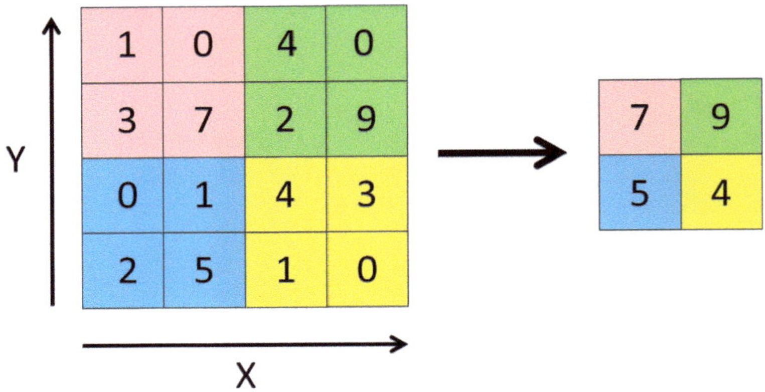

Fig. 6.18 Pooling process example from a 4 × 4 2D matrix of values across X and Y axis. Multiple values in a specified grid size (2 × 2 squares) are summarised by choosing the highest values within the squares

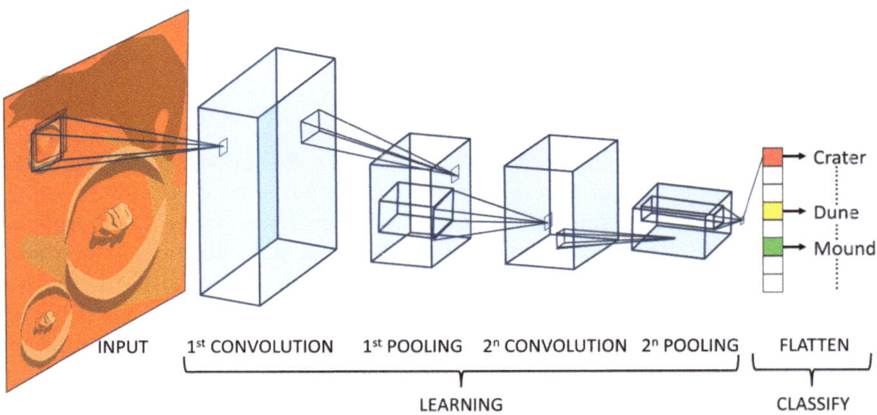

Fig. 6.19 Overview of CNN process. Complexities within an image (input) are sequentially broken down through convolutions and pooling of results. At the conclusion of the learning process, flattening of the results allows classification of image features

operators. A sample image is analysed by an operator and manually extracted and labelled in a process called supervised learning. Additional imagery identifying the absence of features, has also been used to increase the robustness of the training model (Zandri 2022).

Transfer learning can reduce this training overhead through applying outputs from training data used in previous algorithms to a new model. This method requires that previous and new algorithms can be applied to related datasets. An example of this method is the adaptation of pre-trained crater locators developed on older orbiter imagery to newer or combined datasets (Nagle-McNaughton et al. 2022).

One of the challenges of applying CNN or other AI to Martian feature extraction is class imbalance, where the size of a particular feature in most images would skew statistical confidence in the model. An example would be where an algorithm would be 99% correct at guessing the absence of a feature in every sampled image simply because only a minor portion of the image is occupied by that feature. A specific example includes small craters, which might occupy only a few percent of the total image swath and hence reducing statistical confidence. Artificially increasing the number of features in the training dataset is a process called over-sampling; and can improve algorithm accuracy by providing more instances to learn from. Alternatively, under-sampling, or reducing the number of occurrences of where the feature is absent in the training data set can also reduce class imbalance.

Synthetic imagery creation is a third alternative to generate training data and reduce imbalance. While it seems counterintuitive to build "fake" data given the terabytes of imagery returned from Mars to date, examples of specific features may be too few to train on or simply not exist. Generative Adversarial Networks (GAN) uses CNN to generate realistic synthetic data designed to simulate the feature of interest, while a second machine learning network uses differences between real and "fake" data to grow both networks respectively. Martian examples of GANS have been improvement in rover navigation through synthetic lander image generation (Wang et al. 2021), potentially resulting in an increased accuracy of orbital terrain classifiers (Nagle-McNaughton 2023).

Application

Crater detection is one of the most popular applications of CNN processing. Craters are found throughout Mars, and analysis of the number of times particular sized impacts are present is used to infer the age of an area under study (see Chap. 8). Obtaining and using imagery of consistent resolution and coverage is a prime consideration for this work, as coarse resolutions may lead to biases in the training (e.g. class imbalance, above), and make classification of smaller craters difficult. Algorithms may be trained by visually identifying the circular pattern of a crater in imagery, or using elevation data to note the depression and, if present, crater rim.

The latter method requires elevation data of sufficient detail to resolve craters of the desired size (Nagle-McNaughton et al. 2022). Recently, CTX imagery was used to train a CNN for detecting new craters using 1858 manually identified impact site, and 4973 non-impact examples. The algorithm was then able to be deployed over all CTX data collected to that time (112,000 images) and successfully identified fresh impacts ~ 71% of the time without human intervention (Gao et al. 2021). From this example where 29% of fresh impacts were not identified, it should be noted that best practices for AI-assisted feature detection is for researchers to understand how accuracy limitations may impact their analysis. If desired accuracy is not obtained through AI, further human intervention and analysis may be required to verify results.

6.4.2 Edge-Based AI

Edge-based processing (the capability of systems to analyse, process, and store data locally) and algorithm deployment is a promising future field for Martian AI. This would involve including ML algorithm capability on board the spacecraft or surface lander, allowing a mission to process collected data and extract meaningful information on operation rather than having to transmit raw data back to Earth for exploitation. Using AI to extract scientifically useful information for raw data could potentially save communications bandwidth, reducing the need for high-power transmitters (Francis et al. 2017).

For surface missions, live control is impossible due to the lag imposed by speed of light restrictions. Using edge-based machine augmentation significantly increases the distances able to be travelled by Mars rovers. This is critical as surface missions are expensive to plan and operate, with a significant risk of failure (Verma et al. 2023). Therefore, the use of machine-augmentation to maximise vehicle safety and science return is of high interest to mission designers and scientists.

The Sojourner rover, first wheeled vehicle landed on Mars possessed a degree of hazard avoidance systems to ensure the safe operation of the vehicle. Laser-based active distance sensors were able to detect obstacles such as rocks or voids in front of the vehicle, while tilt sensors protected Sojourner from tipping over during slope traverses (Matijevic 1998).

Perseverance, NASA's latest Mars rover uses AI to process data collected from navigation cameras into a 3D obstacle and hazard map, from which a safe navigation path is automatically plotted. The stereo imagery of the navigation cameras is processed into a 3D point cloud, from which an elevation map is produced. Additional processing produces a traverse cost map, where obstacles are ranked as high cost, and trafficable paths as low-cost. The rover then prioritises and selects the most appropriate path, performs the drive and repeats the cycle (McHenry et al. 2020).

This process increases the amount of science that can be performed on the surface by removing communication delays in transmitting raw camera data to Earth for processing, which would then need to be uploaded as a set of move instructions to the rover (Chi et al. 2018). Future Mars missions, such as sample return are expected to create further demand for AI/ML to increase autonomy.

Surface-based ML has also extended to self-determining the most scientifically useful information to be gathered. Mars Curiosity's Autonomous Exploration for Gathering Increased Science (AEGIS) allows the rover to target ground features, such as rocks (Fig. 6.20). The software identifies rocks satisfying preprogrammed criteria such as contrast or shape difference from surroundings. AEGIS prioritises instruments, such as Chemcam to return data on these features without human intervention and to target the rover's laser system (Estlin et al. 2014).

Edge-based processing will be an area of growth in the future, as computer processing boards increase in computing power and become lighter. In contrast, edge-based processing has yet to be deployed significantly on orbital systems. Most orbital missions have returned extensive remotely sensed data from a variety of

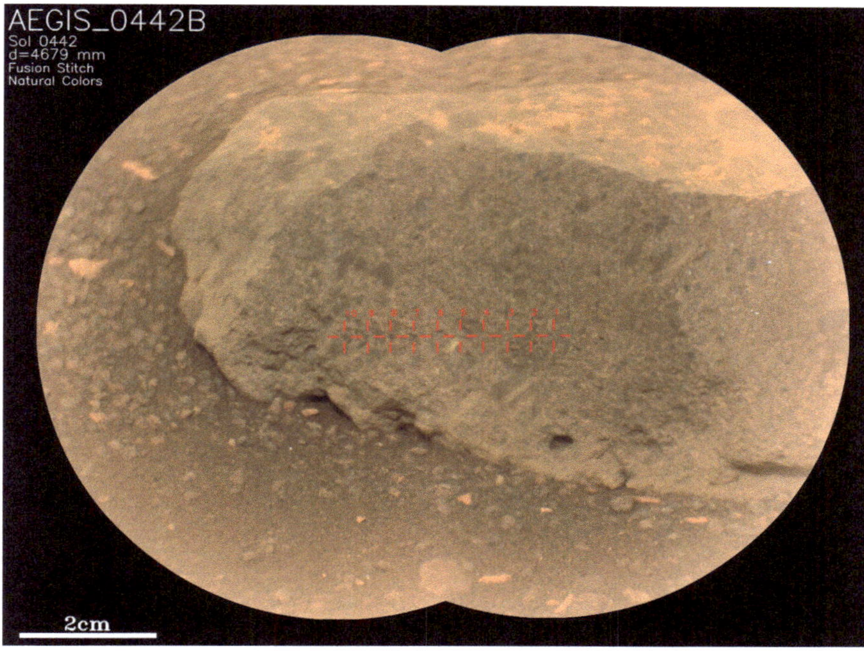

Fig. 6.20 Autonomous Exploration for Gathering Increased Science (AEGIS) results where the Perseverance Rover to autonomously choose and sample rocks using the most appropriate scientific instruments. Minimal human intervention is required during the process (PIA25289, NASA/JPL)

sensors, almost all of which is of scientific interest. This reduces the requirement to preprocess data and return a subset addressing a specific scientific question to Earth as whole datasets that can be generated into near-global coverage are of benefit to the scientific community (Diniega et al. 2021). Spacecraft-based AI/ML will also be of more value to smaller, low-cost "Space 2.0" missions (Chap. 2), where power, storage and transmissions are at a premium. In this case, algorithms able to reduce significant amounts of data, such as from hyperspectral sensors, to return low-bandwidth results, are of interest (Goodwill et al. 2022).

Edge processing may also provide benefit to landing site selection, where recent work is exploring AI/ML techniques on board the spacecraft. This would perform on the fly assessments on a high-performance spaceflight computer to continually update a lander's descent profile for achieving the safest possible landing. Imagery of ever-increasing resolution would be ingested by the segmentation algorithms as the spacecraft approached the surface, to automatically identify mission ending hazards not detectable from orbit (Claudet et al. 2022). Initial deployment of this method was in fact used during the Perseverance Rover descent and landing phases (Farley et al. 2020).

As with terrestrial ML, interest in its application for Mars is increasing as a method to address the supply of returned data that outmatches human's ability to exploit it. By increasing the ability to analyse Martian remotely sensed datasets, AI/ML is helping increase the understanding on processes acting on the planet. ML is increasingly being used for extended missions, as budgetary constraints limit human staffing and the operations rely more on multi-sol planning and ML/AI-determined surface target identification. Furthermore, ongoing generation of standardised datasets able to produce consistent and testable results will greatly increase the utility of Martian ML.

Martian ML applications and interpreting the planet's history and evolution are made possible through the GIS and image processing methods described in this chapter. GIS enables accurately positioning and overlay of multiple, disparate datasets on a planetary surface and extract meaningful data enables training of algorithms, as well as conduct volume analysis, change detection and geomorphic analysis. In addition, the ability of GIS and image processing to support accurate landings on the Martian surface highlights them as a central tool for understanding the planet and mission planning.

References

Albawi S, Mohammed TA, Al-Zawi S (2017) Understanding of a convolutional neural network. In: 2017 international conference on engineering and technology (ICET), pp. 1–6, https://doi.org/10.1109/ICEngTechnol.2017.8308186.

Archinal BA, A'Hearn MF, Bowell E, Conrad A, Consolmagno GJ, Courtin R, Fukushima T, Hestroer D, Hilton JL, Krasinsky GA, Neumann G, Oberst J, Seidelmann PK, Stooke P, Tholen DJ, Thomas PC, Williams IP (2011) Report of the IAU working group on cartographic coordinates and rotational elements: 2009. Celestial Mech Dyn Astron. https://doi.org/10.1007/s10569-010-9320-4

Baltsavias EP (1996) Digital ortho-images—a powerful tool for the extraction of spatial-and geo-information. ISPRS J Photogramm Remote Sens 51:63–77

Batson RM, Ellianson EM (1995) Digital maps of Mars. Photogramm Eng Remote Sens 6:1495–1507

Beyer RA, Alexandrov O, McMichael S (2018) The Ames Stereo Pipeline: NASA's open-source software for deriving and processing terrain data. Earth Space Sci 5:537–548

Boazman SJ, Davis JM, Grindrod PM, Balme MR, Vermeesch P, Baird T (2021) Measuring ripple and dune migration in Coprates Chasma, Valles Marineris: a source to sink Aeolian System on Mars? J Geophys Res Planets. https://doi.org/10.1029/2020JE006608

Bogue R (2012) Mars curiosity: sensors on the red planet. Sens Rev 32:187–193. https://doi.org/10.1108/02602281211233151

Broxton MJ, Edwards LJ (2008) The Ames Stereo pipeline: automated 3D surface reconstruction from orbital imagery. Lunar and Planetary Science Conference 39, abstract # 2419.

Callas JL, Golombek MP, Fraeman AA (2019) Mars exploration rover opportunity end of mission report. No. JPL-CL-19-7647

Callas JL (2015) Mars Exploration Rover Spirit end of mission report (No. JPL-Publ-16-2).

Cassidy LW (1990) Star trackers for spacecraft applications. In: Acquisition, tracking, and pointing IV, 1304, p 58. SPIE, https://doi.org/10.1117/12.2322198

Catling DC, Moore JM (2003) The nature of coarse-grained crystalline hematite and its implications for the early environment of Mars. Icarus 165:277–300. https://doi.org/10.1016/S0019-1035(03)00173-8

Chen A et al (2014) EDL engineering constraints, 1st Landing Site Workshop, https://marsnext.jpl.nasa.gov/workshops/2014_05/05_LSW1_EDL_Eng_Constraints_v6.pdf. Accessed 23 Jul 2023

Chi W, Chien SA, Agrawal J, Rabideau G, Benowitz E, Gaines DM, Fosse E, Kuhn S, Biehl J (2018) Embedding a scheduler in execution for a planetary rover. ICAPS 2018:312–320

Christensen P, Engle AE, Syahrul Dickenshied S, Noss D, Gorelick N, Weiss-Malik M (2009) JMARS—a planetary GIS. AGU Fall Meeting Abstracts.

Claudet T, Tomita K, Ho K (2022) Benchmark analysis of semantic segmentation algorithms for safe planetary landing site selection. IEEE Access 10:41766–41775. https://doi.org/10.1109/ACCESS.2022.3167763

Curkendall D, Hare T, Anderson RC, Dobinson E, Plesea L (2006) Mars GIS landing site suitability models. In: 37th annual lunar and planetary science conference, p 2110

Deuchlera C, Wählischa M, Gehrkeb S, Haubera E, Obersta J, Jaumanna R (2004) Combining Mars data in GRASS GIS for geological mapping. ISPRS, Istanbul

Di K, Yue Z, Liu Z, Wang S (2013) Automated rock detection and shape analysis from mars rover imagery and 3D point cloud data. J Earth Sci 24:125–135. https://doi.org/10.1007/s12583-013-0316-3

Diniega S, Bramson A, Buratti B, Buhler P, Burr D, Chojnacki M, Conway S, Daubar I, Dinwiddie CL, Galofre AG et al (2021) Mars as a "natural laboratory" for studying surface activity on a range of planetary bodies. Bull Am Astron Soc 53:123. https://doi.org/10.3847/25c2cfeb.950513cc

Duxbury TC, Christensen P, Smith DE, Neumann GA, Kirk RL, Caplinger MA, Albee AA, Seregina NV, Neukum G, Archinal BA (2014) The location of Airy-0, the Mars prime meridian reference, from stereo photogrammetric processing of THEMIS IR imaging and digital elevation data. J Geophys Res Planets 119:2471–2486. https://doi.org/10.1002/2014JE004678

Earthdata (2023) https://www.earthdata.nasa.gov/engage/open-data-services-and-software/data-information-policy/data-levels. Accessed 9 Nov 2023

Estlin T, Gaines D, Bornstein B, Schaffer S, Tompkins V, Thompson DR, Altinok A, Anderson RC, Burl M, Castaño R, Blaney D (2014) Automated targeting for the MSL rover ChemCam spectrometer. In: 12th international symposium on artificial intelligence, robotics, and automation in space (i-SAIRAS), pp 17–19

Farley KA, Williford KH, Stack KM, Bhartia R, Chen A, de la Torre M, Hand K, Goreva Y, Herd CD, Hueso R, Liu Y (2020) Mars 2020 mission overview. Space Sci Rev 216:1–41. https://doi.org/10.1007/s11214-020-00762-y

Francis R, Estlin T, Doran G, Johnstone S, Gaines D, Verma V, Burl M, Frydenvang J, Montaño S, Wiens RC, Schaffer S (2017) AEGIS autonomous targeting for ChemCam on Mars Science Laboratory: deployment and results of initial deployment and results of initial science team use. Sci Robot. https://doi.org/10.1126/scirobotics.aan4582

Gao A, Daubar IJ, Wexler D, Wagstaff KL, Bickel V, Doran G (2021) Using machine learning to complement new martial crater inventories. In: Proceedings of the 52nd lunar and planetary science conference, virtual, pp 15–19

Golombek MP, Cook RA, Economou T, Folkner WM, Haldemann AFC, Kallemeyn PH, Knudsen JM, Manning RM, Moore HJ, Parker TJ, Rieder R, Schofield JT, Smith PH, Vaughan RM (1997) Overview of the Mars Pathfinder mission and assessment of landing site predictions. Science 278:1743–1748. https://doi.org/10.1126/science.278.5344.1743

Golombek MP, Grant JA, Parker TJ, Kass DM, Crisp JA, Squyres SW, Haldemann AFC, Adler M, Lee WJ, Bridges NT, Arvidson RE (2003a) Selection of the Mars Exploration Rover landing sites. J Geophys Res Planets. https://doi.org/10.1029/2003JE002074

Golombek M, Grant J, Kipp D, Vasavada A, Kirk R, Fergason R, Bellutta P, Calef F, Larsen K, Katayama Y, Huertas A (2012) Selection of the Mars Science Laboratory landing site. Space Sci Rev 170:641–737. https://doi.org/10.1007/s11214-012-9916-y

Golombek M, Kipp D, Warner N, Daubar IJ, Fergason R, Kirk RL, Beyer R, Huertas A, Piqueux S, Putzig NE, Campbell BA (2017) Selection of the InSight landing site. Space Sci Rev 211:5–95

Golombek M, Williams N, Warner NH, Parker T, Williams MG, Daubar I, Calef F, Grant J, Bailey P, Abarca H, Deen R, Ruoff N, Maki J, McEwen A, Baugh N, Block K, Tamppari L, Call J, Ladewig J, Stoltz A, Weems WA, Mora-Sotomayor L, Torres J, Johnson M, Kennedy T, Sklyanskiy E (2020) Location and setting of the Mars InSight Lander, instruments, and landing site. Earth Space Sci. https://doi.org/10.1029/2020EA001248

Golombek M, Grant J, Parker T, Kass D, Crisp J, Squyres S, Adler M, Haldemann A, Carr M, Arvidson R, Weitz C (2003b) Mars exploration rover landing site selection. In: Sixth international conference on Mars (No. 3222). LPI

Goodwill JS, MacKinnon JP, Sakano K, Wilson CM (2022) Onboard hyperspectral image classification via transfer learning for communication-limited spacecraft. In: 2022 IEEE aerospace conference (AERO), pp 1–13. https://doi.org/10.1109/AERO53065.2022.9843805

Greeley R (2013) Introduction to Planetary geomorphology. Cambridge University Press, New York, p 237

Greeley RM, Batson RM (eds) (1990) Planetary mapping, Cambridge University Press, Cambridge, 296 pp

Gwinner K, Scholten F, Spiegel M, Schmidt R, Giese B, Oberst J, Heipke C, Jaumann R, Neukum G (2009) Derivation and validation of high-resolution digital terrain models from Mars Express HRSC data. Photogramm Eng Remote Sens 75:1127–1142

Hargitai H, Wang J, Stooke PJ, Karachevtseva I, Kereszturi A, Gede M (2017) Map projections in planetary cartography. In: Choosing a map projection, pp 177–202

Hirt C, Claessens SJ, Kuhn M, Featherstone WE (2012) Kilometer-resolution gravity field of Mars: MGM2011. Planet Space Sci 67:147–154. https://doi.org/10.1016/j.pss.2012.02.006

Hord CW, Barth CA, Stewart AI, Lane AL (1972) Mariner 9 ultraviolet spectrometer experiment: photometry and topography of Mars. Icarus 17:443–456

Hu H, Wu B, Chen L (2019) Color balancing and geometrical registration of high-resolution planetary imagery for improved orthographic image mosaicking. Planet Space Sci 178:104719. https://doi.org/10.1016/j.pss.2019.104719

Kirk RL, Mayer DP, Fergason RL, Redding BL, Galuszka DM, Hare TM, Gwinner K (2021) Evaluating stereo Digital Terrain Model quality at Mars rover landing sites with HRSC, CTX, and HiRISE images. Remote Sens 13:3511. https://doi.org/10.3390/rs13173511

Kreslavsky MA, Head JW III (2000) Kilometer-scale roughness of Mars: results from MOLA data analysis. J Geophys Res Planets 105:26695–26711

Li R, Arvidson RE, Di K, Golombek M, Guinn J, Johnson A, Maimone M, Matthies LH, Malin M, Parker T, Squyres SW (2007) Opportunity rover localization and topographic mapping at the landing site of Meridiani Planum, Mars. J Geophys Res Planets. https://doi.org/10.1029/2006JE002776

Li Z, Liu F, Yang W, Peng S, Zhou J (2021) A survey of convolutional neural networks: analysis, applications, and prospects. IEEE Trans

Liang X, Chen W, Cao Z, Wu F, Lyu W, Song Y, Li D, Yu C, Zhang L, Wang L (2021) The navigation and terrain cameras on the Tianwen-1 Mars rover. Space Sci Rev 217:1–20. https://doi.org/10.1007/s11214-021-00813-y

Marco R, Figuera B, Huu P, Rossi AP, Minin M, Flahaut J, Halder A (2018) Online characterization of planetary surfaces: PlanetServer, an open-source analysis and visualization tool. Planet Space Sci 150:141–156. https://doi.org/10.1016/j.pss.2017.09.007

Masursky H, Crabill NL (1976) The Viking landing sites: selection and certification. Science 193:809–812. https://doi.org/10.1126/science.193.4255.809

Masursky H (1981) Viking site selection and certification. National Aeronautics and Space Administration, Washington DC, vol 429

Matijevic J (1998) Autonomous navigation and the sojourner microrover. Science 280:454–455

McEwan AS et al (2007) Mars reconnaissance orbiter's high resolution imaging science experiment (HiRISE). J Geophys Res. https://doi.org/10.1029/2005JE002605

McHenry M, Abcouwer N, Biesiadecki J, Chang J, Sesto TD, Johnson A, Litwin T, Maimone M, Morrison J, Rieber R, Toupet O, Twu P (2020) Mars 2020 autonomous rover navigation. In:

Rieber R, McHenry M, Twu P, Stragier MM (eds) AAS, 2020, Planning for a Martian road trip-the Mars2020 mobility systems design, 2022 IEEE aerospace conference (aero). IEEE, pp 1–18. https://doi.org/10.1109/AERO53065.2022.9843375.

Morris EC, Jones KL, Berger JP (1978) Location of Viking 1 lander on the surface of Mars. Icarus 34:548–555. https://doi.org/10.1016/0019-1035(78)90044-1

Muller JP, Tao Y, Sidiropoulos P, Gwinner K, Willner K, Fanara L, Waehlisch M, van Gasselt S, Walter S, Steikert R, Schreiner B (2016) EU-FP7-iMARS: analysis of Mars multi-resolution images using auto-coregistration, data mining and crowd source techniques: processed results–a first look. ISPRS Int Arch Photogram Remote Sens Spatial Inf Sci 41, 453–458. https://doi.org/10.5194/isprsarchives-XLI-B4-453-2016

Nagle-McNaughton TP (2023) Synthetic Mars. Zenodo. https://doi.org/10.5281/zenodo.6338470

Nagle-McNaughton TP, Scuderi LA, Erickson N (2022) Squeezing data from a rock: machine learning for Martian science. Geosciences 12:248. 10.33910/rs14163883

Neumann GA, Rowlands DD, Lemoine FG, Smith DE, Zuber MT (2001) Crossover analysis of Mars orbiter laser altimeter data. J Geophys Res Planets 106:23753–23768

Paar G, Ortner T, Tate C, Deen RG, Abercrombie P, Vona M, Proton J, Bechtold A, Calef F, Barnes R, Koeberl C (2023) Three-dimensional data preparation and immersive misfsion-spanning visualization and analysis of Mars 2020 Mastcam-Z Stereo image sequences. Earth Space Sci. https://doi.org/10.1029/2022EA002532

Putzig NE, Mellon MT (2007) Apparent thermal inertia and the surface heterogeneity of Mars. Icarus 191:68–94. https://doi.org/10.1016/j.icarus.2007.05.013

Quantin-Nataf C, Lozac'h L, Thollot P, Loizeau D, Bultel B, Fernando J, Allemand P, Dubuffet F, Poulet F, Ody A, Clenet H, Leyrat C, Harrisson S (2018) MarsSI: Martian surface data processing information system. Planet Space Sci 150:157–170. https://doi.org/10.1016/j.pss.2017.09.014

Scholten F, Gwinner K, Roatsch T, Matz KD, Wählisch M, Giese B, Oberst J, Jaumann R, Neukum G (2005) Mars Express HRSC data processing—methods and operational aspects. Photogramm Eng Remote Sens 71:1143–1152

Sidiropoulos P, Muller JP, Watson G, Michael G, Walter S (2018) Automatic co-registration and orthorectification (ACRO) and subsequent mosaicking of NASA high-resolution imagery over the Mars MC11 quadrangle, using HRSC as a baseline. Planet Space Sci 151:33–42. https://doi.org/10.1016/j.pss.2017.10.012

Smith DE, Zuber MT (1998) The relationship between MOLA northern hemisphere topography and the 6.1-Mbar atmospheric pressure surface of Mars. Geophys Res Lett 25:4397–4400

Spiegel M (2007) Improvement of interior and exterior orientation of the three line camera HRSC with a simultaneous adjustment. Int Arch Photogramm Remote Sensing 36(3/W49B):161–166

Sutton SS, Chojnacki M, McEwen AS, Kirk RL, Dundas CM, Schaefer EI, Conway SJ, Diniega S, Portyankina G, Landis ME, Baugh NF (2022) Revealing active Mars with HiRISE digital terrain models. Remote Sens 14:2403. https://doi.org/10.3390/rs14102403

Tao Y, Muller J-P, Sidiropoulos P, Veitch-Michaelis J, Yershov V (2016) An optimised system for generating multi-resolution DTMs using NASA MRO datasets. In: The international archives of the photogrammetry, remote sensing and spatial information sciences, Volume XLI-B3, 2016 XXIII ISPRS Congress, 12–19 July 2016, Prague, Czech Republic

Verma V, Maimone MW, Gaines DM, Francis R, Estlin TA, Kuhn SR, Rabideau GR, Chien SA, McHenry MM, Graser EJ, Rankin AL (2023) Autonomous robotics is driving Perseverance rover's progress on Mars. Sci Robot. https://doi.org/10.1126/scirobotics.adi3099

Walpert M (2023) A risk assessment of potential landing zones on Mars using GIS, machine learning, and stereo-photogrammetry. In: IGARSS 2023–2023 IEEE international geoscience and remote sensing symposium, pp 4206–4209. IEEE. https://doi.org/10.1109/IGARSS52108.2023.10281708

Wang C, Zhang Z, Zhang Y, Tian R, Ding M (2021) GMSRI: a texture-based martian surface rock image dataset. Sensors 21:5410

Way D, Dutta S, Zumwalt C, De León SS (2020) EDL simulation results for the Mars 2020 landing site safety assessment. In: 2020 IEEE Aerospace Conference. IEEE, pp 1–5

Withers P, Jakosky BM (2016) Implications of MAVEN's planetographic coordinate system for comparisons to other recent Mars orbital missions. American Geophysical Union. https://doi.org/10.1002/2016JA023470

Wu B, Dong J, Wang Y, Rao W, Sun Z, Li Z, Tan Z, Chen Z, Wang C, Liu WC, Chen L (2022) Landing site selection and characterization of Tianwen-1 Zhurong rover on Mars. J Geophys Res Planets. https://doi.org/10.1029/2021JE007137

Zandri E.M (2022) Image processing machine learning algorithms for interplanetary small-sats images to support Martian rovers' navigation. Master's Thesis, Politechnic University of Torino, Italy

Chapter 7
Mars Unveiled: Atmosphere and Surface

7.1 Atmosphere

The Martian atmosphere differs greatly from Earth's in that it consists predominantly of CO_2, an attribute it shares with Venus. Additional gasses and water vapour are present in the Martian atmosphere but in vastly different concentrations from that of Earth (Table 7.1; Cox 2000; Franz et al. 2017).

The atmospheric pressure at the surface of Mars averages 700 Pa (< 1% of Earth), near the triple-point of water (611 Pa) where it exists simultaneously as a solid, liquid or gas. The low pressure plus average surface temperatures of 240 K preclude long-term existence of liquid water in the current Martian climate (Carr 2006). Exposed water simultaneously sublimates into the atmosphere and freeze onto the surface. As a consequence of its low pressure, the thin Martian atmosphere is unable to retain much heat, with temperatures rapidly decreasing with altitude, and night-time temperatures being as much as 83 K colder than those of midday (Haberle 2015).

Atmospheric pressure and temperature are highly variable between Martian seasons, with substantial portions of the CO_2 gasses exchanging with the south pole as part of a regular freeze–thaw process. The lowest pressures (decrease of 25–30%) occur during the southern winter, with the highest pressures occurring during northern summer months (polar condensation and sublimation overview, Fig. 7.1 [1, 2]). Owing to the eccentricity of the planet's orbit, the Martian seasons have greater variation in length compared to Earth. Table 7.2 compares the Martian season lengths with those of Earth. Its southern winter is longer than its summer (Appelbaum and Landis 1991).

Unlike Earth, the Martian atmosphere is not governed by heating and cooling of large oceans. Convection processes occur due to solar heating of the terrain, causing warmer gasses to rise and expand, with cooler air sinking to take its place (Wind flow to warmer regions from polar to equatorial regions and night to day side (Fig. 7.1 [3] and [4]). The dusty Martian atmosphere responds rapidly to local solar heating without heatsinks afforded by large oceans. Localised dust devils frequently occur,

S. Hobbs, *Sensing Mars: Remote Sensing to Understand the Red Planet*,
https://doi.org/10.1007/978-3-032-24416-1_7

Table 7.1 Atmospheric constituent comparison between Earth and Mars

Gas	Mars (%)	Earth (%)
Carbon dioxide	95	0.0391
Nitrogen	2.6	78.1
Argon	1.9	0.93
Oxygen	0.16	20.9
Carbon monoxide	0.06	< 200 ppm
Water vapour	210 ppm (variable)	1 (average)
Nitrogen oxide	~ 100 ppm	< 1 ppm
Neon	< 10 ppm	18 ppm
Hydrogen Deuterium Oxygen (HDO)	< 1 ppm	< 1 ppm
Krypton	< 1 ppm	1.1 ppm
Xenon	< 1 ppm	< 1 ppm

more common in the Southern hemisphere than the north, and have been directly observed by orbital and surface missions. These local-scale systems can reach 1–6 km in height on Mars and form frequently, likely enhanced through the hyper-arid near-surface conditions (Thomas and Gierasch 1985). Dust devils are instrumental in replenishing atmospheric dust, with the rarefied atmosphere favouring dust to fine sand. They also change local albedo through scouring the surface of the upper coating of dust, leaving a lower albedo track. (Balme and Greeley 2006). An engineering benefit of dust devils was discovered during the MER operations, where solar panels were inadvertently cleaned of their dust coating through coincidental action of one of these phenomena (Greeley et al. 2010).

7.1.1 Atmospheric Column

Additional space age remote sensing of the Martian atmosphere has consisted of radio occultation experiments dating from Mariner 4, where the Mars-Earth communication between transmitter and ground station is attenuated by the spacecraft passing behind the planet (Fjeldbo et al. 1966). Entry, descent and landing data from surface missions have also contributed to our understanding of how the Martian atmosphere varies with height. Changes in densities and temperatures impact the descent profile of landing missions and provide hard data behind initial atmospheric predictions (Karlgaard et al. 2021). The higher latitudes also subject to polar air mass effects that drive high and low-pressure masses, with fronts between them. Like Earth, the fronts can cause local storms, though with less severity due to the low-pressure environment and minimal water vapour (Fig. 7.1 [5]; Hollingsworth 2007; Read et al. 2015).

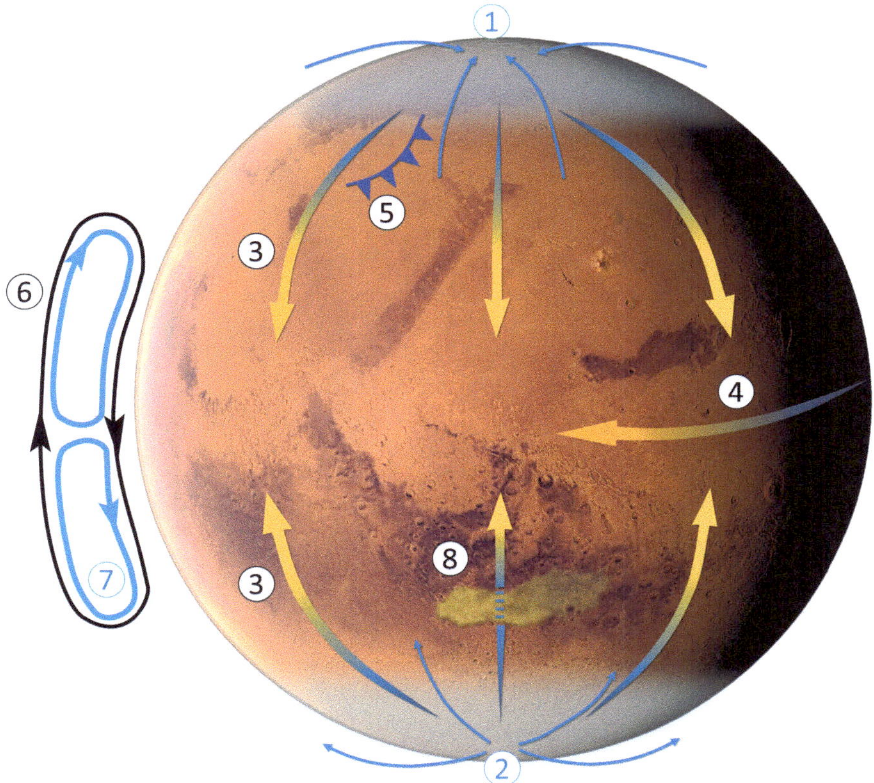

Fig. 7.1 Representative atmospheric cycles of Mars. (1) Condensation of volatiles during the northern polar winter. (2) Sublimation of volatiles during the southern pole summer. (3) Wind flow from higher, colder latitudes to warmer, equatorial regions. (4) Thermal tide from night to day side. (5) Front in northern latitudes. (6) Single Hadley cycle during Martian solstice. (7) Multiple Hadley cycles at equinox. (8) Dust storm

Table 7.2 Terrestrial versus Martian season lengths in days and sols

Season (North Hemisphere)	Terrestrial season length (days)	Martian season length (sols)
Spring	93	194
Summer	93	178
Autumn	90	142
Winter	89	154

The profoundly different gas composition of the Martian atmosphere, in addition to its lower density compared with Earth is also reflected in its unique vertical structure. The atmospheric column is divided into three broad regions: troposphere, mesosphere and thermosphere.

The troposphere extends 40 km from the surface and is dominated by surficial heating and suspended dust. These inputs provide convection and Hadley cycles that drive most of the observed Martian weather. Near the equator Hadley cell motion operates, with warm air flowing northward and southward to 30° latitude, where mid-latitude conditions cause it to cool, sink and return to the equator. Rotation of the planet as well as topographic obstacles, such as the Tharsis volcanoes, serve to redirect the air flow to generally northeast and southeast in the northern and southern hemisphere respectively (Haberle 1999). During season solstice, Hadley cells may reach higher latitudes to the planet's poles, becoming one cell (Fig. 7.1 [6]), with two cells forming at the equinox (Fig. 7.1 [7], Guendelman and Kaspi 2018).

The lower Martian gravity gives rise to an increased scale height (11.2 km) when compared with that of Earth (8.5 km). The decrease in temperature with altitude is only 2.5 °C/km, lower than a modelled 4.3 °C/km, and is caused by suspended dust absorbing solar radiation. This absorption can also cause the lower portion of the atmosphere to expand by up to 10 km and can adversely affect the timing of parachute deployments and rocket firings of descending craft if not accounted for (Bougher et al. 2017).

Thermal heating and convection drive Martian winds (Fig. 7.1 [3, 4]), with measured speeds between 1 and 4 m/s at the surface, and gusts of up to 30 m/s (Nakamura and Anderson 1979). These winds have eroded much of the Martian surface and formed an assortment of aeolian features, such as dunes and yardangs.

Above 15 km altitude, atmospheric temperature is controlled by radiation rather than convection, and temperatures decrease with altitude up to the mesosphere. From 40–100 km, CO_2, which would normally trap heat at lower elevations, acts in the opposite way on Mars, radiating heat into space and making this region—the mesopause—the coldest part of the planet's atmosphere. Unlike Earth, which has two mesopauses at 85 km and 100 km respectively, Mars has a single mesopause at an altitude of approximately 100 km. Temperatures of 100–120 K in this region are sufficient to cause CO_2 cloud formation in the mesosphere (Lewis and Mulholland 2015).

UV aurora glows have also been detected in this region below 100 km by NASA's Mars MAVEN spacecraft. These aurorae are driven by fossilised magnetic fields that are remnant of a global field that operated early in Mars' history. Aurorae likely vary in intensity in response to solar output, with an event observed to be 25 times brighter than previous emissions being associated with a massive solar storm (Schneider et al. 2021, Fig. 7.2).

The thermosphere is controlled by heating from high energy UV radiation and extends to 230 km altitude. As solar UV radiation received by Mars is dependent on its distance from the sun, the planet's increased eccentricity when compared to Earth varies the temperature by 65 K, with ranges between 175 and 240 K and extremes of 390 K reported (Bougher et al. 2017). The high thermosphere temperatures encourage

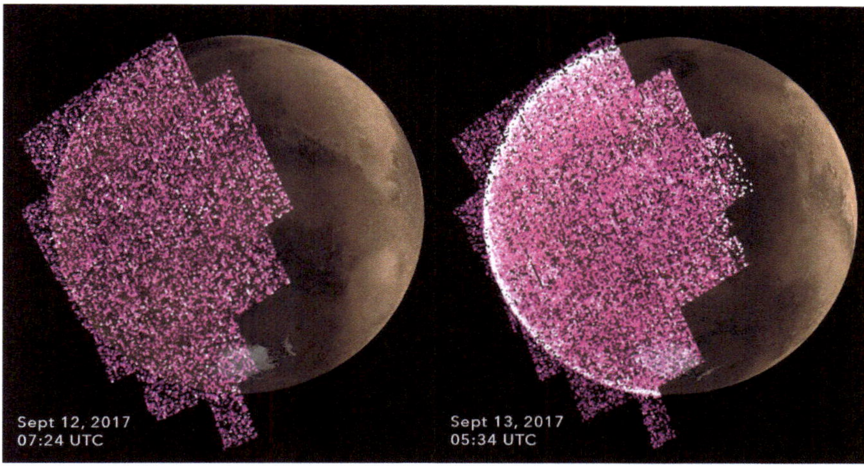

Fig. 7.2 Detection of Aurora around Mars generated by increased solar radiation (PIA21855, NASA/JPL)

loss of hydrogen from the atmosphere, as molecules achieve escape velocity (~5.1 m/s). High energy solar photons break water molecules into hydrogen and oxygen in the lower atmosphere, with the hydrogen molecule decomposing into individual atoms in the exosphere that subsequently leave the planet. Ionic oxygen dissociative recombination with CO_2 also leads to these atoms achieving escape velocity (Leblanc et al. 2017).

Sputtering occurs when heavy ions accelerated by the solar wind collide with atmospheric gasses and send them into deep space. Heavier molecules, such as CO_2 that would normally be bound by Martian gravity can be lost to space in this way. Remote sensing measurements from the MAVEN spacecraft used argon isotope ratio comparison as an indicator for atmospheric loss through sputtering since the early Noachian period. Results and modelling suggest the depletion of up to 65% of its original argon, suggesting an early atmosphere of at least 1 bar (Jakosky et al. 2017). Researchers believe that the effects of sputtering greatly increased following the shutdown of Mars' magnetic field early in its history, with localised magnetic fields, responsible for UV aurorae described above provide some protection against the phenomenon in the present day.

Three localised ozone layers, much thinner than that of Earth, have been discovered around Mars. A layer at 35–70 km height above the south pole was found to increase in concentration until mid-winter. Researchers interpret the formation of this layer through Hadley transport of near-equator air to the pole where UV radiation acts on CO_2 to produce atomic oxygen. In the polar night this combine to form oxygen molecules and ozone (Montmessin and Lefèvre 2013). Two additional ozone layers at 30 km and 40–60 km were discovered at low to mid-latitudes. The latter layer is only detectable during Martian northern spring and summer. Martian ozone formation is favourable when cold temperatures, such as near the poles, inhibit water vapour and subsequent solar generation of hydrogen radicals that can destroy the O_3 molecules through catalytic cycles (Krasnopolsky 2006).

7.1.2 Clouds and Storms

The space age has confirmed telescopic observations of atmospheric disturbances on Mars. Yellow clouds and polar hoods were readily seen and documented with changes to the Martian seasons. Persistent cloud patches were also documented appearing over Nix Olympia, now known to be the largest volcano in the Solar System, Olympus Mons. Additionally, morning hazes, corresponding to low-temperature vapour condensations that form fog on Earth, were seen near Mars' terminator. This type of fog routinely occurs in Valles Marineris (illustrated in Fig. 7.3).

The yellow clouds observed through telescopes since the 1600 s are now identified to be dust storms, caused by a preponderance of surface and atmospheric dust as opposed to water-based clouds due to the scarcity of H_2O on the planet. The abundance of Martian dust is thought to have originated from long-term deflation of the Medusae Fossae Formation (MFF) near the Martian equator (Fig. 7.4a–c). This is supported by chemical similarities between MFF and surface dust spectral measurements (Ojha et al. 2018).

Approximately 2.9×10^{12} kg of dust is thought to exchange between the surface and atmosphere per year (Pollack et al. 1977). One mechanism of dust distribution

Fig. 7.3 Illustration of fog in Valles Marineris

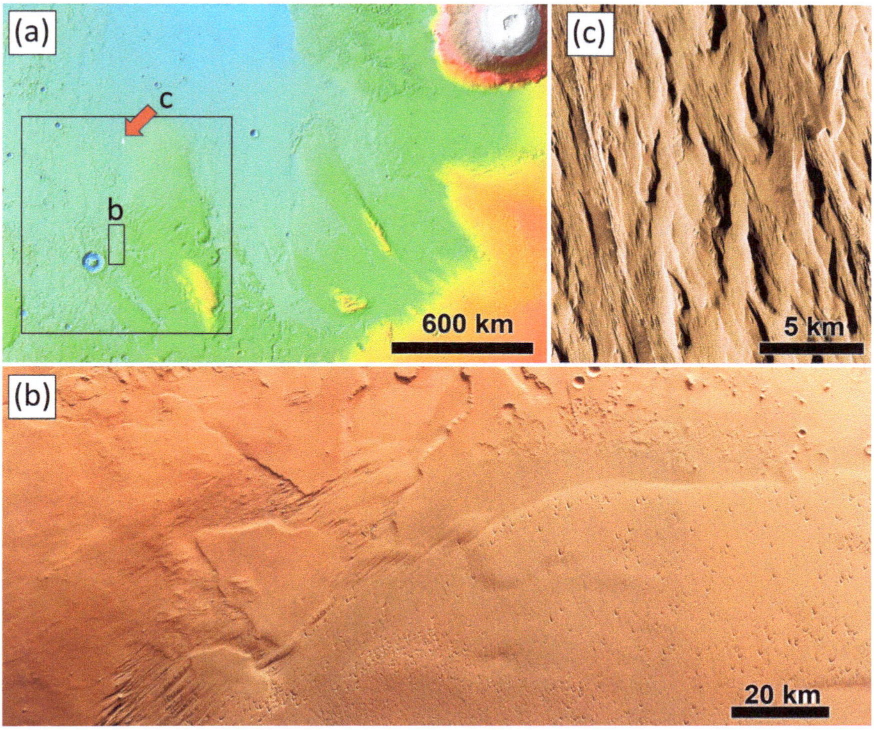

Fig. 7.4 **a** Context MOLA map of the MFF. Labelled insets denote locations of parts (**b**) and (**c**). **b** Mars Express image of MFF (ESA/DLR/FU Berlin, CC BY-SA 3.0 IGO). **c** Colourised HiRISE image of layered material in the MFF (ESP063264-1895, NASA/JPL)

around the planet are dust storms (representative, Fig. 7.1 [8]). These storms can be local or regional events (tens to hundreds of km in size) and usually increase through dust lifting during the southern spring and summer near perihelion. For example, the Hellas Basin, with temperature differences between its the southern edge and the CO_2-covered south pole has been observed to be subject to numerous localised dust storms (Chiu et al. 2018).

As described above, perihelion increases daytime heating, causing strong winds and increased dust devil activity. Local storms, normally lasting three Martian sols or less, may instead persist and grow, causing localised cooling and wind-driving temperature gradients (Fig. 7.5a, b). This self-feeding mechanism raises more dust, and may increase to a planet wide event. Global dust storms have also been observed through Earth-based telescopes, orbital and surface missions, where their effect can be devastating (Fig. 7.5c). The Opportunity rover succumbed to power loss when a planet-wide dust storm had reduced its available solar energy below a critical level to enable recharge of its batteries (Guzewich et al. 2020). After 30–40 sols

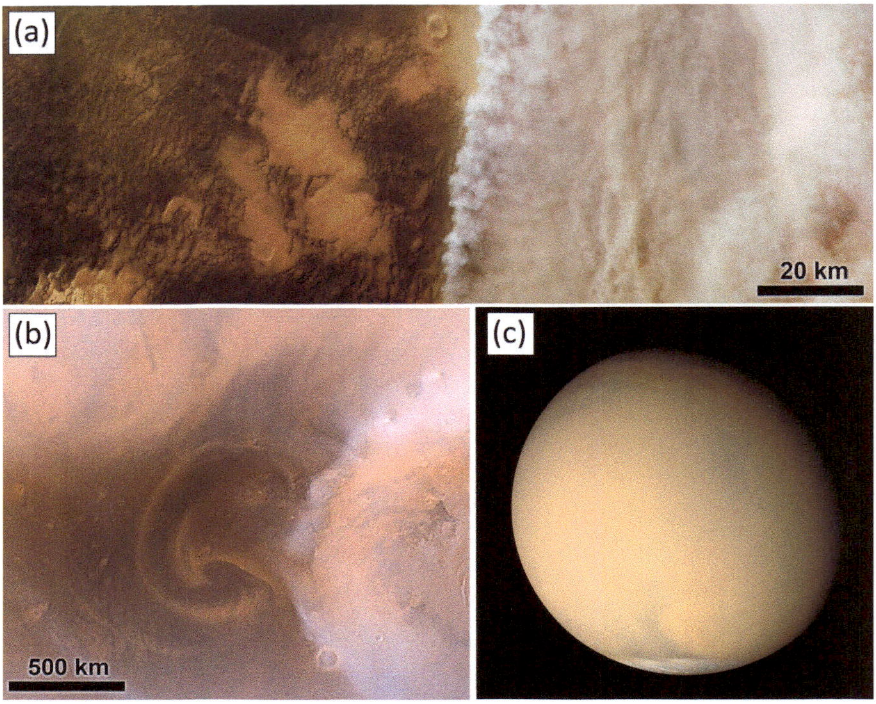

Fig. 7.5 **a** Mars Express image of a dust storm near the north polar ice cap (ESA/DLR/FU Berlin, CC BY-SA 3.0 IGO). **b** Dust storm front from the north pole (PIA02807, NASA/JPL). **c** Global dust storm imaged in 2001 by Mars Global Surveyor (PIA03170, NASA/JPL)

duration, global dust storms may peak and the levels of dust block sunlight, causing homogenous cooling of the surface and subsequent abatement of dust lifting winds.

7.1.3 Clouds

Clouds of water and CO_2 composition occur on Mars, and have been imaged by orbital and surface missions (Curiosity and Perseverance images, Fig. 7.6a, b). Like on Earth, Martian water-based clouds form when a parcel of air rises through convection and cools until the parcel reaches the dew pointwhere the parcel's temperature equals the condensation temperature of water. When the air is at its maximum vapour saturation pressure, the addition of more water vapour will form droplets. A similar process occurs with CO_2 clouds, though the dew point (150 K) is much lower than for water and hence these types of clouds mostly form near the polar caps or at higher altitudes. Non-polar CO_2 clouds were observed to form near the by the Curiosity Rover. They were estimated to be at least 60 km in altitude, where the cold conditions during aphelion caused gaseous CO_2 to condense out (González-Galindo et al. 2011).

As on Earth, Martian clouds mark the influence of temperature gradients, atmospheric waves and winds caused by convection. Ice particle sizes may vary with season between 1 and > 3 μm, with larger particle sizes at the bottom of the cloud layer and smaller at the top (Guzewich and Smith 2019). In contrast, terrestrial cloud droplets are 20 μm in diameter, with actual rain drops averaging 2 mm (De Wolf 2001). Observations at the Phoenix lander site near the Martian south pole suggest that ice condensing out of uplifted atmospheric gasses grew large enough

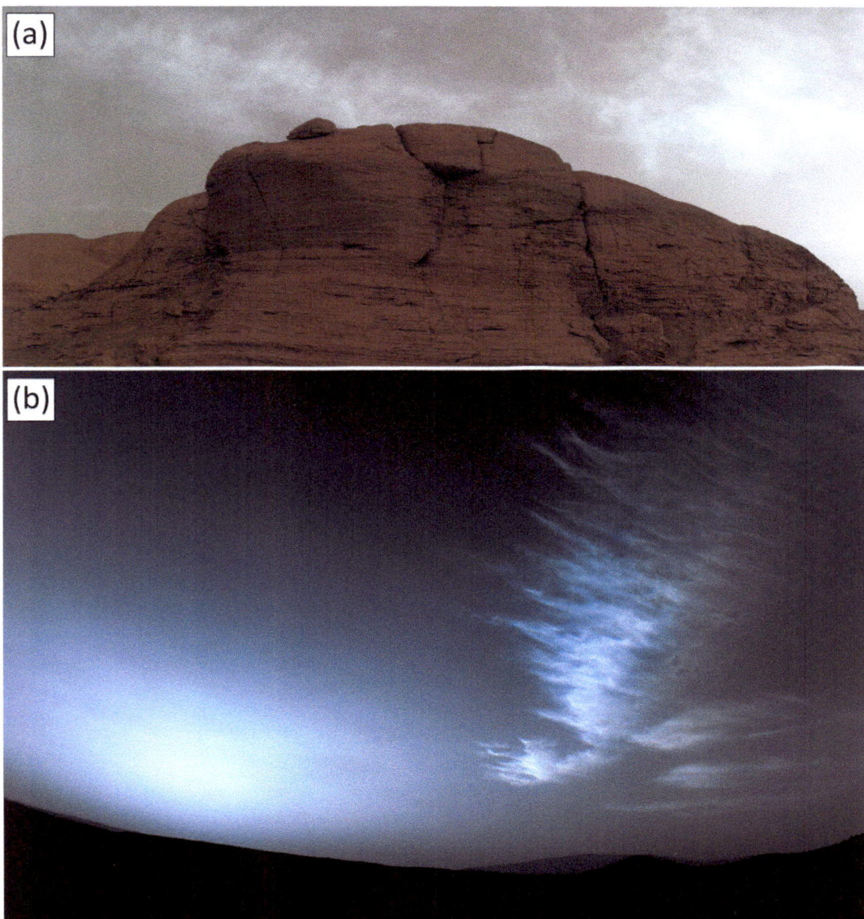

Fig. 7.6 **a** Martian clouds captured by Curiosity after sunset on Sol 3063 (PIA24622, NASA/JPL-Caltech/MSSS). **b** Clouds imaged by Perseverance in 2023 (PIA25821, NASA/JPL-Caltech/Kevin M. Gill)

to precipitate downwards during the planet's night. Sublimation is believed to occur during the local morning, with gasses raised once again through convective turbulence (Whiteway et al. 2009). This activity may become vigorous enough to generate microburst-like snowstorms within the planetary boundary layer (Spiga et al. 2017).

The Tharsis region has commonly produced clouds, with local winds blowing against the elevated topography causing lift and vapour condensation (Benson et al. 2003, Fig. 7.7a).

In 2018, Mars Express discovered an extremely elongated, orographic cloud stretching 1800 km westward of Arsia Mons, one of the large Tharsis volcanoes. Periodic remote sensing identified the cloud commenced around the Southern solstice between solar longitude (L_S) of 220°–230°, with a daily cycle of forming off the

Fig. 7.7 **a** Orographic cloud extending southeast of Arsia Mons (ESA/DLR/FU Berlin, CC BY-SA 3.0 IGO). **b** Terrestrial orographic clouds appearing over a topographic obstacle such as this mountain range near Hanksville, Utah

western flank of the volcano before local sunrise and expanding westward in the mesosphere at 170 m/sec at 45 km altitude. After a 2.5-h period, the cloud detached from the volcano and drifted westward before evaporating (Hernández-Bernal et al. 2021). Modelling of the cloud formation showed it was consistent with seasonally-occurring fast wind speeds causing strong vertical uplift, a subsequent temperature drop and condensation at 40–50 km altitude (Hernández-Bernal et al. 2022). Similar cloud formation regularly occurs on Earth, where mountain topography causes uplift of air and subsequent condensation. The preponderance of water vapour and much denser atmosphere makes cloud formation in this manner much more common than on Mars, occurring at much lower altitudes and being more compact and clearly defined (Holton 1992, Fig. 7.7b).

7.1.4 Stability of Water on Present Mars

The seasonal temperature and pressure variations, in addition to changes caused by elevation differences mean that liquid water can theoretically exist at lower elevation regions, such as the northern plain areas near the Martian equator, Valles Marineris, and Hellas and Argyre impact basins (Kostama et al. 2006). Mars Odyssey GRS mapping did not identify appreciable quantities of near-equator water, which are believed to be unstable at these latitudes.

The discovery of hydrated minerals by CRISM and OMEGA support the likelihood of past liquid water on Mars (Fig. 7.8a). Perchlorates and brines are also of interest to fluvial stability as dissolved salts lower the melting point of water up to 70 K below its nominal freezing point. Calcium perchlorate was discovered at the Phoenix Mars lander site, as well as in Gale Crater by the Curiosity rover. Elemental chlorine also seems abundant on Mars, as indicated by Mars Odyssey, MER and Sojourner observations (Banin 2005). Calcium perchlorate, likely present at 0.5–1% concentrations, is a hygroscopic salt with a natural affinity for water and capable of forming a brine in a process called deliquescence. This process occurs in terrestrial environments where moisture is absorbed from the atmosphere by a substance to form a solution. Recent research was conducted to identify possible Martian sites where deliquescence could occur (Pál and Kereszturi 2022). Mars climate modelling indicated that Acidalia and Utopia Planitia above 30° N and Argyre and Hellas Planitia near 60° S and in warmer seasons are possible areas where deliquescence may occur on Mars.

Discoveries and mapping of aqueous minerals reveal a complex history of multiple fluvially active periods, dominating in the Noachian, but also present in Hesperian and Amazonian periods (Fig. 7.8a). This makes it difficult to correlate the presence of hydrated minerals spanning much of Mars' history with where liquid water is expected to be stable to the present day (Wray 2021). Such water, if it exists, would be transient in nature, with contemporary mineral alteration processes being limited and hard to distinguish from older material. Direct surface observations of the sites identified above are more useful in confirming the actual presence of liquid water.

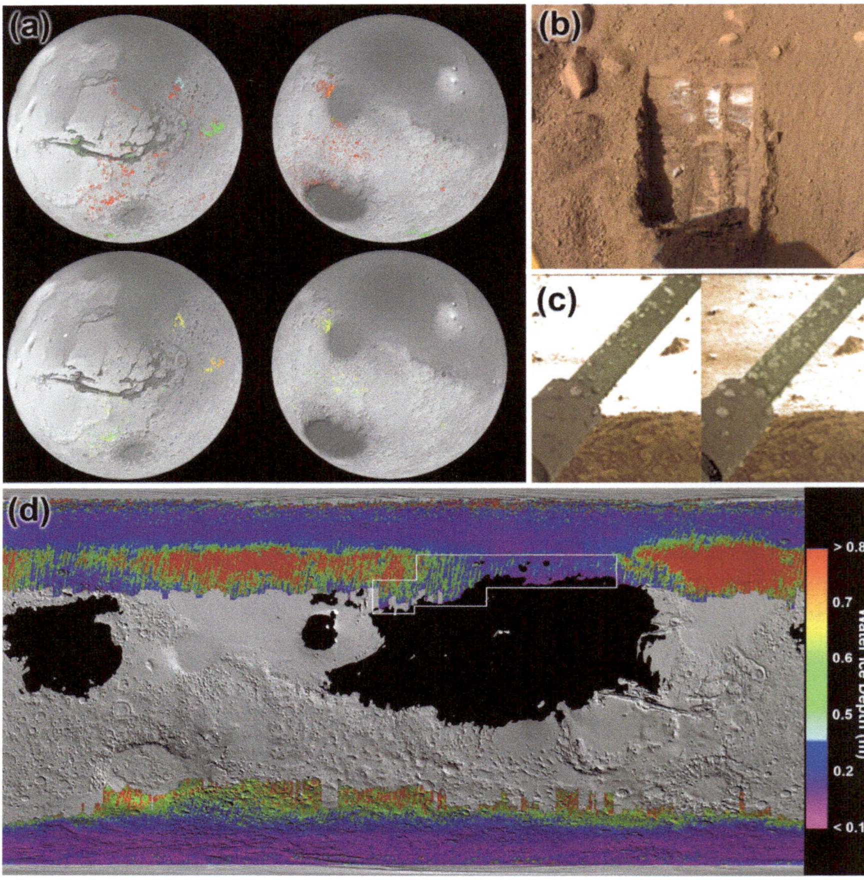

Fig. 7.8 **a** Locations of hydrated minerals (hydrous clays, red, hydrated sulphates, green, carbonate salts, orange and hydrated silica and aluminosilicate clays, blue, ESA/DLR/FU Berlin, CC BY-SA 3.0 IGO). **b** Water ice as revealed by the Phoenix lander trenching tool (PIA10911, NASA/JPL). **c** Colourised view of likely liquid briny water drops on the Phoenix lander legs changing over time (PIA10759, NASA/JPL). **d** Map of underground ices on Mars and highlighted region of a potential human landing site (PIA23514, NASA/JPL)

7.1.5 Present-Day Water Activity

Such a surface-based discovery was made by the Zhorong rover that operated in Utopia Planitia. This mission observed crusts, cracks, aggregates and bright polygonal ridges on the surfaces of hydrated, salt-rich dunes at 25° N. The features are consistent with spectroscopic evidence of hydrated materials and support recent (0.4–1.4 Ma) interactions of water-rich brines with regolith. CO_2 frost related processes are ruled out as possible mechanisms due to its instability of dry ice at these latitudes

(Khuller et al. 2021). Temperatures and pressures at the Zhorong site were found to be consistent with the formation of liquid brines stable enough to cause the dune erosion observed by the rover (Qin et al. 2023).

In 2008 the Phoenix Mars lander discovered direct evidence of liquid water ice just below the surface during arm scoop operations (Fig. 7.8b). The exposed ice deposit sublimed into the thin Martian atmosphere over a period of days (Smith et al. 2009). The mission may also have imaged possible water brines that had splashed onto the lander strut during landing. Comparison of images taken 13 days apart appeared to show two of the possible droplets merging into a single feature under surface tension (Smith et al. 2009, Fig. 7.8c).

While not possible to confirm the photographed features as liquid water, thermodynamic analysis supports formation of brines in colder temperatures in the Martian arctic (Rennó et al. 2009). A thermal and electrical conductivity probe operating on the lander was able to measure a change in electrical capacity corresponding to a rise in daytime temperatures. This result was consistent with progressive melting of salty ice being measured by the Phoenix instrument (Stillman and Grimm 2011). The presence of subsurface water ice is also supported by ground penetrating radar data from MRO, and gamma ray detections from Mars Odyssey, showing shallow ice deposits near the poles (Fig. 7.8d).

Remote sensing by the MARSIS radar instrument on board Mars Express has identified bright basal reflections at Ultimi Scopuli, similar to radar responses at terrestrial ice/liquid water interfaces. The reflections were 1.5 km below the region's surface ice, with modelled pressures and estimations of dissolved salts to allow the existence of briny water at 200 K temperature (Lauro et al. 2019). Additional research on estimations of the dust content using spacecraft motion data were used to support the 300 K temperature model—being near the freezing point of most brines. The water is likely kept liquid through pressure and remnant geothermal heating and held in place beneath ice and dust layers around the Martian south pole (Lauro et al. 2022).

The presence of an underground lake is of high interest to researchers, though the depth of the feature makes it difficult to conceive of an exploratory mission within the coming decades. If confirmed to consist of liquid water, the Ultimi Scopuli feature will be the largest present-day reservoir known to exist on the planet (Fig. 7.9).

7.2 Elemental Composition and Mineralogy

7.2.1 Elemental Abundances

The red colour of Mars was known for thousands of years by the ancients, and early telescopic observations indicated the similarity between this and iron oxide (Chap. 1, global abundance map shown in Fig. 7.10). The advent of the space age dispelled

Fig. 7.9 The location of an underground Martian lake near the Martian south pole is supported by strong radar returns of conductive material (NASA/JPL)

the notion of vegetation and revealed the reddish coating to be surficial deposits of dust, covering underlying materials of different mineralogy.

Subsequent discoveries showed Mars to be a complex planet with terrains of different ages and geology, with recent remote sensors (detailed in Chaps. 3–5) able to directly discriminate between mineralogy species. From 1997, continuous orbital

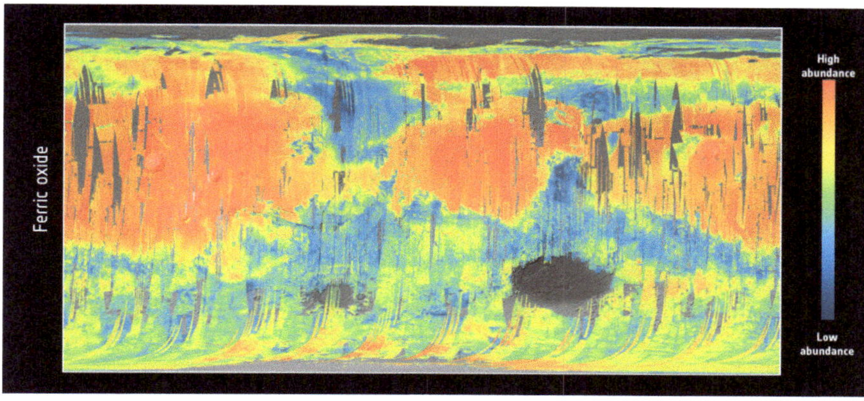

Fig. 7.10 Ferric oxide abundance map as sampled by Mars Express OMEGA (ESA/CNES/CNRS/IAS/Université Paris-Sud, Orsay; background image: NASA MOLA)

observations and multiple surface missions have demonstrated that while global-scale trends exist on Mars, the planet, like Earth, also possesses an abundance of local-scale features.

Understanding the mineralogy of Mars provides important insight into the formation and evolution of the planet. Minerals form in accordance with geophysical processes acting on them, such as volcanic processes, and can be subsequently altered through additional activities. The presence of mafic (volcanically formed, dark-toned minerals) rocks, such as olivine, for example, can be used to identify a formation as primary volcanic in nature, as this mineral is readily altered through the action of water (Váci et al. 2020). Secondary oxides, such as hematite, were used to prioritise the Meridiani Planum landing site for the Opportunity Mars Rover as this mineral is readily formed in the presence of liquid water.

Mineralogy analysis is also used to critically test inferences provided through geomorphology or Earth analogue studies. An example is the Gusev Crater Spirit Rover landing site. This site was chosen as it appeared physically similar to a crater lake, where a channel seemed to have transported liquid water into the crater in Mars' past. Post-landing spectroscopy analysis of the Gusev Crater floor identified the abundance of olivine, suggesting lava intrusion into the crater (Haskin et al. 2005).

Local scale mineralogy analysis can now be conducted in detail, with sophisticated and long-lasting rover missions carrying a suite of remote sensing instruments. Extended investigations in regions specifically chosen for scientific interest has generated a wealth of information regarding formation and evolution of Mars to the present day. Global, as well as site-specific mineralogy will be discussed in the below section.

7.2.2 *Global Mineralogy*

While being a terrestrial planet, Mars' mantle may contain twice the amount of iron, plus more phosphorous and potassium than that of Earth, with a core richer in sulphur. Mineralogical remote sensing from orbit and surficial missions indicate that Sulphur and chlorine are more abundant as volatiles in the Martian crust. Table 7.3 lists the primary and secondary mineral types detected on Mars (Ehlmann and Edwards 2014).

Table 7.3 Martian mineralogy

Distinction	Class	Mineral
Primary	Framework silicates	Olivine
		Orthopyroxenes
		Clinopyroxenes
		Plagioclase Feldspar
		Alkali Feldspars
	Sulphides	Pyrrhotite
		Pyrite/marcasite
	Oxides	Magnetite
		Ilmenite
Secondary	Oxides	Hematite
		Goethite
		Akageneite
	Clays (phyllosilicates)	Fe/Mg smectites
		Al smectites
		Kaolin minerals
		Chlorite
		Serpentine
		Al/K phyllosilicates
	Other hydrated silicates	Prehnite
		Analcime
		Opaline silica
	Carbonates	Mg/Ca/Fe carbonates
	Sulphates	Kieserite, szomolnokite, Mg-polyhydrated sulphates
		Gypsum
		Alunite
		Jarosite
	Chlorides	Chlorides
	Perchlorates	Perchlorates

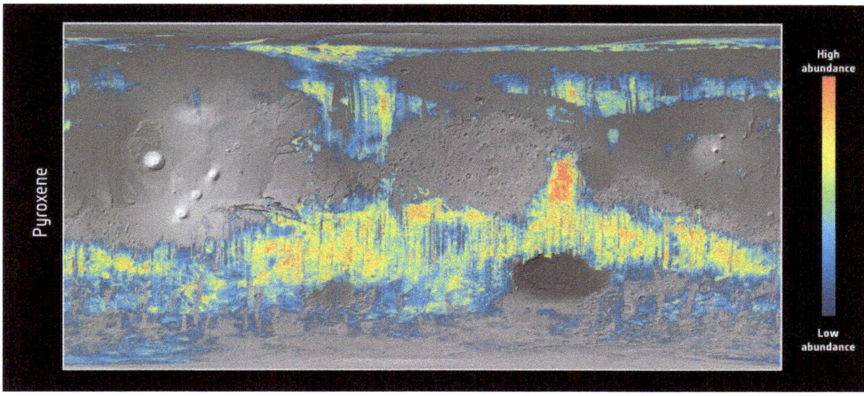

Fig. 7.11 Pyroxene abundance map as sampled by Mars Express OMEGA (ESA/CNES/CNRS/IAS/Université Paris-Sud, Orsay; Background image: NASA MOLA)

Discoveries from the space age identified the crust of Mars to be primarily basaltic, a rock type commonly found on the terrestrial planets and the Moon (Ruzicka et al. 2001; Barlow 2008). Globally, Mars is primarily igneous, with lava-derived minerals such as olivine, feldspars and pyroxenes (Table 7.3, Fig. 7.11).

The southern highlands, the oldest landforms on Mars, possess igneous minerals indicative of high-temperature lava flows, such as low-calcium pyroxenes. This indicates that early Mars had enough internal energy to support more energetic volcanism. Hydrated minerals have also been discovered on Mars, particularly in the northern plains (Carter et al. 2013).

The younger terrain north of the Martian dichotomy (dichotomy description: Chap. 1) is interpreted to be volcanic in nature, consisting of andesitic and basaltic andesitic material, plagioclase and high-silica glasses (Kraft et al. 2003; Ehlmann and Edwards 2014) as evidenced by sharp Si–O absorption and subsurface mafic materials exposed by cratering. Aqueous alteration geology and possible marine sediments, such as phyllosilicate-rich basalt alterations or silica coatings are also suggested by mineralogical analysis, providing some support for a Noachian ocean hypothesised to have been present in these regions (Scheller et al. 2021; described in detail in Chap. 9).

Lavas flowing into and around impact basins is evidenced by discoveries of olivine associated with these features (Ody et al. 2013; Ehlmann and Edwards 2014). High Mg concentrations suggest fracturing of the lower crust by these impacts has caused Noachian-Hesperian magma eruptions that would be fluid enough to flow into topographic lows such as pre-existing water-carved channels or around rims left by the impact basins. The Spirit Mars rover's mineralogical analysis of olivine in Gusev Crater described above, is an example of olivine-rich material subsequently altering an aqueous-formed feature.

Hydrated Minerals

On Earth, the physical plus chemical actions of solvents such as water serve to alter igneous and mafic minerals (Eggleton et al. 1987). Martian mineralogy analysis has also identified alteration of volcanic materials by surface processes, including weathering through dry mass wasting, aeolian, ice or fluvial processes (secondary minerals, Table 7.3). Remote sensing and in situ analysis has identified hydrated minerals on Mars and are of particular interest for studies into past habitability and astrobiological implications. Along with the presence of altered volcanic material in the norther highlands described above, a predominance of hydrated silicate clay material is dominant in the southern highlands, and were also discovered at Gale and Jezero Crater (Bristow et al. 2015; Tarnas et al. 2019) surface sites. Martian clays exist where primary minerals were altered in place by sedimentary processes, such as crater or valley basins. Clays are also associated with craters, where the impact process has excavated this material and deposited it in ejecta. Fe/Mg phyllosilicate-bearing basaltic materials are present in layers hundreds of metres, and deep cratering has exposed early Noachian units of Fe/Mg smectites, and chlorinated minerals. Deep excavation of minerals formed through thermal (up to 400 °C) alteration (e.g. prehnite, opaline silica, serpentine) indicate geothermal activity processes operating during the early Noachian period.

These types of hydrated minerals are generally present in large impact craters in Hesperian or Amazonian regions, meaning that impact into these terrains has excavated older material onto the surface. Based on the rarity of finding hydrated minerals in these younger regions, it is likely that significant aqueous alteration through geothermal activity was dominant during the early periods of Martian history.

Though rarer, Mars exploration has discovered liquid water-based alteration in post-Noachian geology. Endeavour Crater, explored by the Opportunity rover, showed evidence for Fe/Mg clay-altered bedrock and fracturing that occurred after the crater was formed (Arvidson et al. 2014). Hesperian volcanic outbursts, associated heating and catastrophic release of water-rich material such as with chaos terrains has likely provided conditions for localised aqueous alteration. Hydrated minerals such as silica, sulphates, jarosites and smectites were also discovered in the Valles Marineris system (Chojnacki and Hynek 2008), Nili Patera and Gusev Crater. While exploring its Gusev Crater landing site (Fig. 7.12a, b), the Spirit rover excavated silicate deposits through dragging a non-functioning wheel. Subsequent analysis of the Home Plate terrain revealed that this site was modified through superheated water emerging as a geyser, hot spring or fumarole, onto the surrounding plateau, with rover imagery suggesting the presence of silicate deposits in the region (Fig. 7.12c, Ruff et al. 2020).

Mineral analysis of regions interpreted as crater lakes has suggested the erosion and transport of hydrated minerals from uplands, which are likely to have been deposited within the topographic lows of crater basins (Milliken and Bish 2010). Jezero, Eberswelde and Holden Craters are examples where the spectral signature of the hydrated minerals on the crater floors is similar to that of the surrounding terrain. Sulphate minerals such as gypsum and polyhydrated Mg sulphates, are also found in

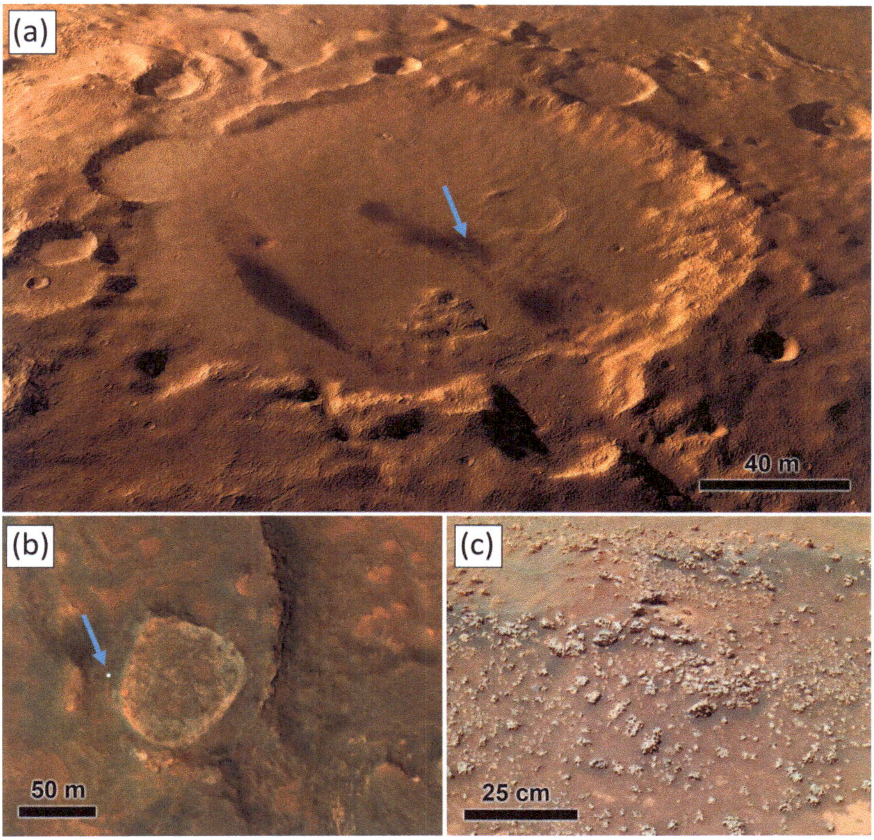

Fig. 7.12 **a** Artist illustration of Gusev Crater with arrow showing the location of Columbia Hills. **b** Home Plate, with the Spirit Rover highlighted (ESP_021925_1650, NASA/JPL). **c** Silicate deposits as imaged by Spirit at Home Plate (PIA2388, NASA/JPL)

many closed basin crater lakes and differ in composition from the surrounding terrain. These are consistent with deposits left over once acidic lake water had evaporated into the atmosphere (Sheppard et al. 2021). The Opportunity Mars rover's analysis of Meridiani Planum indicated Hesperian-aged sedimentary units, including hematite, sulphates and others, overlaid Noachian clay minerals. Episodic fluvial upwelling formed temporary lakes, causing precipitation of minerals (Andrews-Hanna et al. 2010).

As described above, the mineralogy of Mars as explored by decades of remote sensors suggests a complex planet where multiple processes have shaped its geology. Early alteration of igneous material into clays indicates abundant aqueous activity during the Noachian to Hesperian periods. By the Hesperian, volcanic outgassing likely made standing liquid water more acidic, as evidenced by discoveries of chlorine and sulphate mineralogy in these regions. Hydrated mineral types and locations support past surface outflows, groundwater upwelling and local ponding that reduced in frequency over geologic time. The paucity of crystalline or aqueous altered minerals in the Amazonian period point to a marked decline in the availability of liquid water. In this latter period, aqueous alteration likely occurred where localised, as opposed to global, processes operated (Fairén et al. 2009).

The planet's mineralogy alteration was supported by its atmosphere, which despite its low pressure and profoundly different gas abundance when compared to Earth, may enable the existence of liquid water on the planet today. Mars 'atmosphere drives circulation patterns originating from surface thermal heating rather than large oceans. Hadley cells, thermal tides, dust activity and storms distribute the thermal energy and drive Martian weather. Recent discoveries of CO^2 clouds, ozone layers and polar aurorae have also suggested a more dynamic, active planet than previously thought.

References

Andrews-Hanna JC, Zuber MT, Arvidson RE, Wiseman SM (2010) Early Mars hydrology: Meridiani playa deposits and the sedimentary record of Arabia Terra. J Geophys Res Planets. https://doi.org/10.1029/2009JE003485

Appelbaum J, Landis GA (1991) Solar radiation on Mars: Update 1991. In: European Space Power Conference (No. NAS 1.15: 105216)

Arvidson RE, Squyres SW, Bell JF III, Catalano JG, Clark BC, Crumpler LS, de Souza Jr PA, Fairén AG, Farrand WH, Fox VK, Gellert R (2014) Ancient aqueous environments at Endeavour crater, Mars. Science 343:1248097. https://doi.org/10.1126/science.1248097

Balme M, Greeley R (2006) Dust devils on Earth and Mars. Rev Geophys 44:1–22. https://doi.org/10.1029/2005RG000188

Banin A (2005) The enigma of the Martian soil. Science 309:888–890. https://doi.org/10.1126/science.1112794

Barlow N (2008) Mars: an introduction to its interior, surface and atmosphere. Cambridge University Press, New York, NY, p 264

Benson JL, Bonev BP, James PB, Shan KJ, Cantor BA, Caplinger MA (2003) The seasonal behavior of water ice clouds in the Tharsis and Valles Marineris regions of Mars: Mars Orbiter Camera observations. Icarus 165:34–52. https://doi.org/10.1016/S0019-1035(03)00175-1

Bougher SW, Roeten KJ, Olsen K, Mahaffy PR, Benna M, Elrod M, Jain SK, Schneider NM, Deighan J (2017) The structure and variability of Mars dayside thermosphere from MAVEN NGIMS and IUVS measurements: seasonal and solar activity trends in scale heights and temperatures. J Geophys Res Space Phys 122:1296–1313. https://doi.org/10.1002/2016JA023454

Bristow TF, Bish DL, Vaniman DT, Morris RV, Blake DF, Grotzinger JP, Rampe EB, Crisp JA, Achilles CN, Ming DW, Ehlmann BL (2015) The origin and implications of clay minerals from Yellowknife Bay, Gale Crater, Mars. Am Miner 100:824–836. https://doi.org/10.2138/am-2015-5077CCBYNCND

Carr M (2006) The surface of Mars. Cambridge University Press, New York, NY, p 307

Carter J et al (2013) Hydrous minerals on Mars as seen by the CRISM and OMEGA imaging spectrometers: updated global view. J Geophys Res Planets. https://doi.org/10.1029/2012JE004145

Chiu CK, Kwing-lam C, Jing X (2018) Dust activity over the Hellas basin of Mars during the period of southern spring. Icarus 311:306–316. https://doi.org/10.1016/j.icarus.2018.04.011

Chojnacki M, Hynek BM (2008) Geological context of water-altered minerals in Valles Marineris, Mars. J Geophys Res Planets. https://doi.org/10.1029/2007JE003070

Cox AN (ed) (2000) Allen's astrophysical quantities (Fourth ed.). AIP Press, pp 258–259. ISBN 0-387-98746-0

De Wolf DA (2001) On the Laws-Parsons distribution of raindrop sizes. Radio Sci 35:639–642

Eggleton RA, Foudoulis C, Varkevisser D (1987) Weathering of basalt: changes in rock chemistry and mineralogy. Clays Clay Miner 35:161–169. https://doi.org/10.1346/CCMN.1987.0350301

Ehlmann BL, Edwards CS (2014) Mineralogy of the Martian surface. Annu Rev Earth Planet Sci 42:291–315. https://doi.org/10.1146/annurev-earth-060313-055024

Fairén AG, Schulze-Makuch D, Rodríguez AP, Fink W, Davila AF, Uceda ER, Furfaro R, Amils R, McKay CP (2009) Evidence for Amazonian acidic liquid water on Mars—A reinterpretation of MER mission results. Planet Space Sci 57:276–287. https://doi.org/10.1016/j.pss.2008.11.008

Fjeldbo G, Fjeldbo WC, Eshleman VR (1966) Models for the atmosphere of Mars based on the Mariner 4 occultation experiment. J Geophys Res 71:2307–2316

Franz HB, Trainer MG, Malespin CA, Mahaffy PR, Atreya SK, Becker RH, Benna M, Conrad PG, Eigenbrode JL, Freissinet C, Manning HL (2017) Initial SAM calibration gas experiments on Mars: quadrupole mass spectrometer results and implications. Planet Space Sci 138:44–54. https://doi.org/10.1016/j.pss.2017.01.014

González-Galindo F, Määttänen A, Forget F, Spiga A (2011) The Martian mesosphere as revealed by CO_2 cloud observations and general circulation modelling. Icarus 216:10–22. https://doi.org/10.1016/j.icarus.2011.08.006

Greeley R, Waller DA, Cabrol NA, Landis GA, Lemmon MT, Neakrase LD, Pendleton Hoffer M, Thompson SD, Whelley PL (2010) Gusev Crater, Mars: observations of three dust devil seasons. J Geophys Res Planets. https://doi.org/10.1029/2010JE003608

Guendelman I, Kaspi Y (2018) An axisymmetric limit for the width of the Hadley cell on planets with large obliquity and long seasonality. Geophys Res Lett 45:13–213. https://doi.org/10.1029/2018GL080752

Guzewich SD, Smith MD (2019) Seasonal variation in Martian water ice cloud particle size. J Geophys Res Planets 124:636–643. https://doi.org/10.1029/2018JE005843

Guzewich SD, Fedorova AA, Kahre MA, Toigo AD (2020) Studies of the 2018/Mars year 34 planet-encircling dust storm. J Geophys Res Planets. https://doi.org/10.1029/2020JE006700

Haberle RM (1999) Modelling the atmospheric structure and dynamics of the Martian atmosphere. In: The Fifth international conference on Mars

Haberle RM (2015) Solar system/sun, atmospheres, evolution of atmospheres|Planetary Atmospheres: Mars. In: North GR, Pyle J, Zhang F (eds) Encyclopedia of atmospheric sciences, 2nd edn. Academic Press, pp 168–177. https://doi.org/10.1016/b978-0-12-382225-3.00312-1

Haskin LA, Wang A, Jolliff BL, McSween HY, Clark BC, Des Marais DJ, McLennan SM, Tosca NJ, Hurowitz JA, Farmer JD, Yen A (2005) Water alteration of rocks and soils on Mars at the Spirit rover site in Gusev crater. Nature 436:66–69. https://doi.org/10.1038/nature03640

Hernández-Bernal J, Sánchez-Lavega A, del Río-Gaztelurrutia T, Ravanis E, Cardesín-Moinelo A, Connour K, Tirsch D, Ordóñez-Etxeberria I, Gondet B, Wood S, Titov D (2021) An extremely elongated cloud over Arsia Mons volcano on Mars: I. Life cycle. J Geophys Res Planets 126. https://doi.org/10.1029/2020JE006517

Hernández-Bernal J, Spiga A, Sánchez-Lavega A, del Río-Gaztelurrutia T, Forget F, Millour E (2022) An extremely elongated cloud over Arsia Mons volcano on Mars: 2 Mesoscale modelling. J Geophys Res Planets. https://doi.org/10.1029/2022JE007352

Hollingsworth JL (2007) Cyclogenesis and frontal waves on mars. In: Seventh international conference on Mars. http://www.lpi.usra.edu/meetings/7thmars2007I

Holton JR (1992) Dynamic meteorology. Academic, San Diego, CA

Jakosky BM, Slipski M, Benna M, Mahaffy P, Elrod M, Yelle R, Stone S, Alsaeed N (2017) Mars' atmospheric history derived from upper-atmosphere measurements of 38Ar36. Science 355:1408–1410. https://doi.org/10.1126/science.aai7721

Karlgaard CD, Korzun AM, Schoenenberger M, Bonfiglio EP, Kass DM, Grover MR (2021) Mars InSight entry, descent, and landing trajectory and atmosphere reconstruction. J Spacecr Rocket 58:865–878

Khuller AR, Christensen PR, Harrison TN, Diniega S (2021) The distribution of frosts on Mars: links to present-day gully activity. J Geophys Res Planets. https://doi.org/10.1029/2020JE006577

Kostama V-P, Kreslavsky MA, Head JW (2006) Recent high-latitude icy mantle in the northern plains of Mars: characteristics and ages of emplacement. Geophys Res Lett. https://doi.org/10.1029/2006GL025946

Kraft MD, Michalski JR, Sharp TG (2003) Effects of pure silica coatings on thermal emission spectra of basaltic rocks: considerations for Martian surface mineralogy. Geophys Res Lett. https://doi.org/10.1029/2003GL018848

Krasnopolsky VA (2006) Photochemistry of the Martian atmosphere: seasonal, latitudinal, and diurnal variations. Icarus 185:153–170. https://doi.org/10.1016/j.icarus.2006.06.003

Lauro SE, Soldovieri F, Orosei R, Cicchetti A, Cartacci M, Mattei E, Cosciotti B, Di Paolo F, Noschese R, Pettinelli E (2019) Liquid water detection under the south polar layered deposits of Mars—A probabilistic inversion approach. Remote Sensing 11:2445. https://doi.org/10.3390/rs11202445

Lauro SE, Pettinelli E, Caprarelli G, Baniamerian J, Mattei E, Cosciotti B, Stillman DE, Primm KM, Soldovieri F, Orosei R (2022) Using MARSIS signal attenuation to assess the presence of South Polar Layered Deposit subglacial brines. Nat Commun 13:5686. https://doi.org/10.1038/s41467-022-33389-4

Leblanc F, Modolo R, Curry S, Luhmann J, Lillis R, Chaufray JY, Hara T, McFadden JP, Halekas JS, Eparvier F, Larsen D (2017) On Mars' atmospheric sputtering after MAVEN first two years. In: European Planetary Science Congress, vol 11, EPSC2017-58. https://doi.org/10.1002/2018GL077199

Lewis SR, Mulholland DP (2015) The physics of Martian weather and climate: a review. Rep Prog Phys 78:125901

Milliken RE, Bish DL (2010) Sources and sinks of clay minerals on Mars. Philos Mag 90:2293–2308. https://doi.org/10.1080/14786430903575132

Montmessin F, Lefèvre F (2013) Transport-driven formation of a polar ozone layer on Mars. Nat Geosci 6:930–933. https://doi.org/10.1038/ngeo1957

Nakamura Y, Anderson DL (1979) Martian wind activity detected by a seismometer at Viking Lander 2 site. Geophys Res Lett 6:499–502

Ody A, Poulet F, Bibring J-P, Loizeau D, Carter J et al (2013) Global investigation of olivine on Mars: insights into crust and mantle compositions. J Geophys Res Planets 118:234–262. https://doi.org/10.1029/2012JE004149

Ojha L, Lewis K, Karunatillake S, Schmidt M (2018) The Medusae Fossae Formation as the single largest source of dust on Mars. Nat Commun 9:2867. https://doi.org/10.1038/s41467-018-05291-5

Pál BD, Kereszturi Á (2022) Deliquescence probability maps of Mars and key limiting factors using GCM model calculations. Icarus 376:114856. https://doi.org/10.1016/j.icarus.2021.114856

Pollack JB, Colburn D, Kahn R, Hunter J, Van Camp W, Carlston CE, Wolf MR (1977) Properties of aerosols in the Martian atmosphere, as inferred from Viking Lander imaging data. J Geophys Res 82:4479–4496

Qin X, Ren X, Wang X, Liu J, Wu H, Zeng X, Sun Y, Chen Z, Zhang S, Zhang Y, Chen W (2023) Modern water at low latitudes on Mars: Potential evidence from dune surfaces. Sci Adv. https://doi.org/10.1126/sciadv.add8868

Read PL, Lewis SR, Mulholland DP (2015) The physics of Martian weather and climate: a review. Rep Prog Phys 78:125901

Rennó NO, Bos BJ, Catling D, Clark BC, Drube L, Fisher D, Goetz W, Hviid SF, Keller HU, Kok JF, Kounaves SP (2009) Possible physical and thermodynamical evidence for liquid water at the Phoenix landing site. J Geophys Res Planets. https://doi.org/10.1029/2009JE003362

Ruff SW, Campbell KA, Van Kranendonk MJ, Rice MS, Farmer JD (2020) The case for ancient hot springs in Gusev crater, Mars. Astrobiology 20:475–499. https://doi.org/10.1089/ast.2019.2044

Ruzicka A, Snyder GA, Taylor LA (2001) Comparative geochemistry of basalts from the Moon, Earth, HED asteroid, and Mars: implications for the origin of the Moon. Geochim Cosmochim Acta 65:979–997

Scheller EL, Ehlmann BL, Hu R, Adams DJ, Yung YL (2021) Long-term drying of Mars by sequestration of ocean-scale volumes of water in the crust. Science 372:56–62. https://doi.org/10.1126/science.abc7717

Schneider NM, Milby Z, Jain SK, Gérard JC, Soret L, Brain DA, Weber T, Girazian Z, McFadden J, Deighan J, Jakosky BM (2021) Discrete aurora on Mars: insights into their distribution and activity from MAVEN/IUVS observations. J Geophys Res Space Phys. https://doi.org/10.1029/2021JA029495

Sheppard RY, Thorpe MT, Fraeman AA, Fox VK, Milliken RE (2021) Merging perspectives on secondary minerals on mars: a review of ancient water-rock interactions in Gale Crater inferred from orbital and in-situ observations. Minerals 11:986. https://doi.org/10.3390/min11090986

Smith PH, Tamppari LK, Arvidson RE, Bass D, Blaney D, Boynton WV, Carswell A, Catling DC, Clark BC, Duck T, DeJong E (2009) H_2O at the Phoenix landing site. Science 325:58–61. https://doi.org/10.1126/science.1172339

Spiga A, Hinson DP, Madeleine JB, Navarro T, Millour E, Forget F, Montmessin F (2017) Snow precipitation on Mars driven by cloud-induced night-time convection. Nat Geosci 10:652–657. https://doi.org/10.1038/ngeo3008

Stillman DE, Grimm RE (2011) Dielectric signatures of adsorbed and salty liquid water at the Phoenix landing site, Mars. J Geophys Res Planets. https://doi.org/10.1029/2011JE003838

Tarnas JD, Mustard JF, Lin H, Goudge TA, Amador ES, Bramble MS et al (2019) Orbital identification of hydrated silica in Jezero Crater, Mars. Geophys Res Lett 46:12771–12782. https://doi.org/10.1029/2019GL085584

Thomas P, Gierasch PJ (1985) Dust devils on Mars. Science 230:175–177

Váci Z, Agee CB, Herd CD, Walton E, Tschauner O, Ziegler K, Prakapenka VB, Greenberg E, Monique-Thomas S (2020) Hydrous olivine alteration on Mars and Earth. Meteorit Planet Sci 55:1011–1030. https://doi.org/10.1111/maps.13479

Whiteway JA, Komguem L, Dickinson C, Cook C, Illnicki M, Seabrook J, Popovici V, Duck TJ, Davy R, Taylor PA, Pathak J (2009) Mars water-ice clouds and precipitation. Science 325:68–70. https://doi.org/10.1126/science.1172344

Wray JJ (2021) Contemporary liquid water on Mars? Annu Rev Earth Planet Sci 49:141–171. https://doi.org/10.1146/annurev-earth-072420-071823

Chapter 8
Mars Unveiled: Cratering, Volcanism and Aeolian Activity

8.1 Crater Processes

Impact craters were the first definitive features to be identified on Mars in the space age, surprising astronomers and scientists by their discovery by Mariner 4 (Chap. 1). This find placed Mars within the family of almost every other solid surfaced body in the Solar system as possessing craters (Earth and Io's energetic surface processes have actively erased most impact craters from these worlds, Lopes-Gautier et al. 1999; Johnson and Bowling 2014).

Impact craters form when comets or asteroids collide with a solid surface at velocities up to tens of km/sec and exceeding the speed of sound for the underlying regolith (Fig. 8.1 [1]). The initial contact compresses both the front of the projectile and the regolith immediately beneath, accelerating it to significant percentages of the impact velocity. Shockwaves developed by the impact far exceed the cohesive strength of the projectile and underlying material, laterally discharging material in jets. Excavation commences once the shockwave reaches the rear of the projectile, a rapid process lasting a tenth of a second or less. The shockwave is reflected back into the planetary surface, forcing decompression, expansion downward then upward and outward ejection, in some cases material achieving escape velocity to leave Mars (Fig. 8.1 [2, 3]).

In larger impacts, ejected material may form secondary craters of their own, as high velocity objects fall back in a ballistic trajectory to the surface. This crater ejecta grows along with the crater, and is deposited on the surrounding terrain. Most impact shockwaves are sufficient to melt the impactor, and in large enough collisions large portions of the target surface. Eventually the shockwave pressure forces fall below that which is needed to sustain ejection of materials and the crater ceases to grow.

In larger craters the centre may rebound to form a central peak (Fig. 8.1 [4]). The completed feature generally possesses pulverised rock (ejecta deposits (Fig. 8.1 [5]), and breccia that has flowed to the centre of the crater (Fig. 8.1 [6]), and localised marginal slumping of the inner crater walls (Fig. 8.1 [7]).

S. Hobbs, *Sensing Mars: Remote Sensing to Understand the Red Planet*,
https://doi.org/10.1007/978-3-032-24416-1_8

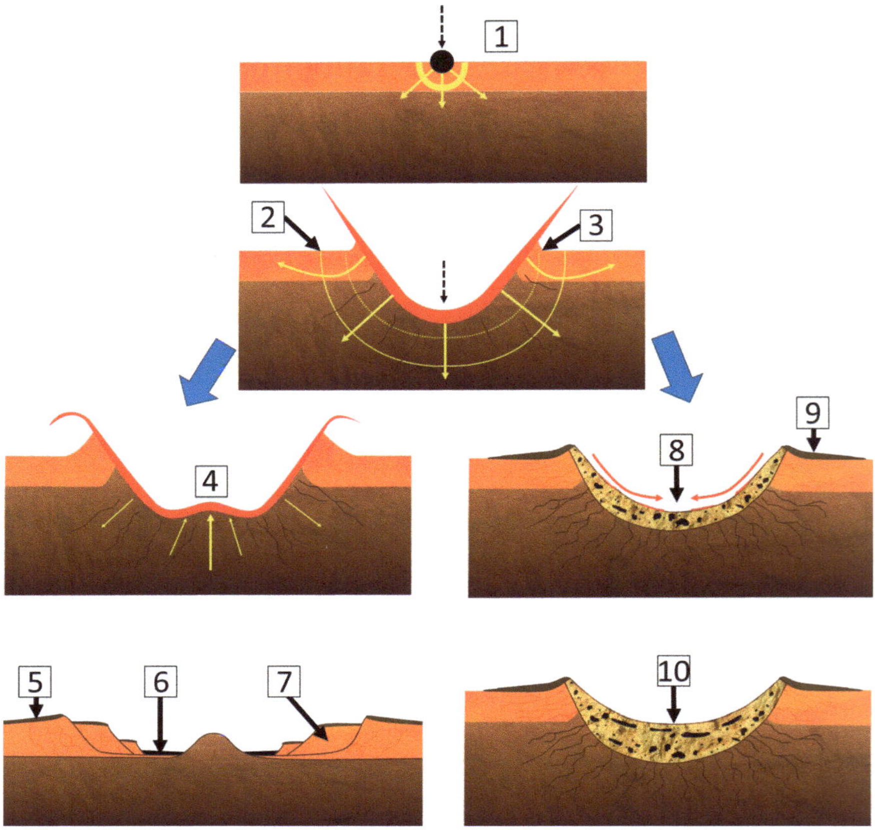

Fig. 8.1 Crater formation process on Mars. (1) An object strikes the surface creating shockwaves. The shockwave and surface response causes (2) compression and (3) rarefaction waves that excavate, shatter and shape the surface. In complex crater formation the centre of the crater rebounds (4) forming a central peak. The completed crater possesses deposited ejecta (5), and solidified breccia melt (6). Subsequent alteration through marginal collapse of the inner walls (7) may reshape the crater. In simple crater formation, breccia material flow downslope (8) and ejecta deposit around the crater rim (9). The final simple crater consists of an uplifted rim, breccia deposits (10) and fractured subsurface rock

Smaller craters do not possess a central peak and the molten breccia flow back down toward the centre of the crater to form a bowl-like feature, and ejecta depositing at smaller distances from the crater rim (Fig. 8.1 [8, 9]). The completed simple crater with solidified ejecta deposits (Fig. 8.1 [10]) may remain stable in this state.

The transient width and depth morphology for larger craters is not stable. The centre of larger craters resists compression and moves back to form a central peak. Subsequent slumping activity of the crater rim and peak transition the impact to their final morphology. Residual heat formed by the impact cools over time, and may cause additional slumping and morphological changes.

Studies of the interior of craters and ejecta blankets are of interest to researchers as impacts provide a "free' method of excavating tonnes of Martian regolith to reveal a view of the subsurface. The Spirit and Opportunity Mars Rovers were tasked to visit craters within their traverses for this reason, analysing Martian layers exposed by the impacts (Arvidson et al. 2011). Martian crater erosion and ejecta patterns are different from those of the Moon or other airless bodies (e.g. Fig. 8.2a, b). Lunar ejecta smoothly blankets the surrounding regolith, with radial ridges or grooves extending approximately one crater radius from the rim. Loops or radial series of secondary craters follow this region, with the lack of atmosphere causing all lofted material to follow ballistic trajectories. Ejecta rays can form bright features such as rays that darken over exposure to solar radiation and erosion by micrometeorites; Copernicus Crater being a prominent example (Fig. 8.2c, Pieters et al. 1985).

Diameter and depth of excavation for complex Martian craters are related by (Boyce et al. 2005):

$$d = aD^b. \tag{8.1}$$

where:

- d = excavation depth;
- D = crater diameter;
- a = constant;
- b = constant.

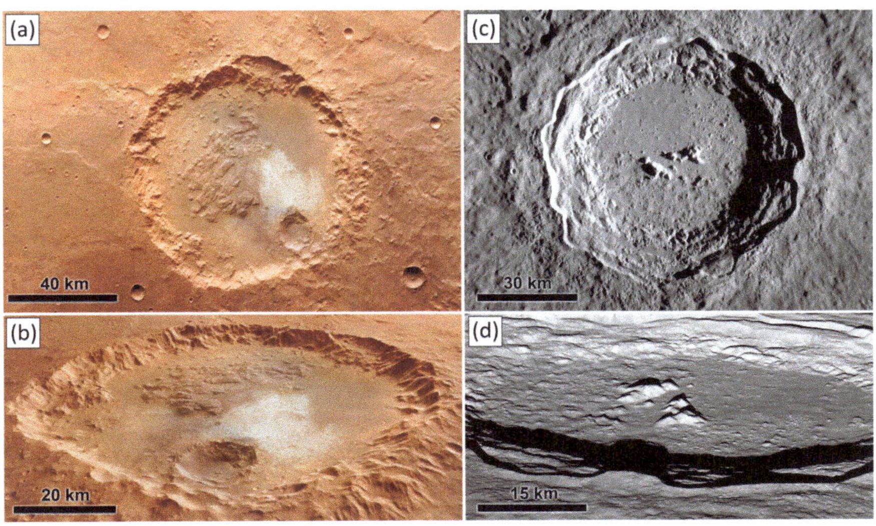

Fig. 8.2 **a** A 70 km diameter crater in Erythraeum Chaos exhibiting eroded, slumped walls and exposure of light toned deposits (ESA/DLR/FU Berlin, CC BY-SA 3.0 IGO). **b** Perspective view of the same crater (ESA/DLR/FU Berlin, CC BY-SA 3.0 IGO). **c** Copernicus Crater and surrounding ejecta on the Moon. What little erosion has occurred since formation is driven by solar darkening and micrometeoroid impacts (NASA/JPL). **d** Perspective view of Copernicus Crater (NASA/JPL)

Values for "a" and "b" are dependent on crater type and degradation and may vary between 0.32–0.35 and 0.49–0.54 respectively (Garvin et al. 2003; Tornabene et al. 2018).

The more diverse Martian geology is reflective of the nature of considerable variations between Martian crater ejecta. One example is the rampart crater, where ejecta surrounding the impact appears fluidised, ending in lobate terminations (e.g. Fig. 8.3a, b). These craters do not exist on the Moon and have attracted interest by through postulations that ejecta may have contained volatiles, such as water ice. This is suggested by flow around obstacles at greater distances than would be expected from dry material (Mouginis-Mark and Baloga 2006). Fluidized shearing moves larger particles to the front of the ejecta flow, increasing localised friction, slowing down and forming a rampart. Solidification of the ejecta occurs when the meltwater heated from the impact sublimes, leaving the solid debris behind. Smaller rampart craters occur at higher latitudes, where ice-rich material residing closer to the surface may have been fluidized by lower energy impacts. Additionally, ejecta mobility from higher-latitude rampart craters appears higher than that of those nearer the equator. This suggests that the increased abundance of ice-rich material provided greater ejecta fluidity. Conversely, larger rampart craters may have required more energy to excavate deeper volatiles for the lobate ejecta to form (Weiss and Head 2017).

Rampart craters may have single (Fig. 8.3c), double (Fig. 8.3d) or multiple ejecta layers. Single layer features are more commonly located near the Martian equator, average 10 km in diameter with ejecta terminating 1–1.5 crater radii from the rim. A water-based interpretation of single layer rampart crater formation is that the impactor failed to entirely penetrate through an ice-rich subsurface layer. Double or multiple rampart craters may form when a higher energy impactor was able to completely pass through the ice-rich layer to penetrate rocky material beneath. Thinner ice-rich layers near the Martian equator are suggested by a higher concentrations multiple rampart craters compared to higher latitude areas (Weiss and Head 2017).

Atmospheric interactions have been modelled as a potential placement mechanism for rampart craters, where intense winds post-impact generated vortices would emplace fine-grained material. In addition, localised heating from the impact thermal pulse would cause localised volatile melting, resulting in an erosion resistant surface from which a pedestal crater could form (Wrobel and Crawford 2006). Similar craters discovered on ice-rich, though airless bodies such as Charon and Ganymede but not on ice-poor worlds such as the Moon or Mercury suggest a greater influence of ice-rich material than atmosphere in the formation of these features (Barlow and Robbins 2015). The competing roles of subsurface and atmospheric volatile movement remains an active research area (Niu et al. 2022).

Measurements of rampart crater width/depth ratios (depth of approximately 0.1 crater diameter), and comparing the differences between single (incomplete penetration) and multiple (full penetration) have provided some estimates of subsurface ice-rich layer (cryosphere) thickness range from 1.3 to 3.3 km from equator to poles respectively (Weiss and Head 2017). It should be noted that the Mars Insight lander, designed to investigate the planet's subsurface, failed to find evidence of a substantial

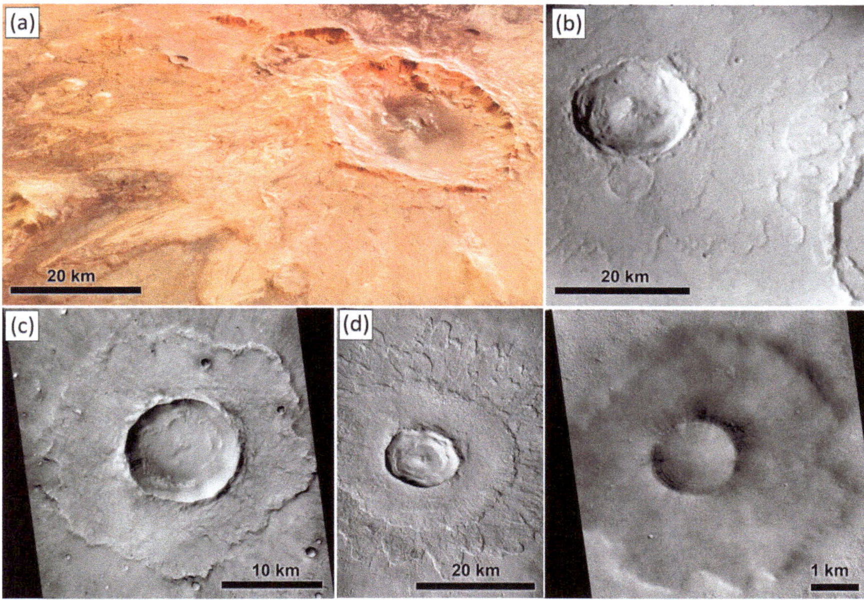

Fig. 8.3 **a** Example of rampart crater impact into fluidised debris north of Hellas Basin (ESA/DLR/FU Berlin, CC BY-SA 3.0 IGO). **b** Yuty rampart crater showing similar fluidised ejecta (Viking 3A07, NASA/JPL). **c** Punsk rampart crater in the Martian northern hemisphere showing a Single ejecta blanket (B21_017766_2007_XN_20N04, NASA/JPL). **d** Steinheim Crater located in Arcadia Planitia is an example showing two distinct ejecta lobes (G03_019221_2327_XN_52N169W NASA/JPL). **e** Pedestal crater in Casius quadrangle (ESP_046487_2345, NASA/JPL)

cryosphere at its landing site to a depth of 300 m. This suggests that current subsurface ice and aquifers may be disconnected and more localised in the Martian crust (Kilburn et al. 2022). Additional research has also suggested temporal cryosphere changes.

Remote sensing analysis of single and multiple ejecta rampart craters in Thaumasia Minor, near the Martian equator, was conducted to calculate excavation depths and infer possible past cryosphere thickness. This work suggested the Martian cryosphere has evolved over time, becoming thinner from the Mid (2 Ga) to later Amazonian periods (0.5 Ga, Dasgupta et al. 2022). Other evidence exists for the presence of a subsurface Martian cryosphere and is covered in more detail in Chap. 9.

Pedestal craters are small (0.5–200 m) impacts located in the lower Martian latitudes with ejecta sitting 20–200 m above the surrounding surface (e.g. Figure 8.3e). Related features are excess ejecta and perched (shallower impact bowl, sometimes filled with superposed material) craters, with larger diameters than pedestal craters (Barlow 2006). These craters have likely been modified through removal of softer surrounding terrain by wind action. Like rampart craters, pedestal craters are believed to have formed through impacts into ice-rich material, with rocky debris and dissolved salts deposited post-impact protecting the crater and ejecta from subsequent erosion.

Analysis of pedestal height suggests that the thickness of the underlying ice-rich material, and not the impact itself, apart from depositing protective rocky material, that determines the degree of inverted relief and hence pedestal height (Pain et al. 2007; Kadish et al. 2010). The ice-rich hypothesis is supported by latitude dependency on crater morphology, where ice is deposited during high obliquity periods, and subsequently altered during the impact. Proximity of pedestals and their placement on surfaces of different ages suggest multiple episodes of mid-latitude ice-rich material deposition and removal (Kadish et al. 2010). Inferences of the amount of ice-rich material has been supported through using ejecta mobility (EM) as a method for quantifying ejecta runout distance. It is a ratio determined by:

$$\text{EM} = \text{average extent of continuous ejecta layer/crater radius} \quad (8.2)$$

Average EM is reasonably consistent across different geologic terrain but increases with latitude (1.43 near the equator and 2.0 between 40° and 50° N, Barlow et al. 2000). Recent research conducted in the Utopia-Isidis Planitia and southern Chryse Planitia suggest regional dependence on past presence of ice-rich material that affects EM ratios. Measured EM ratios in Utopia-Isidis Planitia were consistent, however values in southern Chryse Planitia decreased over time (Niu et al. 2022). This work highlights that importance of regional context in influencing cratering processes, and that the planet has experienced a complex landscape evolution to the present day.

8.1.1 Degradation

Like Earth, Mars exhibits abundant evidence of craters degrading, through to complete erasure, over time (Fig. 8.4a, b). A combination of factors, such as fluvial erosion from rainfall or subsurface flow, plus vegetation or mass wasting serve to degrade terrestrial craters (e.g. Grant and Schultz 1993). Martian craters are also subject to geologic processes occurring after impact include volcanic, fluvial aeolian or ice-related activity. Craters on Tharsis, Syrtis and Hesperia volcanic provinces, have been subject to substantial modification and removal from lava flow-based activity (Breton et al. 2022). Studying the nature of crater degradation has been an important indication of past processes acting on Mars, as the age and amount of erosion provide indicators of the strength of these processes occurring in Mars' past (Weiss and Head 2015).

The type and degree of crater modification is subject to local, as well as global, climate and geology. Hesperian and Amazonian-aged craters residing near the equator have mostly experienced little modification except for those near the MFF. The readily erodible deposit has infilled craters, with wind-blown deposits (Carr

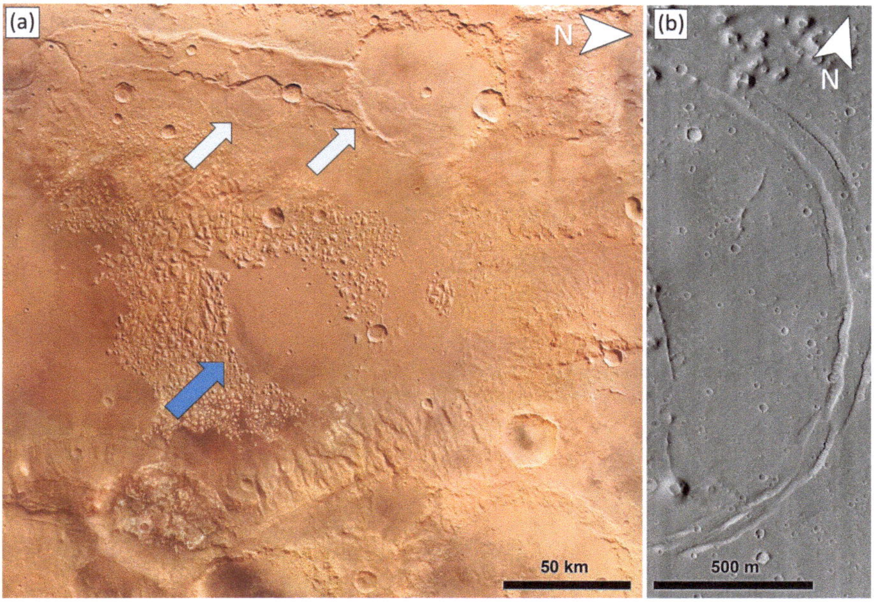

Fig. 8.4 **a** The Sirenum Terrae region in the Martian southern hemisphere possesses heavily degraded craters (white arrows). Atlantis Basin, an impact feature formed during the early in Mars history has almost been completely erased (ESA/DLR/FU Berlin, CC BY-SA 3.0 IGO). **b** HiRISE example of a ghost crater where only the outline of the rim is visible (ESP069243-2030, NASA/JPL)

2006). Partial removal of these deposits has left behind pedestal craters, while post-Noachian mid-latitude modification has also generated pedestal craters as described above.

Craters on Mars' oldest terrains have experienced the most significant erosion, with the removal of ejecta, internal features such as wall terraces and central peak, leaving shallow, flat-floored features with eroded rim (Quantin-Nataf et al. 2019, Fig. 8.4a). It should be noted that smaller craters are degraded faster than larger ones, being more prone to be entirely erased, having implications for surface dating.

So-called "ghost craters" are present in this terrain and represent the last stages of modification, where the central depression is almost completely infilled and only the rim outline remains (Fig. 8.4b). Flat-floored craters may be 1.6–5 times shallower than uneroded impacts, with older Noachian craters being deeper than those on younger Noachian terrain (Weiss and Head 2015). Some modelling on degradation has suggested rainfall-based erosion, pointing to the diffuse nature of the erosion, while interactions of ice in an early cold-wet Mars scenario may also produce craters similar to those observed by remote sensing (Craddock and Howard 2002; Palumbo et al. 2020).

The rate of erosion declined from 3.8 to 3 Ga on transition from the Noachian to the Hesperian, suggesting the strength of erosion also reduced over this period in response to a change in Martian climate to colder, arid conditions (Quantin-Nataf et al. 2019). This corresponds to a period of reduction of fluvial erosion either from postulated warm, wet rainfall or cold, meltwater-based erosion, as evidenced by the presence of other water-eroded features and hydrated minerals in this older terrain (see Chap. 9 for more detail on Martian fluvial processes).

8.1.2 Crater Count Dating

Impact crater analysis also provides essential information on estimating the age of a planetary surface, including Mars. In absence of direct radiological methods, crater counts and stratigraphy remain the only two methods for dating features on Mars. The older the surface, the longer it has been subject to impacts. This normally results in older surfaces possessing more craters; the ancient and heavily cratered Noachis Terra terrain being a prime example. The rate of cratering has not been constant throughout Martian history, and post-impact processes have modified or erased previous impacts, complicating dating through crater counting.

Higher cratering rates of 100–500 times the present rate occurred early in Mars' history, likely caused by material left over from terrestrial planet formation from the Solar nebula (Fig. 8.5). This period occurred at least when Mars' surface was solid enough to preserve the impacts and evidence of the event is also observed on Mercury and the Moon (Hartmann 2019). The duration of this period is estimated to have extended between 4.2–3.5 Ga, spanning the Noachian period and forming a major crater population in Mars' past. A second, post-high bombardment crater population has formed from predominantly asteroid and some cometary impacts (Strom et al. 1992).

From the above periods, crater statistical analysis, greatly enhanced through machine learning techniques (Chap. 6), are used to infer age through crater size frequency distributions. Radiometric calibration of crater age dating has occurred on the Moon, providing a degree or reference for Mars (Benedix et al. 2020).

Crater size frequency analysis has indicated an inverse power law between the size of the crater (measured by diameter D) and the number of times, N, at which it occurs (Hartmann 1981):

$$N = aD^{-b} \tag{8.3}$$

where a and b are Y intercepts (higher values equate to older surfaces) and trendline slope respectively. From early bombardment periods above, values for b are not constant, with large craters sharply decreasing post this period, and steepening below D of 1 km, potentially representing increased abundance of smaller impactors or secondary craters.

Figure 8.6a shows isochron trendlines which are intended to accommodate for the Y intercept (Formula 8.3) and trendline slope changes as described above (Hartmann

Fig. 8.5 Cratering rates, likely caused by an abundance of comet and asteroid debris left over from the formation of the Solar System, were higher for early Mars than today, as shown in this illustration

2005). As an example of crater counting usage, Fig. 8.6b illustrates a count of 30 craters of average diameter of 14 km (blue rings) over Mawrth Vallis, plus six craters of 62 km diameter (green rings). Each crater count and size yields 10^{-4} and 10^{-5} craters/km^2 respectively (blue and green triangles, Fig. 8.6a). These values provide an age estimate of the Noachian period for this area of Mars. Higher resolution imagery focussing on specific features within this landscape has provided more specific detail on its evolution. (e.g. Loizeau et al. 2012).

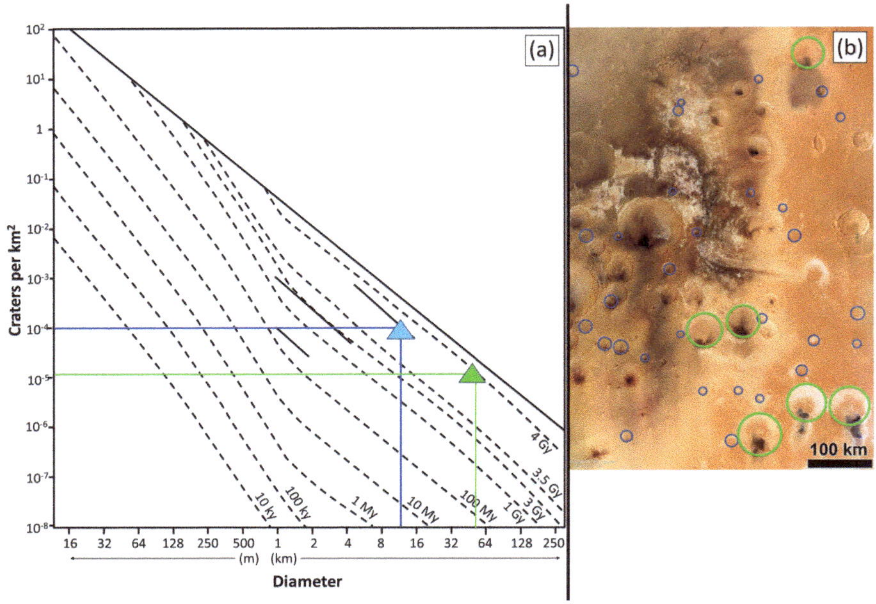

Fig. 8.6 **a** Crater counting isochron diagram with result from Mawrth Vallis example **b** (ESA/DLR/FU Berlin, CC BY-SA 3.0 IGO)

Improved remote sensors have increased image resolution, allowing smaller craters to be detected. The latter attribute is particularly important as crater formation can often produce secondary impacts. These generate additional craters that may falsely increase the apparent age of the surface through incorrect inference of extended impact rates. The ability to identify smaller craters as primary or secondary impacts can reduce falsely increase age estimates (McEwen and Bierhaus 2006) and will provide a better estimate for "b" (Formula 8.3 above).

Impact cratering is also occurring on Mars today. Decades of continuous orbital observations have confirmed Mars is subject to impacts of 1.65×10^{-6} impacts at least 3.9 m in diameter for each square kilometre of the planet's surface per year (Daubar et al. 2013). As with many crater impacts throughout Martian history, clusters of contemporary craters may form when the impactor breaks apart during travel through the Martian atmosphere. The thickness of the Martian atmosphere helps govern the frequency of fragmentation; higher pressures provide more resisting force and encourage breakup.

As impact crater discoveries are subject to the collection areas and frequencies of the orbital remote sensor, observation bias may be a factor in the predicted versus observed frequency discrepancy. High imaging rates over an area by a spacecraft leads to greater chance of contemporary impact detection. In addition, craters are easy to detect if occurring in visibly bright, loosely consolidated areas where high-contrast ejecta rays are produced. Impacts occurring in rocky terrain often lack the prominent ejecta, making them harder to identify in imagery analysis. Techniques using machine learning are working to remove these biases in observation data to provide a more complete catalogue of contemporary impacts. CNN training development on a large CTX images increased classifier confidence and detected fresh impacts in rocky terrain not readily apparent to human observation (Wagstaff et al. 2022).

8.2 Volcanism

Volcanic activity has featured extensively throughout the Solar System, from Mercury, the innermost planet to the dwarf planet Pluto. That Mars possessed volcanoes was only identified during the Mariner 9 mission, when four dark splotches appeared above a globe-circling Martian dust storm in 1971. These were identified as the peaks of four large volcanoes, the biggest, Olympus Mons, remains the largest volcano discovered in the Solar System (Chap. 1).

Basalt occurs across much of the Martian surface, found in extensive lava flows spanning multiple time periods, and lava plains. Up to 60% of the Martian surface has been modified through volcanism. Volcanic features are present from the oldest Noachian terrain (3.7 Ga or older), with a trend of decreasing volcanic activity until the youngest lava flows dated to the late Amazonian, < 500 Ma ago (Werner 2009). Present day volcanism may be possible on Mars, though has yet to be confirmed.

Analysis and interpretation of Martian volcanic activity has relied on knowledge gained from terrestrial volcanoes. On Earth, studies of features analogous to those found on Mars, such as Hawaii, have been conducted far beyond what is possible with current generations of robotic missions (Hughes et al. 2019). Like Earth, Martian volcanism was likely driven by thermal energy left over from planetary formation being transferred by convection to the planet's surface. This energy produced lower density magma that rose until it either reached the density of the host rock, forming a subsurface magma chamber, or continued to the surface as an eruption or outflow (Greeley et al. 1990).

The eruption may be explosive (examples being Arabia Terra mounds, Fig. 8.7a), since it is usually caused by gasses being released from the magma, or are effusive, where molten rock in the form of lava flows on the surface. Explosive pyroclastic flows can release tonnes of ash and cinders into the atmosphere and on the surface. Pyroclastic flow activity is believed to responsible for a series of Noachian highland volcanoes that possess easily erodible and shallow flanks (Hadriacus Mons, Apollinaris Patera, Fig. 8.7b and c, Williams et al. 2007). The ease of erosion of these features is believed to originate from loosely consolidated airfall and pyroclastic deposits, such as those found on Mt St Helenes on Earth (Papale and Dobran 1994).

Planetary remote sensing has suggested differences in volcanic activity between Mars and Earth, generated by differences in their evolution. Reduced gravity on Mars is thought to enable growth of larger, deeper magma chambers due to magma's reduced buoyancy. These feed large scale eruptions with longer effusive lava runouts

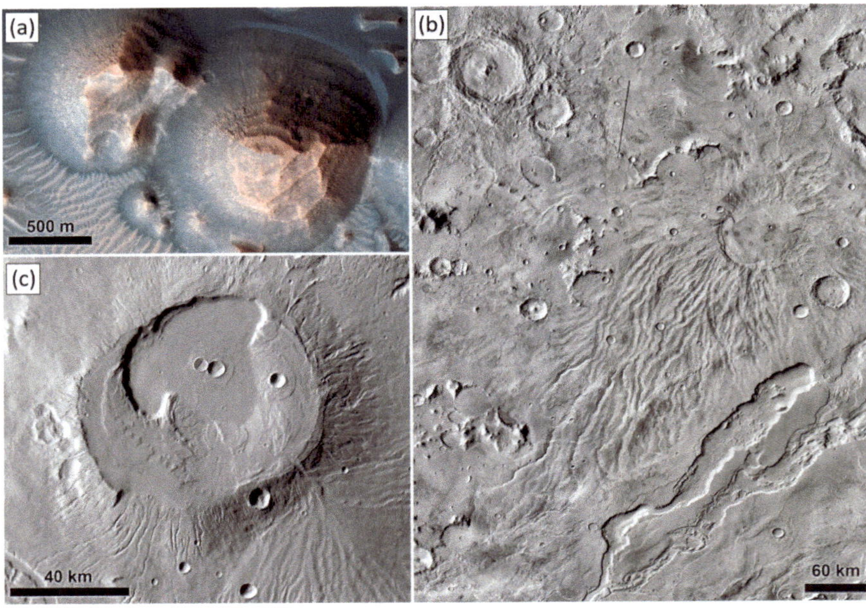

Fig. 8.7 **a** Layered rock mounds in Arabia Terra, formed as the result of explosive super volcanism (PIA24383, NASA/JPL). **b** Hadriacus Mons, a heavily eroded volcano (THEMIS VIS, NASA/JPL). **c** The 80 km wide caldera of Apollinaris Patera whose flanks consist of readily erodible material, likely ash fall (THEMIS VIS, NASA/JPL)

(Brož et al. 2021). In addition, gravity-induced crustal compression that limits mountain growth on Earth is reduced for Mars, in part enabling the growth of the Olympus Mons shield volcano to be 2.5 times higher than Mount Everest.

A lack of tectonic plate drift on Mars also facilitates growth in place of volcanoes that remain in position over their respective magma chambers. In contrast, most terrestrial volcanoes occur at the intersection between tectonic plates, forming linear chains along the plate boundaries. An example of this is Hawaii, where plate movement has shaped volcanic activity into the formation of an island chain while restricting growth as the magma source is moved (Hughes et al. 2019).

Tharsis Montes hosts the most obvious examples of this activity on Mars, with shield volcanoes Olympus Mons, Ascraeus Mons, Pavonis Mons, and Arsia Mons (illustration, Fig. 8.8). This region represents the highest elevation on the planet, and is believed to have formed through tectonic uplift and accumulation of lava flows from a region of anomalously high temperature magma, a "hotspot" (Black and Manga 2016; Mouginis-Mark et al. 2022). The Tharsis bulge dates from the end of the Noachian (3.7 Ga), with the massive volume of material causing extensive grabens and rift valleys. These were caused through the weight of volcanic deposits overloading the crust and causing failures, consisting of lowered elongated blocks of crust between two faults. Multiple lava resurfacing events have occurred since

Fig. 8.8 Illustration of the Tharsis volcanoes, the largest of which, Olympus Mons, is in the foreground

then to shape Tharsis to its present form. The lengths and volumes of the numerous lava flows in this region (up to 310 km long, 440 km^3 volume) are much larger than terrestrial equivalents. This is likely due to the region being active for long periods as opposed to exotic composition or abnormally high eruption rates (Peters et al. 2021).

Much of the past Martian volcanism was produced by low viscosity lava flows, with basaltic rocks forming flat lava plains. Shield volcanoes, such as those in Tharsis Montes, are produced through these types of flows, where the low viscosity prevents development of slopes steeper than 5° (Hauber et al. 2009). The presence of gasses or silicates, plus variation in crystallisation within magma may increase flow viscosity. This is thought to have occurred during later stage eruptions from differentiation within volcanic magma chambers (Chevrel 2013).

Lava tubes have been found on Mars, being up to three orders of magnitude larger than those found on Earth, though likely created through analogous processes of a longer-term lava flow (weeks to months) experiencing outer cooling and solidification. They may also form through lava generating tributary flows beneath a solid crust. The collapse of lava tubes may form sinuous rills, related to similar features found on the Moon. Lava tubes have been identified near Pavonis Mons, Arsia mons and Alba Mons (Fig. 8.9a, b, Léveillé and Datta 2010).

Volcanism has also been postulated by some researchers to be a dominant process for shaping the Martian surface, arguing against significant role of past or present water action. There is an abundance of volcanically-derived minerals such as olivine (Table 7.3) at many regions where geomorphology suggests past water action (Stopar et al. 2006). In addition, low-viscosity lava flows have been suggested to act like

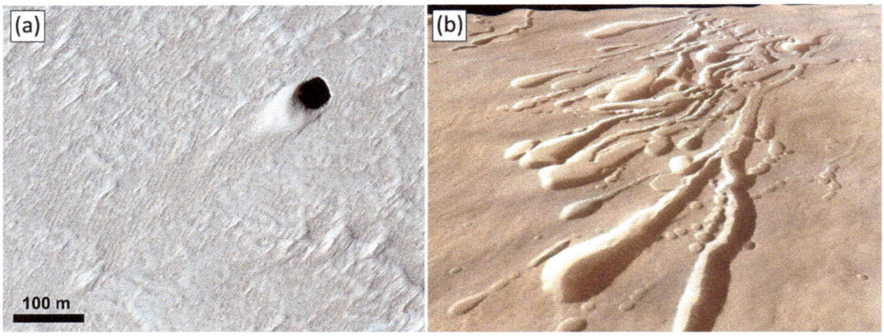

Fig. 8.9 **a** Collapsed roof of a lava tube in the Arsia Mons region imaged by HiRISE (PIA 24149, NASA/JPL). **b** Exposed lava tubes, ranging from less than 280 m to 1.9 km wide, on the flanks of Pavonis Mons (ESA/DLR/FU Berlin, CC BY-SA 3.0 IGO)

water (Leone 2014), carving meandering channels such as those situated at Tharsis (Fig. 8.10).

A more complex interaction of multiple processes is also consistent with lava emplacement and mineralogy found on the planet (described above and in Table 7.3). This is suggested through studies of features believed to have formed through interactions between Martian magma and water. Outgassing of water from magma flows, particularly through the heightened volcanic activity during the Hesperian period, have led to the formation of outflow channels and other fluvial features (Perron et al. 2007). Mapping of thorium, a trace element believed to exist in magma containing

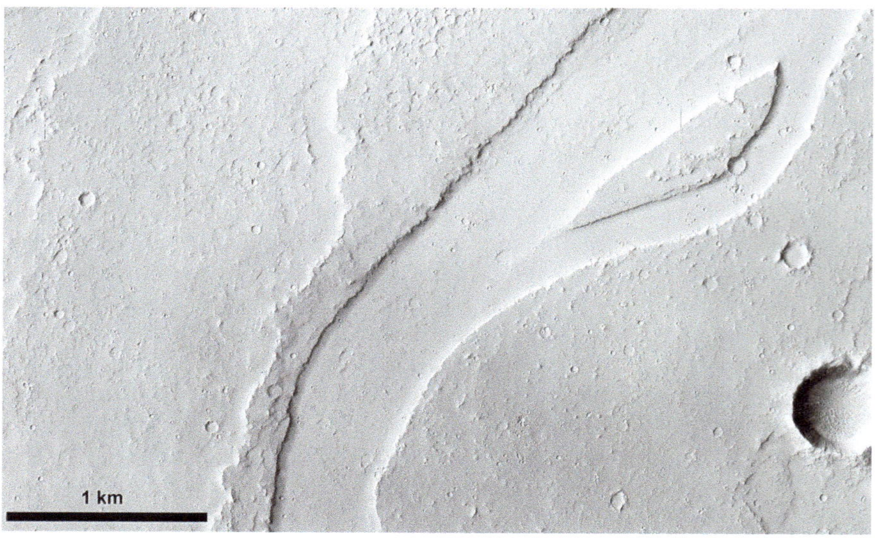

Fig. 8.10 Smooth, flat-bottomed channel within a lava flow at Tharsis (PIA21104, NASA/JPL)

water but left behind during a volcanic release, also supports the past water content within some Martian lavas (Black et al. 2022). Geologic examples are volcanic rootless cones found throughout the northern lowlands, thought to have formed through explosive interactions between magma and subsurface water (Hamilton et al. 2011).

Analysis of channels east of Olympus Mons, particularly Olympica Fossae, provided further support to a complex erosional history of water and lava (Fig. 8.11a, b). The features exhibit a mixture of erosional characteristics typical of fluid flow, such as sinuous channels and streamlined forms, along with pits and sharp walled valleys more typical of localised faulting and tectonic activity. Research has suggested this area was modified by a combination of fluidised lava, with viscosity low enough to carve the features observed, as well as liquid water. Stratigraphic analysis suggests original channel erosion by water, and subsequently used to transport later lava flows (Hargitai and Gulick 2018).

Cerberus Fossae, a long fracture system located on lowland plains within Elysium Planitia and 11.5° north of the equator, is another prominent example of lava/water interaction (Fig. 8.12a, b). This region possesses the most recent volcanic activity on Mars with a small eruption occurring 53–210,000 years ago (Hovarth et al. 2021). A hypothesised mantle plume has likely uplifted the terrain in this region, forming extensional fissures and leading to catastrophic lava and subsurface water release and erosion. The recent activity at Cerberus Fossae, supported by Mars Insight seismic activity findings, also suggests Mars is much more volcanically active than once thought (Broquet and Andrews-Hanna 2023).

This is also suggested from seismic data returned by the Mars Insight lander that operated at Elysium Planitia and was equipped with an instrument buried into the Martian regolith designed to detect Mars quakes. The spacecraft was sent to determine the interior structure of Mars, with an open question as to whether the planet's core remained liquid or had lost most of its thermal energy and solidified. Analysis of seismic waves that passed from the far side of the planet to Insight's location revealed a 1780–1810 km diameter liquid core consisting primarily of iron, with sulphur, oxygen, hydrogen and carbon. The core is surrounded by a rocky mantle up to 1880 km thick, overlaid by a 24–72 km thick crust with a deep lithosphere of up to 500 km (Stähler et al. 2021). The remnant interior thermal energy of the planet may possibly power contemporary volcanism, and likely generate many present-day Mars-quakes, whose seismic waves may reflect off the inner core (Fig. 8.13a). Mars quakes were detected by Insight throughout its mission, the largest being magnitude 4.7 (Fig. 8.13b) and believed to have formed through crustal movement (Fernando et al. 2023; Kawamura et al. 2023). Most of these Mars quakes originated from the Cerberus Fossae region, supporting theories of this potentially remaining an active site to current times.

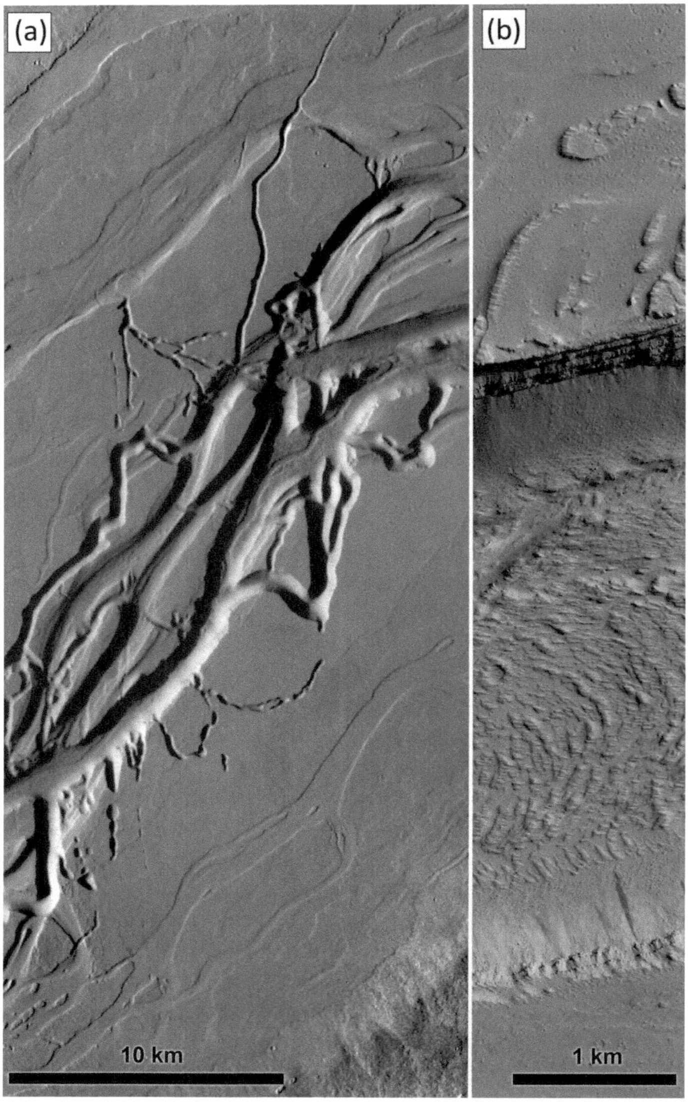

Fig. 8.11 **a** Olympica Fossae channels showing multiple erosive processes (PIA20782, NASA/JPL). **b** HiRISE closeup of the channel northeast of part (**a**) showing layered walls (PSP_003528_2050, NASA/JPL)

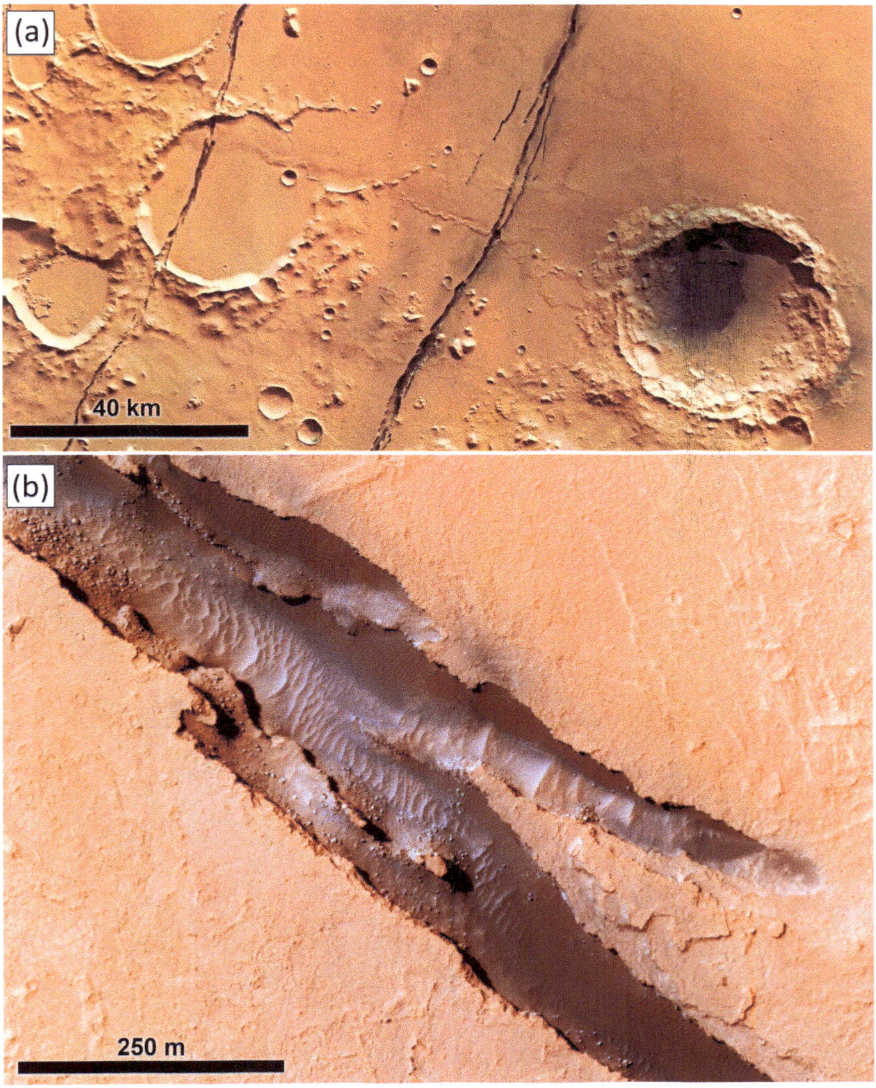

Fig. 8.12 **a** Mars Express view of the fractured terrain of Cerberus Fossae (ESA/DLR/FU Berlin, CC BY-SA 3.0 IGO). **b** HiRISE closeup view of a branching lava channel partially infilled with sand (PSP_009003_1890, NASA/JPL)

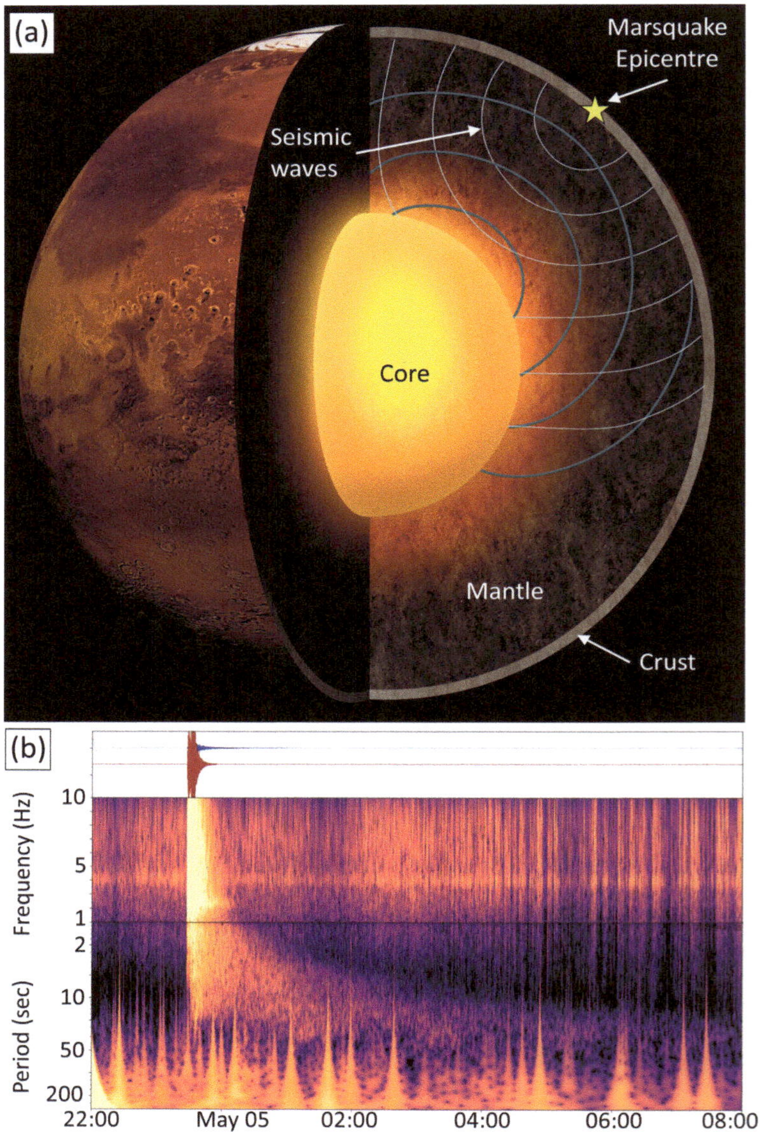

Fig. 8.13 **a** Hypothesised view of seismic waves travelling through the Martian interior and reflecting back to the crust. **b** On 4 May 2022, the InSight lander detected a magnitude 4.7 Mars quake, the largest quake ever observed off Earth (InSight, NASA/JPL)

8.2.1 Mud Volcanism

Remote sensing research has suggested mud volcanism occurring on Mars within the past 100 Ma. On Earth, mud volcanoes differ from their lava-emanating counterparts by releasing pressurised fluid-rich sediment to the surface (Mazzini and Etiope 2017). Pitted cones and knob-like protrusions located in multiple regions of Mars, including Utopia Planitia, Valles Marineris, Acidalia Planitia, Arabia Terra and others possess morphology attributed to mud volcanic activity (Fig. 8.14a, b, Dapremont and Wray 2021). Geologic settings for likely Martian mud volcanism are fissure or fracture related, or from sedimentary basins. Thermal heating, seismic activity or other processes where volatile-rich material is compressed to produce a mud eruption are proposed formation mechanisms (Skinner and Mazzini 2009). Laboratory studies have suggested the low atmospheric pressures solidify the outer sides of the mud eruption, while the centre remains liquid and able to flow. This allows the possibility of fluidised sediment transport in Amazonian periods, which would occasionally break through solid mud to form lobate terminations (Brož et al. 2020).

In the absence of direct surface measurements and sampling at the likely mud volcano sites, both multispectral and hyperspectral analysis from HiRISE and CRISM (Chap. 4) were employed to analyse these features. CRISM analysis identified water hydration spectra correlating with the location of the likely mud volcanoes near Valles Marineris, Acidalia/Chris Planitia, Arabia Terra and Terra Sirenum, but was unable to positively identify specific minerals such as carbonates, phyllosilicates or sulphates associated with terrestrial mud volcanism (Dapremont and Wray 2021).

As noted in Chap. 4, hyperspectral analysis is traditionally used for detailed mineralogical discrimination, though multispectral band ratios have also provided useful insight into material discrimination. This latter technique is particularly useful if there are limitations in hyperspectral coverage or spatial resolution, or if further datasets may reduce uncertainty in previous analysis. This was the case for recent mud volcano research where HiRISE imagery used band ratios for analysis. Results supplemented the CRISM data, suggesting association with past water-based activity (Dapremont and Wray 2023). Future surface missions, guided by the remote sensing research performed above, may be required to confirm the existence of mud volcanism on Mars.

8.2.2 Valles Marineris, Planet-Scale Faulting and Relationship to Volcanism

Valles Marineris, discovered by the Mariner 9 mission, is the largest canyon system in the Solar System (Fig. 7.3). The 4000 km long and up to 11 km deep system occupies one quarter of the planet's circumference. It commences with a series of

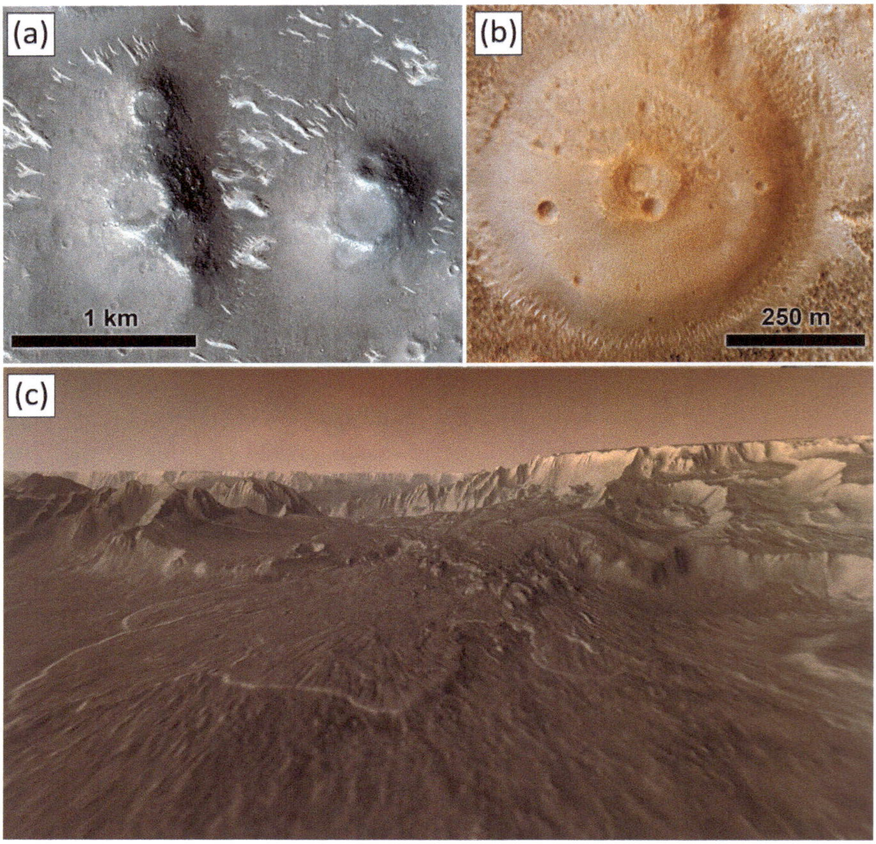

Fig. 8.14 **a** Possible ~ 100 m wide mud volcano on the Martian northern plains (PIA25898, NASA/JPL). **b** Candidate mud volcano in Utopia Planitia (histogram-stretched, PSP_008561_2205, NASA/JPL). **c** Perspective view of landslide alteration in Ganges Chasma, eastern Valles Marineris (PIA02894, NASA/JPL)

networked valleys east of the Tharsis volcanoes (Noctis Labyrinthus) and terminates in a series of chaotic terrains and outflow features that connect with the northern plains. Valles Marineris was probably formed by rift faults, originating as a direct result of crustal stresses caused by Tharsis. Elevation of Tharsis likely caused fracture faults and sinkage through the Martian crust being unable to sustain the weight of the volcanic material added to it. This faulting is postulated to have liberated trapped subsurface water that further undermined and enlarged the fracture zone. Evidence for this water flow is exhibited in Kasei Valles to the north and outflow valleys to the east of Valles Marineris (detailed in Chap. 9). Sub troughs thought to have formed through this phase linked together to form the complete valley system by the late Hesperian (3.7–3.0 Ga, Chap. 1). Since its formation, Valles Marineris was modified through slope failure, including landslides (Fig. 8.14c). Many of these landslides may have occurred through seismic activity destabilising the system's wall slopes to the point of failure (Brunetti et al. 2014).

8.2.3 Case Study Alba Mons—A Unique Volcano on Mars

Alba Mons (Fig. 8.15a) is on the boundary between Hesperian and Amazonian geology. Possessing a variety of volcanic-related morphology, it is an example of a distinctive Martian volcano unique to the planet and inner Solar System. Alba Mons has some of the longest long lava flows found on Mars, spreading up to 1350 km and generating very low slopes (average 0.5°) to a 200 m deep caldera near the centre (Schneeberger and Pieri 1991). The long runouts and low slopes indicated the lava flows were very fluid, able to create morphologies 350 km beyond the central edifice and caldera of the volcano where there are very low and reverse slopes.

Impacts and Tectonic Influence

Alba Mons also resides on the opposite side of the planet to Hellas Basin, the massive impact structure in the Southern highlands that occurred 4.1–3.8 Ga ago (Barlow 2008). The size and energy of the impact would have generated shockwaves that travelled around and through the planet, coming to a focus at the antipode, the opposite side of Hellas. A prominent example of antipodal modifications in the Solar System are the Badlands of hummocky hills opposite the Caloris Basin impact on Mercury, formed by shockwaves generated by the force of the impact (Watts et al. 1991).

Seismic modelling was used to analyse the antipodal effects of the Hellas impact. It suggested the Alba Mons and Hellas region was subject to uplift from interior shockwaves and fracturing from the convergence of surface waves. The effect of these changes likely preconditioned the region to experience subsequent volcanic activity, with the prime driver for volcanism being mantle convection to drive crustal melting and volcanism. The research also highlighted the importance of localised geology affecting results, with differences in crustal thicknesses, the location of Tharsis and availability of partially melted material shifting the volcano's location 2° (~119 km) from the theoretical antipodal centre (Zhang et al. 2022).

The grabens described above were most likely formed through extensional stretching of the terrain. A series of north–south graben south of the central caldera extend for up to 1000 km in length. Pit craters, indicating deep tension cracks, were identified in many of Alba's grabens. Towards the caldera, divergent grabens form a horseshoe shape of faults, surrounding the central edifice. Positioning of these graben suggests that Alba Mons was impacted by deformation of the Tharsis structure. The volcano likely commenced as massive eruptions that formed the ultra-long lava flows, with subsequent eruptions depositing the central edifice and finally the summit dome. This volcanism reduced in intensity over time (Ivanov and Head 2006).

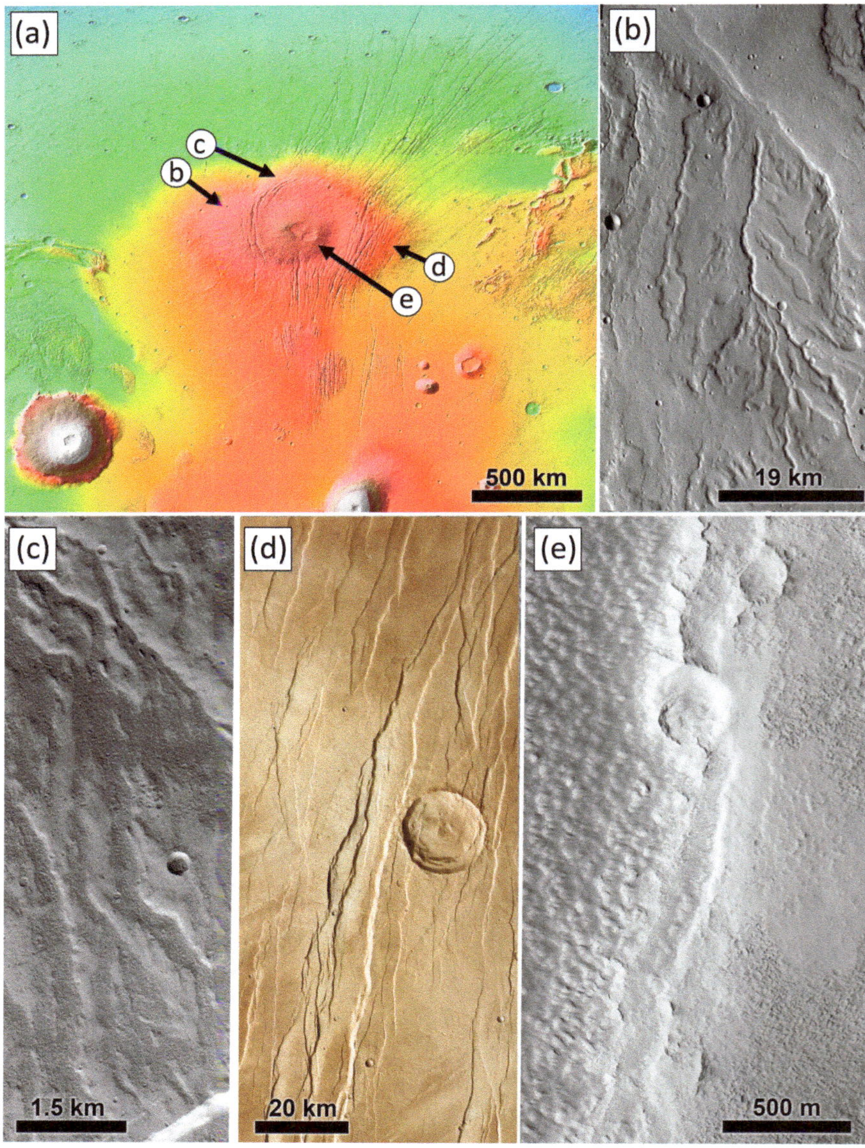

Fig. 8.15 **a** MOLA overview of Alba Mons with location of subsequent figure parts shown. **b** Fluidised lava channels on the northern flank (PIA 17710, NASA/JPL). **c** partially infilled valley network on the north-western flank (PIA06965, NASA/JPL). **d** Graben faulting (ESA/DLR/FU Berlin, CC BY-SA 3.0 IGO). **e** Dust deposits near the caldera (PSP001510-2195, NASA/JPL)

Alba Mons also demonstrates where multiple processes have operated on Mars to shape the geomorphology seen today. Such processes include fluidised lava channels (Fig. 8.15b), valley networks (Fig. 8.15c), faulting (Fig. 8.15d) and dust deposition (Fig. 8.15e). Measured low thermal inertia results supported HiRISE observations of extensive dust deposits on much of the volcano (Fig. 8.15e). This represents post-formation deposition and subsequent erosion by aeolian processes and dry mass wasting. Higher thermal inertia results from the northern flanks and apron suggest better consolidation of rocky material, with some research suggesting the presence of ices. This latter inference is supported by Mars Odyssey Neutron Spectrometer results indicating up to 7.6% of ice, and also within a region thought to have trapped near-surface ice from higher obliquity excursions (Feldman et al. 2008).

Potentially related to fluvial activity are the valley networks discovered on the northern flanks of Alba Mons (Fig. 8.15c). These are similar in appearance to terrestrial water-carved channels that occur within readily erodible material on low slope areas. Like with many other Martian features, interpretation of the valleys demonstrates the limitations of orbital remote sensing in the absence of "ground truth" data. Crater age dating suggests that these valleys have formed during the Amazonian period, post-dating Noachian-aged features in the Southern Highlands and postulated warm wet conditions which favour Martian rainfall. Terrestrial rainfall has also produced analogous features on Earthly volcanoes, but the Amazonian period is thought to have been too cold and dry for rainfall. An alternative hypothesis for the valley formation includes volcanic origins, where lava tube collapse has led to further modification through channelled lava flow (Bahia 2022). The discovery of lobate deposits on Alba Mon's northern facing slopes, which also face the north pole, suggest an alternative hypothesis involving ice-related processes. This is supported by the neutron spectrometer results described above, and the poleward-facing orientation may have favoured ice deposition during periods of Martian obliquity excursions. Latent volcanic heating over the Amazonian period could then have mobilised these deposits, carving valley networks into the easily erodible material (Schiff and Gregg 2022).

This case study of Alba Mons reveals a volcanic feature that has evolved through interactions with multiple surface processes over its billions of years-long histories. As with most of the geologic features identified in this book, researchers work within the limitations of remote sensing's ability to infer the past. These include subsequent geologic events, erasing earlier ones, spectral signatures being hidden under layers of dust, or data constraints from the limited sensor set currently in orbit. Nevertheless, Alba Mons remains a site of active research.

8.3 Aeolian Features

Wind has blown on Mars for billions of years and still acts on the planet in the present day. Remote sensing has yielded substantial results when it comes to aeolian investigation and analysis. An additional benefit of continuous orbital observations of the planet since 1997 is monitoring and quantifying changes over time, such as transport of fine-grained material, yardangs, dune migration and interactions with underlying geomorphology. This sheds light on present day wind activity and also indicators of climate change, where geomorphology of fossilised aeolian features is used to identify the action of a past wind regime that no longer operates.

The present-day thin atmosphere is capable of moving dust particles, the large-scale effects of which cause albedo changes as removal of dust exposes darker-toned basaltic plains beneath. These changes, often linked with seasonal convection and circulation patterns, plus requiring the movement of small amounts of dust to be visible (Christensen 1988) were mistaken for vegetation growth and senescence during the telescopic era of observation (Chap. 1). Fragmentation of bedrock to smaller particles has occurred on Mars through crater impacts, volatile-related, chemical and mechanical erosion. The lack of geologic recycling through tectonic forces or hydrologic activity since the late Noachian and Hesperian periods has produced an abundance of fragmented particles for the Martian winds to redistribute (Carr 2006).

8.3.1 Wind Streaks, Ripples, Yardangs and Ventefacts

Wind streak formation and modification are directly related to albedo change observations and were observed as early as Mariner 9 (Chap. 1). They occur downwind of topographic obstacles, such as craters, where deflection and local vortices can erode or deposit fine grained material. Mars wind streaks can consist of bright deposition, dark deposition or erosional or frost streaks. Dark erosional streaks occur through the removal of overlying dust, revealing darker material underneath. They can themselves be removed through dust-storms depositing fresh layers of ochre dust. Dark depositional streaks can occur down-wind of mafic sand dunes, where fine dark material has blown into the lee side of these features (Thomas et al. 1981).

Near the north pole wind streaks were found to evolve in response to heating and the Martian northern summer. This activity suggests a more complex mechanism of aeolian transport of water frost particles and darker dust or sand (Drake et al. 2006; Smith and Spiga 2018). Wind streaks are also useful indicators for prevailing wind direction (e.g. Fig. 8.16).

Ripples are small scale aeolian features that form on Earth and Mars, with distances between crests (wavelength) of Martian features ranging from centimetres to metres in size (Fig. 8.17a). Observed ripples on Mars are standalone or overprint larger features, such as dunes. Larger ripples of 1–5 m in wavelengths have been studied by surface rover missions (Lapôtre et al. 2018). A suggested formation mechanism based on terrestrial research suggests evolution through saltated particles eroding a wind-facing surface, while lee slopes are unaffected (Bagnold 1941). The wavelength would be set by the length of the saltation bounce of the sand particle. Research modelling has suggested lower atmospheric pressure on Mars

Fig. 8.16 Wind streaks imaged in Syrtis Major by Mars Express. Prevailing wind direction is from the bottom of the image to the north-east (ESA/DLR/FU Berlin, CC BY-SA 3.0 IGO)

favours generation of longer wavelength ripples as compared with terrestrial equivalents as an equilibrium point. This is supported by ripples at higher elevations possessing longer wavelengths than features at higher pressure areas, such as within low lying craters and basins (Lapôtre et al. 2021).

Yardangs are found on Mars and on Earth where extensive aeolian erosion has existed. These features commonly form through unidirectional winds carrying abrasive material that over time carve readily-erodible terrain into parallel ridges and shapes that resemble the base of ships (Fig. 8.17b). The Medusae Fossae Formation, west of Olympus Mons, is readily erodible and yardangs are common in this area. The orientation of yardangs provides an indication for past Martian climate through inferring prevailing wind direction (Liu et al. 2020).

Aeolian abrasion also affects harder material, such as rocks. Ventifacts on Mars are common on many rocks that have been exposed to the prevailing winds (e.g. Figure 8.17c). Wind erosion through constant impacts by airborne material may polish or reshape the surface of the rock, or produce elongated pits, flutes and grooves. The thinner Martian atmosphere has suggested that sand grains accelerated on the windward sides of Martian rocks, possess the energy required to erode them. Dust particles are likely too small to produce the ventifacts so far observed on Mars (Laity and Bridges 2009).

8.3.2 Dunes

Dunes are among the most common aeolian feature found on Mars (Fig. 8.18a–c). Unlike Earth, where most dunes are formed from silicate sand, Martian dunes are

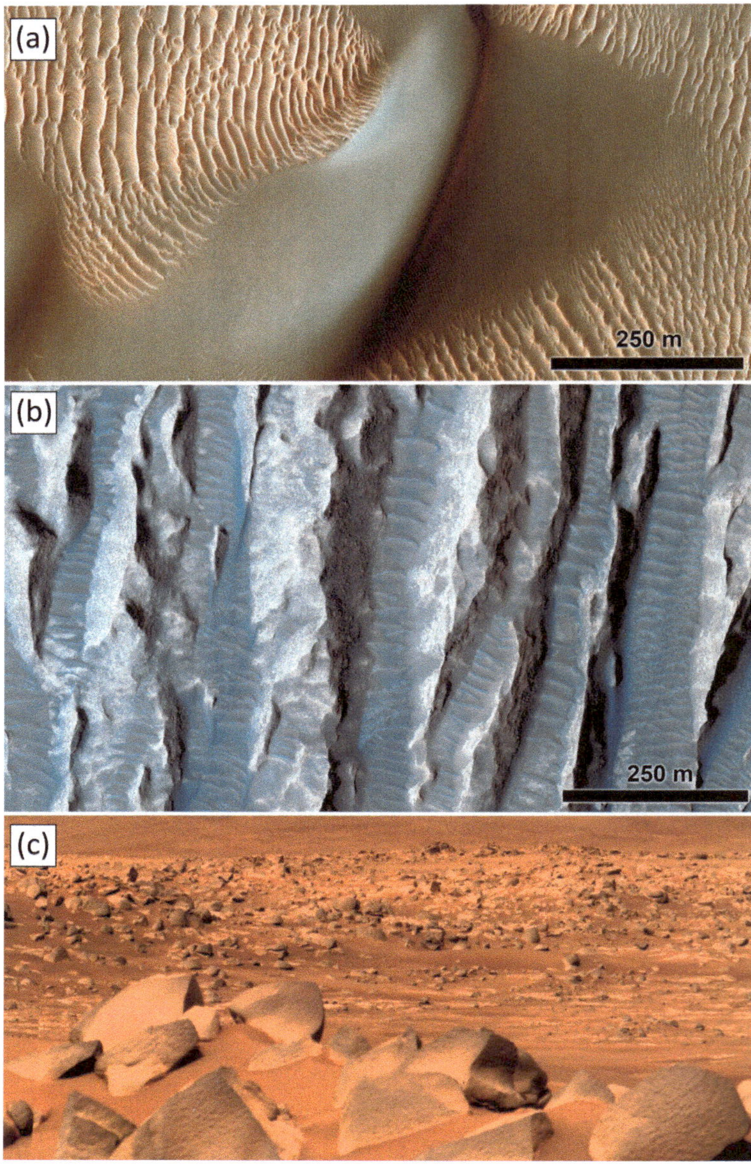

Fig. 8.17 **a** HiRISE image of ripples surrounding mafic dune material (PIA24036, NASA/JPL). **b** Yardangs in Arsinoes Chaos in the far eastern portion of Valles Marineris (PIA19291, NASA/JPL). **c** Rocks in the foreground exhibiting ventifacts at the Santa Cruz site explored by the Perseverance Rover (PIA25075, NASA/JPL)

mainly composed of mafic material. This reflects the fact that volcanic basalt is the most common rock type on Mars, and its erosion has produced an abundance of dark-toned sand particles (e.g. Fig. 8.18c, Cornwall et al. 2015). Coatings of dust overlay some dunes, producing a lighter tone in orbital images and suggesting that they are no longer active and have remained in place long enough for the overlay to form.

Martian dune morphology is driven by the prevailing wind. Mars' thin atmosphere is also capable of transporting sand-sized grains in a process called saltation (Fig. 8.19a–c). Lower velocity winds may cause particles to creep, where they roll along the surface. Higher wind speeds lift some particles from the surface, where they travel in an arc towards the ground and eject other particles into saltation (Fig. 8.18a). This process is the primary mechanism governing the formation and modification of Martian sand dunes by forming sand mounds over time (Fig. 8.18b) that may grow to create dunes (Fig. 8.18c).

Post-Viking mission environment modelling suggested that the ultra-thin Martian atmosphere would require high velocity wind gusts to initiate sand grain saltation (Iversen and White 1982). However, such wind speeds are rare and do not correspond with the frequency of dune alteration identified through periodic remote sensing. More recent work has identified, that while comparison with terrestrial wind erosion is useful, conditions on Mars are unique and limit what terrestrial studies alone can tell us. On Mars, lower than expected wind speeds are sufficient to cause a small proportion of sand-size particles to saltate with increasing energy. Each contact with other sand particles causes more collective saltation, resulting in low-rate changes to sand features such as ripples or dunes (Sullivan and Kok 2017). Topography may also cause localised wind acceleration, as occurs on Earth. Physical elevation such as mounds or crater rims constrains the air into a smaller space, increasing its velocity by up to 30% to initiate saltation in local areas (Bourke et al. 2004). Near-continuous observations of specific dune sites by MRO HiRISE suggest a planetary advancement of 1 m per Martian year. Typical sand movement values of 1–20 m^3/yr, with occasional 35 m^3/yr on Mars is comparable to values of slower moving terrestrial dunes (Chojnacki et al. 2019).

Observations of Martian dunes can thus be used to infer past and present planetary wind regimes. Barchan dunes (Fig. 8.20a) are the most common type found on Mars, forming under a single wind direction that saltates particles up a 5–15° slope and avalanche down a 30–32° lee slope. Barchans may merge to form barchanoids (Fig. 8.20b) or transverse dunes (Fig. 8.20c) with wavy or straight ridges forming at right angles to the prevailing wind.

Two or more slip-faces may form where there is more than one dominant wind direction. Linear or seif dunes (Fig. 8.20d) possess dual slip faces and form from wind directions intersecting at less than right angles. If the prevailing wind blows from multiple directions, star dunes may develop (Fig. 8.20e). Martian star dunes are commonly found at the base of craters, where turbulence caused by the surrounding crater walls produces conditions favourable for the formation of these features (Breed et al. 1979; Hayward et al. 2007).

Fig. 8.18 **a** Perspective view of the vast Martian dune field in Chasma Boreale near the north pole (vertically exaggerated 2.5 times, THEMIS NASA/JPL). **b** Frost-covered dunes within a 5 km crater in the Martian high latitudes (PIA24420, NASA/JPL). **c** Direct observations of the mafic Bagnold dune field by the Curiosity Rover (PIAS20316, NASA/JPL)

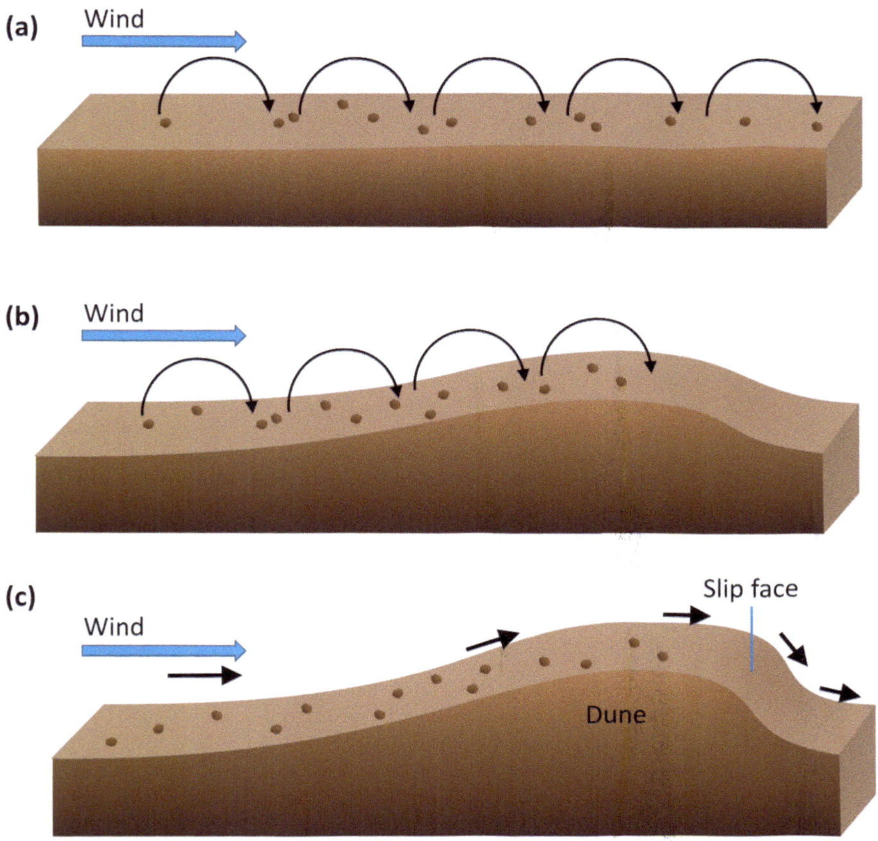

Fig. 8.19 **a** Saltating sand particles bounce along the surface, driven by the prevailing wind. **b** Over time the sand grains may pile up. **c** Should this process continue, the pile of sand may grow to a dune, with sand falling down a slip face on the lee side of the feature

Dunes are present on the Martian surface from the equator to the poles, with a location preference south of the Martian dichotomy. Martian dunes exist as intra and inter-crater types, where features were discovered residing inside and outside of craters respectively.

Intra-crater Dunes

Intra-crater dune fields are very common on Mars. Craters, being topographic lows, provide ideal locations for wind-blown sediment to become trapped and coalesce over time. An alternative mechanism of crater dune formation entails reworking of crater floor sediments eroded from material already present within the crater. Remote sensing analysis with spectral sensors enables inference about which mechanism, or what combination of mechanisms, has been active. Correlating mineralogy between dunes and potential source material would assist in inferring whether the dune sand

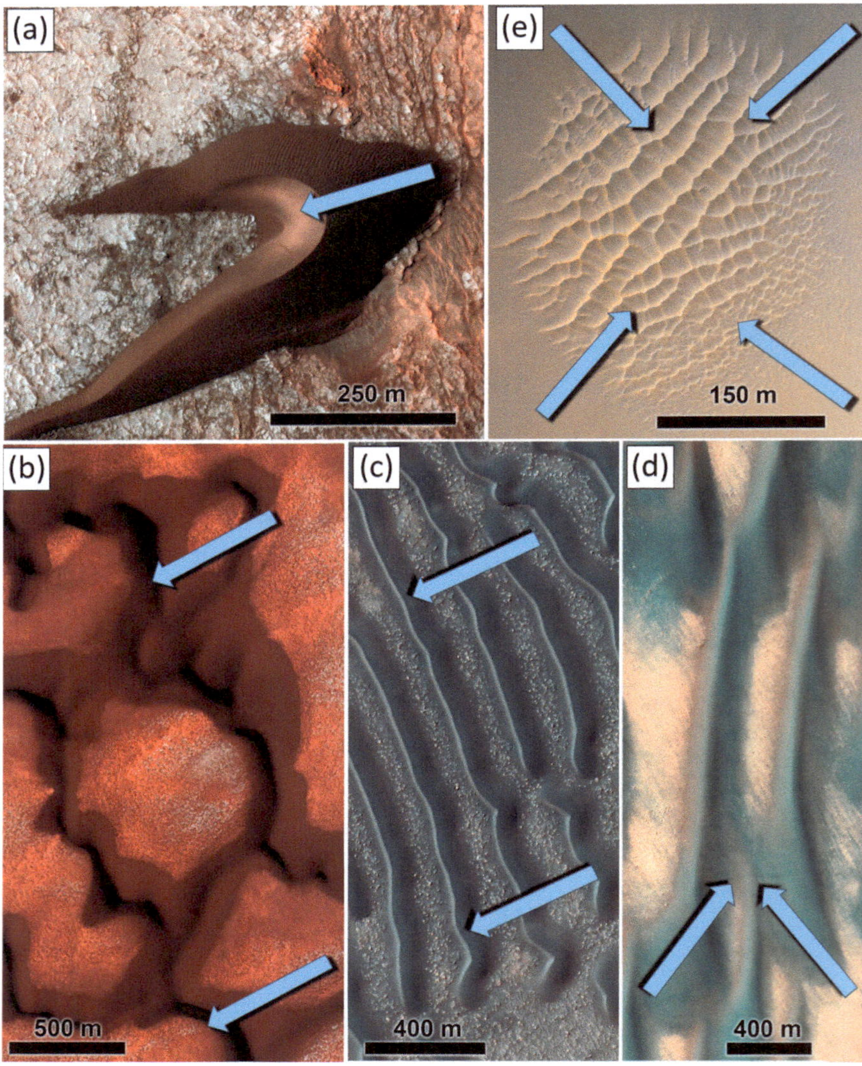

Fig. 8.20 Martian dune examples with prevailing winds marked by blue arrows (all images: NASA/JPL). **a** Example of Martian barchan dune (ESP042223-1890). **b** Barchanoid dunes located near the north pole (ESP018920-2585). **c** Transverse dunes west of Hellas Planitia (ESP06745-1380). **d** Longitudinal dunes in Noachis Terra. **e** Star dune field located in the bottom of Victoria Crater and explored by the Opportunity Mars rover (PIA08813)

originated inside or outside the crater. Analysis conducted this way indicated that layers exposed within many crater walls had eroded downslope to feed dune fields (Tirsch et al. 2009). Discoveries of multiple mafic layers of this type suggest a near-global deposition of volcanic ash during the Noachian period (Werner 2009). This material was also observed emerging from crater floors, providing a possible source for dune formation in addition to sediment from higher elevations becoming trapped inside craters (Tirsch et al. 2011).

A dune field within Proctor Crater, a 150 km diameter flat floored impact feature in the southern highlands, was one of the first to be discovered on Mars (Fig. 8.21a). This crater has been the focus of extensive remote sensing and analysis. Its 36 × 35 km dune field consists of transverse and star dunes of mafic material, providing evidence of a complex interaction of three opposing wind directions of different strengths (Fig. 8.21b). Proctor Crater may be an example where the source material was saltated into the crater from an external origin, long since eroded or buried. A strong westerly wind during the Southern hemisphere autumn and winter has dominated the morphology of the Proctor Crater dunes. Spring and summer katabatic (dense air flowing down slope under the force of gravity) winds blow in the opposite direction, trapping the dune field in its present position (Fenton et al. 2005).

The Gale Crater dune fields are among the most studied features of this type, and provide an example of where increasingly sophisticated remote sensing, and ultimately direct sampling, can quantify how dunes operate on Mars. Gale Crater is similar in size to Proctor but is dominated by a 5 km high central peak (Fig. 8.21c). Viking images were able to identify dark-toned material that surrounds the peak to the west, south and east. MGS MOC imagery provided a step change in resolving power, allowing slip face position and orientation to be measured in the mafic dune fields and sand sea of the crater (Pelkey and Jakosky 2002). Analysis of HiRISE and CTX imagery showed that dominant northerly winds are accelerated and channelled by the topographic influence of the central mound (Fig. 8.21c). Nighttime temperatures power katabatic winds flowing downhill from Mount Sharp and the crater rim. Dunes have formed around this mound, and over geologic time periods, winds have transported much of the sand and eroded sediment out of the crater. This is evidenced by dark material extending to the south, likely sourced from the dune fields within the crater have blown over the southern rim to deposit on the lee side (Hobbs et al. 2010). Recent analysis suggests most of the sediment deflation occurred early in Mars' geologic history (Day and Kocurek 2016).

Quantitative analysis of the dunes by the Curiosity Rover confirmed a chemical makeup dominated by basaltic minerals, with small quantities of quartz, hematite and anhydrite. These results provided ground truthing of inferences made from orbit, a rare opportunity given the limited number of surface missions that have operated on the planet (Fig. 8.18c; Achilles et al. 2017). Curiosity was also able to measure dune modification through small ripple migration of ~ 2.8 cm/sol, an observation only made possible through the presence of the rover at Gale Crater (Bridges et al. 2017).

An additional, side benefit of intra-crater dune studies was demonstrated through the landing site selection process for the Curiosity Mars Rover. Modelling of atmospheric interactions with topography were used to infer wind strength and direction

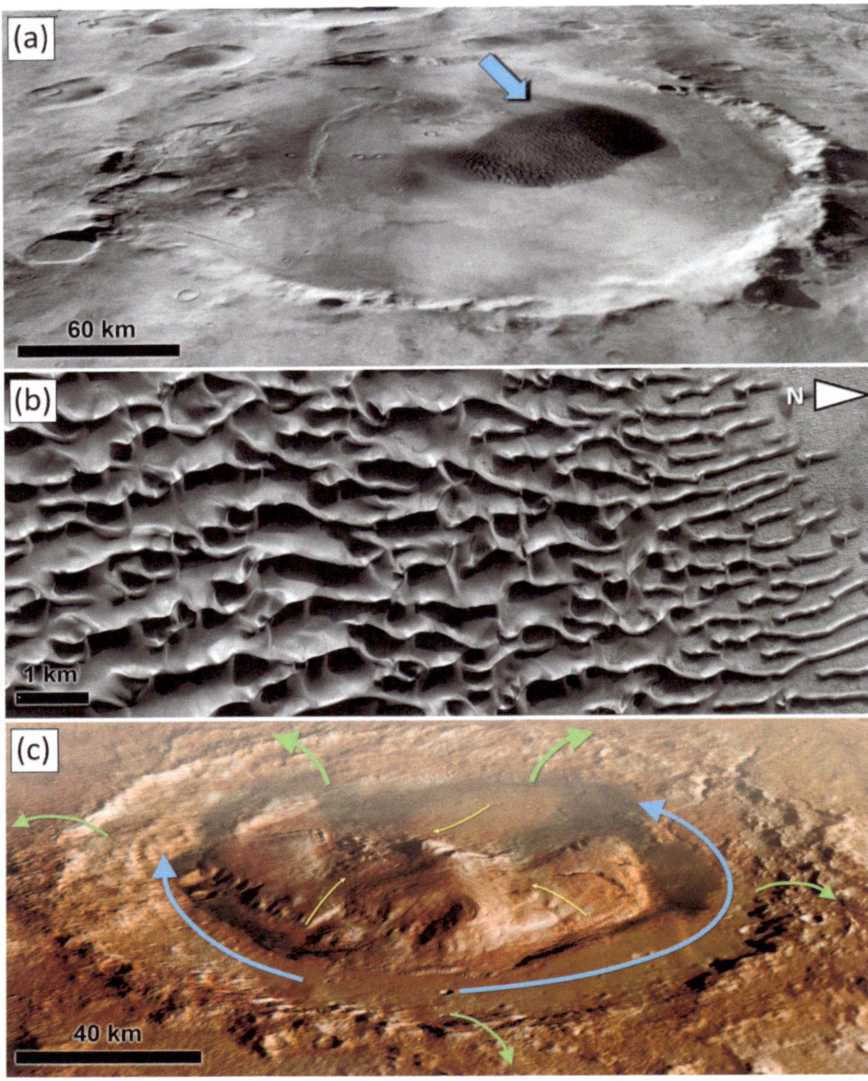

Fig. 8.21 **a** Perspective view of Proctor Crater with the associated intra-crater dune field highlighted by the blue arrow (CTX composite, NASA/JPL). **b** HiRISE imagery of part of the dune field showing a transition in morphology from the edge to the centre of the field (ESP018990-1325, NASA/JPL). **c** Illustration of Gale Crater showing the mafic dune field surrounding the central mound. The blue arrows denote prevailing wind flowing around the mound, green arrows denote flow outside the crater, yellow arrows show flow over the mound. At nighttime the directions are reversed

corresponding to a particular Martian season. This was important to understand air pressure, as well as drift for a surface payload descending by parachute as they affected the landing (Chap. 2). Additionally, mapping of the dunes identified a traversable path for the rover on its journey up Mount Sharp to avoid becoming immobilised in low-cohesion material (Grotzinger et al. 2012).

Inter-crater Dunes

Dune fields are also present outside craters on Mars. A massive sand sea surrounding the Martian north pole represents the most substantial dune field of this type on Mars. Olympia Undae is between 78 and 83° latitude, covers 470,000 km^2 and consists of barchan and transverse dunes (Fig. 8.18a). As with other dune fields, the Olympia Undae sand sea consist of weathered basalt, but more complex geologic interactions are suggested by concentrations of gypsum discovered along dune crests in the eastern region. As gypsum is a hydrated sulphate mineral requiring liquid water to form, its discovery was of high interest to planetary scientists. Recent research suggests an origin from early-Amazonian regolith, indicating the presence of at least transient liquid water near the north pole at this period (Szynkiewicz and Bishop, 2021; Das et al. 2022).

Olympia Undae also has some of the most active dunes on the planet, despite frost and ice deposition constraining migration during the north polar winter. Dune morphology in this area is dominated by the seasonal heating and cooling of the north polar cap, and the winds that it generates. Downslope winds from the pole are rotated by Coriolis forces funnelled through a series of chasms. Dunes spiral southward to merge with the main erg during the short summer period when temperatures are high enough to free ice-cemented or CO_2-buried sand, allowing saltation to resume (Fishbaugh and Head 2005). Despite the limited sand availability due to ice cementing and frost overlaying, ripples in this region were observed to migrate at a rate of 10 $\pm$ 6 m/yr over 22 days during the northern summer (Chojnacki et al. 2021).

Fossilised Dunes

Other inter-crater dune types exist where, while sharing the morphology of sand features described above, are degraded and frozen in place. These dunes are considered relict or "fossilised", and are very useful in preserving information about a wind regime operating early in Martian history. Examples of these features, exhumed during the Late Amazonian period, lie within Melas and Coprates Mensa, Valles Marineris (Fig. 8.22a). The preservation of such features is not as common on Earth, with its more energetic weathering processes and degradation by vegetation. Slip face analysis of the fossilised dunes within Valles Marineris inferred wind directions consistent with the present day. This suggests that variations in climate in this region over the Late Amazonian period were not sufficient to significantly alter the convection and Coriolis force-driven wind cycles seen today (Chojnacki et al. 2020).

In contrast, analysis of other fossilised dunes has recorded evidence of changing wind direction through climate change. Hesperian-aged fossilised dunes in the Apollinaris Sulci region indicated a previous climate driven by northerly winds (Fig. 8.22b, Hunt et al. 2022). In Utopia Planitia, the Chinese Zhurong rover imaged barchan

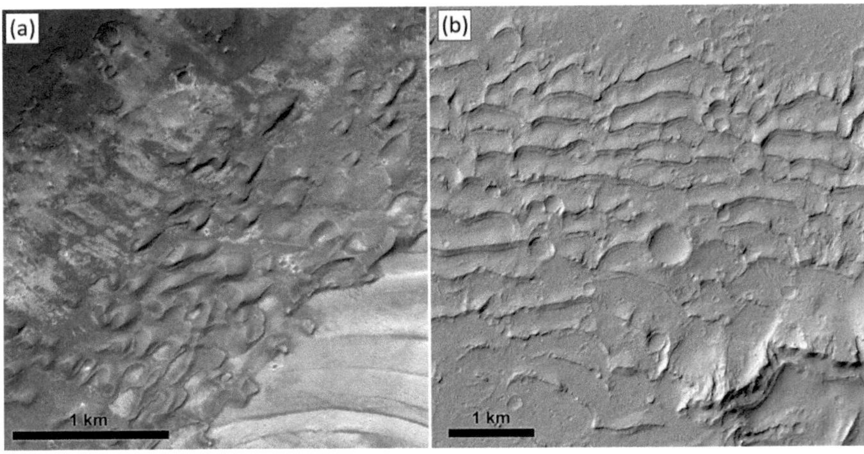

Fig. 8.22 **a** Fossilised dunes in Valles Marineris (ESP025666-1685, NASA/JPL). **b** Fossilised transverse ridges in the Apollinaris Sulci region (PSP010453-1675, NASA/JPL)

dunes whose slip faces indicated that were originally formed with north-easterly prevailing winds. Crater dating and mineralogy evidence of hydrous chemical alterations suggested that these dunes formed during a past Martian ice age of 0.4–201 Ma and corresponded with a global obliquity change (Laskar et al. 2002). Cementation of these dunes preserved the record of this previous wind direction. A subsequent change of wind regime from the north-west marking the likely end of the ice age was evidenced by post-sedimentation erosion of the barchans into longitudinal dunes, and superposition of dark dunes (Liu et al. 2023).

The above and other fossilised dune examples show how study of aeolian features can inform the evolution of the Martian climate. Aeolian features remain an active area of research, being one of the surface processes currently active on Mars and monitored regularly through remote sensing. Active cratering is also of interest to researchers, not only providing access to otherwise hidden sub-surface features, but insights into crater age dating methods and understanding how present and past impact cratering has operated on Mars and the Solar System.

Whether volcanism remains active to the present day remains an open question. Certainly, volcanism and lava flows have changed the landscape of most of the planet, often in combination with water-based erosion. Remote sensing is essential in placing these processes in context, helping understand geological, climatic and environmental history at a global scale. Direct surface missions can then be targeted to specific Martian sites to provide additional detail and place localised lander measurements into a broader planetary context.

References

Achilles CN, Downs RT, Ming DW, Rampe EB, Morris RV, Treiman AH, Morrison SM, Blake DF, Vaniman DT, Ewing RC, Chipera SJ (2017) Mineralogy of an active eolian sediment from the Namib dune, Gale Crater, Mars. J Geophys Res Planets 122:2344–2361. https://doi.org/10.1002/2017JE005262

Arvidson RE, Ashley JW, Bell JF III, Chojnacki M, Cohen J, Economou TE, Farrand WH, Fergason R, Fleischer I, Geissler P, Gellert R (2011) Opportunity Mars Rover mission: overview and selected results from Purgatory ripple to traverses to Endeavour crater. J Geophys Res Planets. https://doi.org/10.1029/2010JE003746

Bagnold RA (1941) The physics of blown sand and desert dunes. London: Methuen. Prog Phys Geogr 18:91–96

Bahia (2022) Morphological and hydrological analysis of volcanic flank valleys—evidence for a volcanic origin. Planet Space Sci. https://doi.org/10.1016/j.pss.2022.105592

Barlow NG (2006) Impact craters in the northern hemisphere of Mars: layered ejecta and central pit characteristics. Meteor Planet Sci 41:1425–1436. https://doi.org/10.1111/j.1945-5100.2006.tb00427.x

Barlow N (2008) Mars: an introduction to its interior, surface and atmosphere. Cambridge University Press, New York, p 264

Barlow NG, Boyce JM, Costard FM, Craddock RA, Garvin JB, Sakimoto SEH, Kuzmin RO, Roddy DJ, Soderblom LA (2000) Standardizing the nomenclature of Martian impact crater ejecta morphologies. J Geophys Res 105:26733–26738. https://doi.org/10.1029/2000JE001258

Barlow NG, Robbins S (2015) Layered ejecta. In: Hargitai H, Kereszturi Á (eds) Encyclopedia of planetary landforms. Springer, New York, NY. https://doi.org/10.1007/978-1-4614-3134-3_475

Benedix GK, Lagain A, Chai K, Meka S, Anderson S, Norman C, Bland PA, Paxman J, Towner MC, Tan T (2020) Deriving surface ages on Mars using automated crater counting. Earth Space Sci. https://doi.org/10.1029/2019EA001005

Black BA, Manga M (2016) The eruptibility of magmas at Tharsis and Syrtis Major on Mars. J Geophys Res Planets 121:944–964. https://doi.org/10.1002/2016JE004998

Black BA, Manga M, Ojha L, Longpré M-A, Karunatillake S, Hlinka L (2022) The history of water in Martian magmas from thorium maps. Geophys Res Lett. https://doi.org/10.1029/2022GL098061

Bourke MC, Bullard J, Barnouin-Jha O (2004) Aeolian sediment transport pathways and aerodynamics at troughs on Mars. J Geophys Res. https://doi.org/10.1029/2003JE002155

Boyce JM, Mouginis-Mark P, Garbeil H (2005) Ancient oceans in the northern lowlands of Mars: evidence from impact crater depth/diameter relationships. J Geophys Res Planets. https://doi.org/10.1029/2004JE002328

Breed CS, Grolier MJ, McCauley JF (1979) Morphology and distribution of common 'sand' dunes on Mars: comparison with the Earth. J Geophys Res Solid Earth 84:8183–8204

Breton S, Quantin-Nataf C, Pan L, Mandon L, Volat M (2022) Insight into Martian crater degradation history based on crater depth and diameter statistics. Icarus 377:114898

Bridges NT, Sullivan R, Newman CE, Navarro S, Van Beek J, Ewing RC, Ayoub F, Silvestro S, Gasnault O, Le Mouélic S, Lapotre MGA (2017) Martian aeolian activity at the Bagnold Dunes, Gale Crater: the view from the surface and orbit. J Geophys Res Planets 122:2077–2110. https://doi.org/10.1002/2017JE005263

Broquet A, Andrews-Hanna JC (2023) Geophysical evidence for an active mantle plume underneath Elysium Planitia on Mars. Nat Astron 7:160–169. https://doi.org/10.1038/s41550-022-01836-3

Brož P, Krýza O, Wilson L, Conway SJ, Hauber E, Mazzini A, Raack J, Balme MR, Sylvest ME, Patel MR (2020) Experimental evidence for lava-like mud flows under Martian surface conditions. Nat Geosci 13:403–407. https://doi.org/10.1038/s41561-020-0577-2

Brož P, Bernhardt H, Conway SJ, Parekh R (2021) An overview of explosive volcanism on Mars. J Volcanol Geoth Res 409:107125. https://doi.org/10.1016/j.jvolgeores.2020.107125

Brunetti MT, Guzzetti F, Cardinali M, Fiorucci F, Santangelo M, Mancinelli P, Komatsu G, Borselli L (2014) Analysis of a new geomorphological inventory of landslides in Valles Marineris, Mars. Earth Planet Sci Lett 405:156–168. https://doi.org/10.1016/j.pss.2004.07.016hal-00102264f

Carr M (2006) The Surface of Mars. Cambridge University Press, New York, NY, p 307

Chevrel MO (2013) Rheology of Martian lava flows: an experimental approach. Dissertation, LMU München, Faculty of Geosciences. https://doi.org/10.5282/edoc.15783

Chojnacki M, Banks ME, Fenton LK, Urso AC (2019) Boundary condition controls on the high-sand-flux regions of Mars. Geology 47:427–430. https://doi.org/10.1130/G45793.1

Chojnacki M, Fenton LK, Weintraub AR, Edgar LA, Jodhpurkar MJ, Edwards CS (2020) Ancient Martian aeolian sand dune deposits recorded in the stratigraphy of Valles Marineris and implications for past climates. J Geophys Res Planets. https://doi.org/10.1029/2020JE006510

Chojnacki M, Vaz DA, Silvestro S, Silva DC (2021) Widespread megaripple activity across the north polar ergs of Mars. J Geophys Res Planets. https://doi.org/10.1029/2021je006970

Christensen PR (1988) Global albedo variations on Mars: implications for active aeolian transport, deposition, and erosion. J Geophys Res Solid Earth 93:7611–7624

Cornwall C, Bandfield JL, Titus TN, Schreiber BC, Montgomery DR (2015) Physical abrasion of mafic minerals and basalt grains: application to Martian aeolian deposits. Icarus 256:13–21

Craddock RA, Howard AD (2002) The case for rainfall on a warm, wet early Mars. J Geophys Res Planets 107:21–31

Dapremont AM, Wray JJ (2023) Mars mud volcanoes in color: a new approach to the study of subsurface sediment mobilization. Icarus 115733. https://doi.org/10.1016/j.icarus.2023.115733

Dapremont AM, Wray JJ (2021) Insights into Mars mud volcanism using visible and near-infrared spectroscopy. Icarus 359:114299. https://doi.org/10.1016/j.icarus.2020.114299

Das E, Mustard JF, Tarnas JD, Pascuzzo AC, Kremer CH (2022) Investigating the origin of gypsum in Olympia Undae: characterizing the mineralogy of the basal unit. Icarus 372:114720

Dasgupta D, Kundu A, Dasgupta N (2022) An insight to the cryospheric level in Mars: case study from the Thaumasia Minor. Icarus 372:114725. https://doi.org/10.1016/j.icarus.2021.114725

Daubar IJ, McEwen AS, Byrne S, Kennedy MR, Ivanov B (2013) The current Martian cratering rate. Icarus 225:506–516. https://doi.org/10.1016/j.icarus.2013.04.009

Day M, Kocurek G (2016) Observations of an aeolian landscape: from surface to orbit in Gale Crater. Icarus 280:37–71. https://doi.org/10.1016/j.icarus.2015.09.042

Drake NB, Tamppari LK, Baker RD, Cantor BA, Hale AS (2006) Dust devil tracks and wind streaks in the North Polar Region of Mars: a study of the 2007 Phoenix Mars Lander Sites. Geophys Res Lett. https://doi.org/10.1029/2006GL026270

Feldman WC, Mellon MT, Gasnault O, Maurice S, Prettyman TH (2008) Volatiles on Mars: scientific results from the Mars Odyssey Neutron spectrometer. In: Bell III JF (ed) The Martian surface: composition, mineralogy, and physical properties. Cambridge University Press, Cambridge, 636

Fenton LK, Toigo AD, Richardson MI (2005) Aeolian processes in Proctor Crater on Mars: mesoscale modeling of dune-forming winds. J Geophys Res Planets. https://doi.org/10.1029/2004JE002309.

Fernando B, Daubar IJ, Charalambous C, Grindrod PM, Stott A, Ateqi A, Atri D, Ceylan S, Clinton J, Fillingim M, Hauber E (2023) A tectonic origin for the largest marsquake observed by InSight. Geophys Res Lett. https://doi.org/10.1029/2023GL103619

Fishbaugh KE, Head JW (2005) Origin and characteristics of the Mars north polar basal unit and implications for polar geologic history. Icarus 174:444–474. https://doi.org/10.1016/j.icarus.2004.06.021

Garvin JB, Sakimoto SEH, Frawley JJ (2003) Craters on Mars: global geometric properties from gridded MOLA topography. In: Sixth international conference on Mars (No. 3277). Lunar and Planetary Institute.

Grant JA, Schultz PH (1993) Degradation of selected terrestrial and Martian impact craters. J Geophys Res Planets 98:11025–11042. https://doi.org/10.1029/93JE00121

Greeley RM, Batson RM (eds) (1990) Planetary mapping. Cambridge University Press, Cambridge, 296 pp

Grotzinger JP, Crisp J, Vasavada AR et al (2012) Mars Science Laboratory mission and science investigation. Space Sci Rev 170:5–56. https://doi.org/10.1007/s11214-012-9892-2

Hamilton CW, Fagents SA, Thordarson T (2011) Lava-ground ice interactions in Elysium Planitia, Mars: geomorphological and geospatial analysis of the Tartarus Colles cone groups. J Geophys Res Planets. https://doi.org/10.1029/2010JE003657

Hargitai HI, Gulick VC (2018) Late Amazonian–aged channel and island systems located East of Olympus Mons, Mars. In: Dynamic Mars. Elsevier, pp 121–154. https://doi.org/10.1016/B978-0-12-813018-6.00004-2

Hartmann WK (2005) Martian cratering 8: isochron refinement and the chronology of Mars. Icarus 174:294–320. https://doi.org/10.1016/j.icarus.2004.11.023

Hartmann WK (2019) History of the terminal cataclysm paradigm: epistemology of a planetary bombardment that never (?) happened. Geosciences. https://doi.org/10.3390/geosciences9 070285

Hartmann WK (1981) Chronology of planetary volcanism by comparative studies of planetary cratering. In: Basaltic Volcanism on the Terrestrial Planets. Pergamon, New York, pp 1049–127

Hauber E, Bleacher J, Gwinner K, Williams D, Greeley R (2009) The topography and morphology of low shields and associated landforms of plains volcanism in the Tharsis region of Mars. J Volcanol Geoth Res 185:69–95. https://doi.org/10.1016/j.jvolgeores.2009.04.015

Hayward RK, Mullins KF, Fenton LK, Hare TM, Titus TN, Bourke MC, Colaprete A, Christensen PR (2007) Mars global digital dune database and initial science results. J Geophys Res Planets. https://doi.org/10.1029/2007JE002943

Hobbs SW, Paull DJ, Bourke MC (2010) Aeolian processes and dune morphology in Gale Crater. Icarus 210:102–115. https://doi.org/10.1016/j.icarus.2010.06.006

Horvath DG, Moitra P, Hamilton CW, Craddock RA, Andrews-Hanna JC (2021) Evidence for geologically recent explosive volcanism in Elysium Planitia, Mars. Icarus 365:114499. https://doi.org/10.1016/j.icarus.2021.114499

Hughes SS, Haberle CW, Kobs Nawotniak SE, Sehlke A, Garry WB, Elphic RC, Payler SJ, Stevens AH, Cockell CS, Brady AL, Heldmann JL (2019) Basaltic terrains in Idaho and Hawai 'i as planetary analogs for Mars geology and astrobiology. Astrobiology 19:260–283. https://doi.org/10.1089/ast.2018.1847

Hunt AR, Day M, Edgett KS, Chojnacki M (2022) The lithified aeolian dune field adjacent to the Apollinaris Sulci, Mars: geological history and paleo-wind record. Icarus 373:114788. https://doi.org/10.1016/j.icarus.2021.114788

Ivanov MA, Head JW (2006) Alba Patera, Mars: topography, structure, and evolution of a unique late Hesperian–early Amazonian shield volcano. J Geophys Res Planets. https://doi.org/10.1029/2005JE002469

Iversen JD, White BR (1982) Saltation threshold on earth, mars and Venus. Sedimentology 29:111–119

Johnson BC, Bowling TJ (2014) Where have all the craters gone? Earth's bombardment history and the expected terrestrial cratering record. Geology 42:587–590. https://doi.org/10.1130/G35 754.1

Kadish SJ, Head JW, Barlow NG (2010) Pedestal crater heights on Mars: a proxy for the thicknesses of past, ice-rich, Amazonian deposits. Icarus 210:92–101. https://doi.org/10.1016/j.icarus.2010.06.021

Kawamura T, Clinton JF, Zenhäusern G, Ceylan S, Horleston AC, Dahmen NL, Duran C, Kim D, Plasman M, Stähler SC, Euchner F (2023) S1222a—the largest marsquake detected by InSight. Geophys Res Lett. https://doi.org/10.1029/2022GL101543

Kilburn R, Dasent J, Wright V, Manga M (2022) Lithology, pore-filling media, and pore closure depth beneath InSight on Mars inferred from shear wave velocities. J Geophys Res Planets. https://doi.org/10.1029/2022je007539

Laity JE, Bridges NT (2009) Ventifacts on Earth and Mars: analytical, field, and laboratory studies supporting sand abrasion and windward feature development. Geomorphology 105:202–217. https://doi.org/10.1016/j.geomorph.2008.09.014

Lapôtre MGA, Ewing RC, Weitz CM, Lewis KW, Lamb MP, Ehlmann BL, Rubin DM (2018) Morphologic diversity of Martian ripples: implications for large-ripple formation. Geophys Res Lett 45:10229–10239. https://doi.org/10.1029/2018GL079029

Lapôtre MG, Ewing RC, Lamb MP (2021) An evolving understanding of enigmatic large ripples on Mars. J Geophys Res Planets. https://doi.org/10.1029/2020JE006729

Laskar J, Levrard B, Mustard JF (2002) Orbital forcing of the Martian polar layered deposits. Nature 419:375–377

Leone G (2014) A network of lava tubes as the origin of Labyrinthus Noctis and Valles Marineris on Mars. J Volcanol Geoth Res 277:1–8. https://doi.org/10.1016/j.jvolgeores.2014.01.011

Léveillé R, Datta S (2010) Lava tubes and basaltic caves as astrobiological targets on Earth and Mars: a review. Planet Space Sci 58:592–598. https://doi.org/10.1016/j.pss.2009.06.004

Liu J, Di K, Gou S, Yue Z, Liu B, Xiao J, Liu Z (2020) Mapping and spatial statistical analysis of Mars Yardangs. Planet Space Sci 192:105035. https://doi.org/10.1016/j.pss.2020.105035

Liu J, Qin X, Ren X, Wang X, Sun Y, Zeng X, Wu H, Chen Z, Chen W, Chen Y, Wang C, Sun Z, Zhang R, Ouyang Z, Guo Z, Head JW, Li C (2023) Martian dunes indicative of wind regime shift in line with end of ice age. Nature 620:303–309. https://doi.org/10.1038/s41586-023-06206-1

Loizeau D, Werner SC, Mangold N, Bibring JP, Vago JL (2012) Chronology of deposition and alteration in the Mawrth Vallis region, Mars. Planet Space Sci 72:31–43. https://doi.org/10.1016/j.pss.2012.06.023

Lopes-Gautier R, McEwen AS, Smythe WB, Geissler PE, Kamp L, Davies AG, Spencer JR, Keszthelyi L, Carlson R, Leader FE, Mehlman R (1999) Active volcanism on Io: global distribution and variations in activity. Icarus 140:243–264. https://doi.org/10.1006/icar.1999.6129

Mazzini A, Etiope G (2017) Mud volcanism: an updated review. Earth Sci Rev 168:81–112. https://doi.org/10.1016/j.earscirev.2017.03.001

McEwen AS, Bierhaus EB (2006) The importance of secondary cratering to age constraints on planetary surfaces. Annu Rev Earth Planet Sci 34:535–567. https://doi.org/10.1146/annurev.earth.34.031405.125018

Mouginis-Mark PJ, Baloga SM (2006) Morphology and geometry of the distal ramparts of Martian impact craters. Meteorit Planet Sci 41:1469–1482

Mouginis-Mark PJ, Zimbelman JR, Crown DA, Wilson L, Gregg TK (2022) Martian volcanism: current state of knowledge and known unknowns. Geochemistry 82:125886

Niu S, Zhang F, Di K, Gou S, Yue Z (2022) Layered ejecta craters in the candidate landing areas of China's first Mars mission (Tianwen-1): implications for subsurface volatile concentrations. J Geophys Res Planets. https://doi.org/10.1029/2021JE007089

Pain CF, Clarke JDA, Thomas M (2007) Inversion of relief on Mars. Icarus 190:478–491. https://doi.org/10.1016/j.icarus.2007.03.017

Palumbo AM, Head JW, Wilson L (2020) Rainfall on Noachian Mars: nature, timing, and influence on geologic processes and climate history. Icarus 347:113782. https://doi.org/10.1016/j.icarus.2020.113782

Papale P, Dobran F (1994) Magma flow along the volcanic conduit during the Plinian and pyroclastic flow phases of the May 18, 1980, Mount St. Helens eruption. J Geophys Res Solid Earth 99:4355–4373

Pelkey SM, Jakosky BM (2002) Surficial properties of Gale Crater and Melas Chasma, Mars: integration of remote-sensing data. Icarus 160:228–257

Perron JT, Mitrovica JX, Manga M, Matsuyama I, Richards MA (2007) Evidence for an ancient Martian ocean in the topography of deformed shorelines. Nature 447:840–843. https://doi.org/10.1038/nature05873

Peters SI, Christensen PR, Clarke AB (2021) Lava flow eruption conditions in the Tharsis volcanic province on Mars. J Geophys Res Planets. https://doi.org/10.1029/2020JE006791

Pieters CM, Adams JB, Mouginis-Mark PJ, Zisk SH, Smith MO, Head JW, McCord TB (1985) The nature of crater rays: the Copernicus example. J Geophys Res Solid Earth 90:12393–12413

Quantin-Nataf C, Craddock RA, Dubuffet F, Lozac'h L, Martinot M (2019) Decline of crater obliteration rates during early Martian history. Icarus 317:427–433. https://doi.org/10.1016/j.icarus.2018.08.005

Schiff NL, Gregg TK (2022) Probable ice-rich deposits on north-facing slopes in Alba Patera, Mars. Icarus 383:115063. https://doi.org/10.1016/j.icarus.2022.115063

Schneeberger DM, Pieri DC (1991) Geomorphology and stratigraphy of Alba Patera, Mars. J Geophys Res Solid Earth 96:1907–1930. https://doi.org/10.1029/90JB01662

Skinner JA Jr, Mazzini A (2009) Martian mud volcanism: terrestrial analogs and implications for formational scenarios. Mar Pet Geol 26:1866–1878. https://doi.org/10.1016/j.marpetgeo.2009.02.006

Smith IB, Spiga A (2018) Seasonal variability in winds in the north polar region of Mars. Icarus 308:188–196. https://doi.org/10.1016/j.icarus.2017.10.005

Stähler SC, Khan A, Banerdt WB, Lognonné P, Giardini D, Ceylan S, Drilleau M, Duran AC, Garcia RF, Huang Q, Kim D (2021) Seismic detection of the Martian core. Science 373:443–448. https://doi.org/10.1126/science.abi773

Stopar JD, Taylor GJ, Hamilton VE, Browning L (2006) Kinetic model of olivine dissolution and extent of aqueous alteration on Mars. Geochim Cosmochim Acta 70:6136–6152. https://doi.org/10.1016/j.gca.2006.07.039

Strom RG, Croft SK, Barlow NG (1992) The Martian impact cratering record. Mars 383–423

Sullivan R, Kok JF (2017) Aeolian saltation on Mars at low wind speeds. J Geophys Res Planets 122:2111–2143

Szynkiewicz A, Bishop JL (2021) Assessment of sulfate sources under cold conditions as a geochemical proxy for the origin of sulfates in the circumpolar dunes on Mars. Minerals. https://doi.org/10.3390/min11050507

Thomas P, Veverka J, Lee S, Bloom A (1981) Classification of wind streaks on Mars. Icarus 45:124–153. https://doi.org/10.1016/0019-1035(81)90010-5

Tirsch D, Jaumann R, Pacifici A, Poulet F, Roach LH, Mustard JL, Bibring JP, Neukum G (2009) Dark layers as local sources for the dark intra-crater dunes on Mars. Lunar Planet Sci XL 1468:1004

Tirsch D, Jaumann R, Pacifici A, Poulet F (2011) Dark aeolian sediments in Martian craters: composition and sources. J Geophys Res Planets. https://doi.org/10.1029/2009JE003562

Tornabene LL, Watters WA, Osinski GR, Boyce JM, Harrison TN, Ling V, McEwen AS (2018) A depth versus diameter scaling relationship for the best-preserved melt-bearing complex craters on Mars. Icarus 299:68–83. https://doi.org/10.1016/j.icarus.2017.07.003

Wagstaff KL, Daubar IJ, Doran G, Munje MJ, Bickel VT, Gao A, Pate J, Wexler D (2022) Using machine learning to reduce observational biases when detecting new impacts on Mars. Icarus 386:115146. https://doi.org/10.1016/j.icarus.2022.115146

Watts AW, Greeley R, Melosh HJ (1991) The formation of terrains antipodal to major impacts. Icarus 93:159–168. https://doi.org/10.1016/0019-1035(91)90170-X

Weiss DK, Head JW (2015) Crater degradation in the Noachian highlands of Mars: assessing the hypothesis of regional snow and ice deposits on a cold and icy early Mars. Planet Space Sci 117:401–420. https://doi.org/10.1016/j.pss.2015.08.009

Weiss DK, Head JW (2017) Evidence for stabilization of the ice-cemented cryosphere in earlier Martian history: implications for the current abundance of groundwater at depth on Mars. Icarus 288:120–147. https://doi.org/10.1016/j.icarus.2017.01.018

Werner SC (2009) The global Martian volcanic evolutionary history. Icarus 201:44–68. https://doi.org/10.1016/j.icarus.2008.12.019

Williams DA, Greeley R, Zuschneid W, Werner SC, Neukum G, Crown DA, Gregg TKP, Gwinner K, Raitala J (2007) Hadriaca Patera: insights into its volcanic history from Mars Express High Resolution Stereo Camera. J Geophys Res. https://doi.org/10.1029/2007JE002924

Wrobel K, Schultz P, Crawford D (2006) An atmospheric blast/thermal model for the formation of high-latitude pedestal craters. Meteorit Planet Sci 41:1539–1550

Zhang L, Zhang J, Mitchell RN (2022) Dichotomy in crustal melting on early Mars inferred from antipodal effect. The Innovation. https://doi.org/10.1016/j.xinn.2022.100280

Chapter 9
Mars Unveiled: Ice and Water

9.1 Ice

Ice has played a significant role in shaping the surface of Mars throughout its history. Its increased distance from the Sun led pre-space-age astronomers to expect that colder conditions would be favourable to the formation of ice on the planet (Chap. 1). Ice on Mars consists of surficial deposits, as well as subsurface material. Under current conditions of an axial obliquity of 25°, Martian ice to a depth of 1–2 m tends to diffuse outward into the atmosphere below an approximate latitude of 50°, but be stable past these latitudes (Schorghofer and Aharonson 2005). In addition to the north and south poles, remote sensing over the past decades has discovered an abundance of landform features related to past and present ice activity, including direct surface-based observation and excavation of ice by the Phoenix lander in 2008. These will now be discussed.

9.1.1 Poles

The north and south poles of Mars were among the first confirmed physical features to be discovered (Chap. 1). Polar layered deposits consisting of dusty water ice dominate these regions. The current polar caps are likely a remnant of much more extensive ice deposits in Martian history. Remote sensing evidence has suggested that polar layered deposits may have been much larger during earlier periods. MOLA analysis uncovered evidence of an extensive Hesperian-aged southern icesheet that may have covered up to 2% of Mars (Head and Pratt 2001). Both polar caps contain sufficient ice to bend the underlying lithosphere through accumulated weight. Direct observations of ice in the near-polar regions were made through clearing of the top few centimetres of regolith during the Phoenix Mars landing, as well as deliberate excavations of the landing site (Mellon et al. 2009).

S. Hobbs, *Sensing Mars: Remote Sensing to Understand the Red Planet*,
https://doi.org/10.1007/978-3-032-24416-1_9

The 1000 km diameter northern polar cap is elevated 3 km above the surrounding terrain and contains 1.6 million km^3 of water ice (Carr et al. 2003; Fig. 9.1a, b). Observable North Polar Layered Deposits (NPLD) at Planum Boreum, the northern plain of Mars, provide a record of episodic ice accumulation through tens to hundreds of thousands to millions of years. They consist of water ice, with up to 5% atmospheric dust (compared to 10% at the South Pole), and may contain up to 10% by weight of mixed CO_2 ice (Broquet et al. 2020). The northern polar cap has 400–1000 m deep spiral troughs radiating clockwise from the pole. Accumulation of the NPLD causes trough migration poleward, in the opposite direction of the prevailing wind. An early theory suggested Coriolis forces from Mars' rotation and thermal disequilibrium between the colder poles and warmer surrounding environment generates strong katabatic winds. These winds would have enhanced erosion in a spiral pattern (Howard 2000). The 20–70 km-spaced features may have formed in their current location through localised variations in thermal absorption by dust deposits, ice accumulation and ablation (Pelletier 2004). Preferential solar heating through insolation (the amount of solar radiation reaching a given area) causes local sublimation, deepening the troughs. Left over dust deposits cause localised heating events, further modifying the troughs (Bramson et al. 2019). Research using GIS methods supported localised erosion events occurring within segmented troughs that grew to intersect each other in a spiral arrangement. Not only is trough migration study important for understanding the history of the Martian climate, but estimates reveal that ice lost from the North Polar troughs may have accounted for up to 25% of ice-rich material found in the Martian mid-latitudes. This estimate suggests some of the ice-rich material eroded from katabatic wind action migrated equatorward and was preserved during high-obliquity events (Rodriguez et al. 2021).

The 400 km diameter south polar cap in the highlands, 6 km higher in elevation than the northern cap, contains 1.6 million km^3 of water ice (similar in volume to the north polar cap) consisting of two distinct units that were deposited at different periods (Fig. 9.1c). The early unit was deposited and then eroded prior to the superposition of a later polar unit, and both old and younger unit are eroding at 3.6 m/Mars year and 2.2 m/Mars year respectively (Barlow 2008). South Polar Layered Deposits are the equivalent layered features to the NPLD, but have been estimated to have accumulated over 10–30 Myr (Becerra et al. 2019). As with the NPLD, these features present a record of Martian climate change in the southern hemisphere.

South Pole Influence on Martian Atmosphere

The South polar cap is also overlaid by 16,500 km^3 of CO_2 ice, in places up to 1 km thick (Putzig et al. 2018; Alwarda and Smith 2021). Seasonal variations in the CO_2 ice approximate 16–45 m elevation difference (Xiaoting et al. 2022), exchanging volume with the Martian atmosphere. The sublimation and redeposition of CO_2 ice between the Martian southern summer and winter respectively causes atmospheric pressure variations of up to 15% (Buhler and Piqueux 2021). Martian atmospheric pressure variations measured at Gale Crater add to decades of lander mission pressure observations commencing from the first Vikings 1 and 2 landers, where pressure was highest (~ 9.7 mBar) during autumn and lowest (< 8 mBar) near the end of Martian

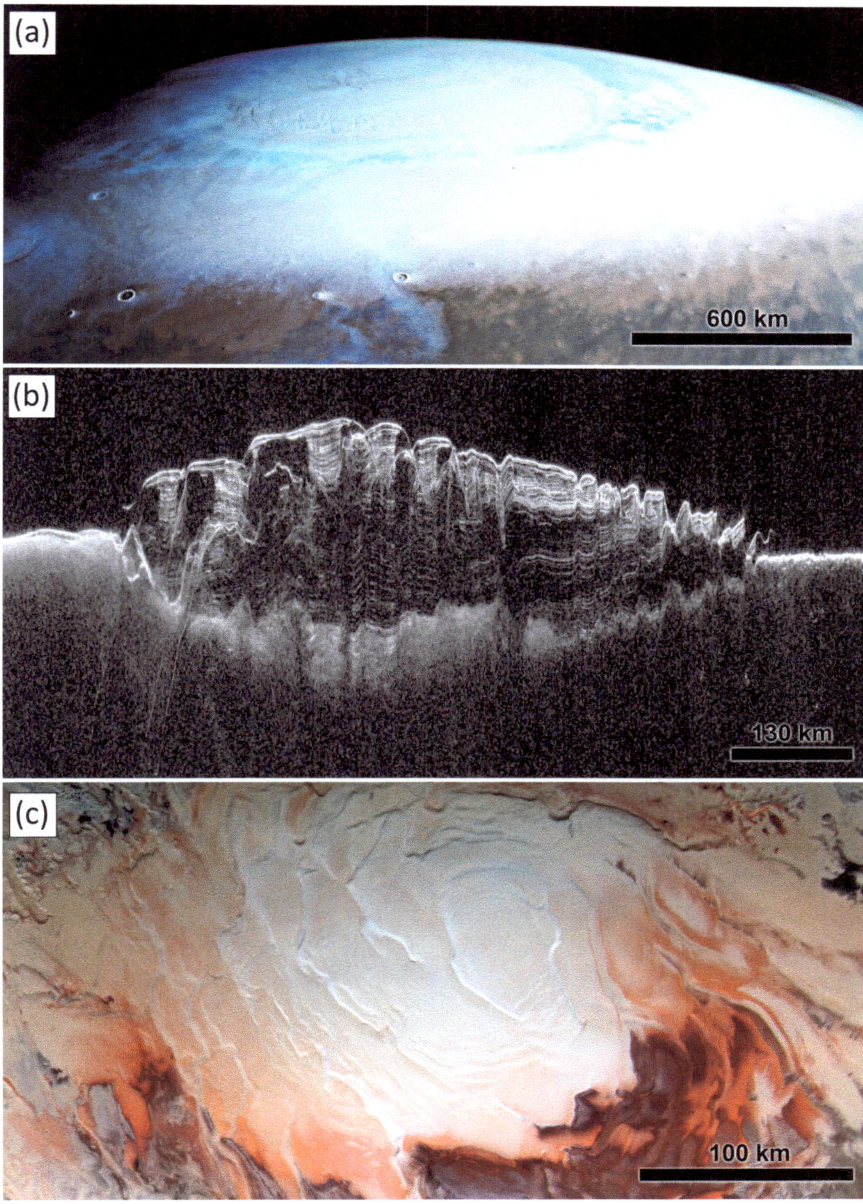

Fig. 9.1 **a** The northern Martian polar cap surrounded by subliming ices in the local spring (ESA/DLR/FU Berlin, CC BY-SA 3.0 IGO). **b** MRO SHARAD cross sectional view of the north pole showing layers of ice deposits, and the depression of the underlying regolith (NASA/JPL). **c** Mars Express image of the Martian south pole (ESA/DLR/FU Berlin, CC BY-SA 3.0 IGO)

winter (Scariah et al. 2023). A perennial deposit identified as the South Pole Residual Cap is observed to endure over the Martian summer. This feature is of importance to atmospheric modellers as its erosion was expected to thicken the Martian atmosphere over time. Decades of observation have failed to identify significant reduction of this residual deposit, and variations of Martian atmospheric pressure have been consistent throughout observed Martian seasons (Lange et al. 2022). This cycle is of importance to mission designers as Martian atmospheric density determination is critical for successful entry, descent and landing operations (Chap. 2).

Remote sensing evidence has demonstrated that H_2O ice in the Martian poles also interacts with the rest of the planet. These exchanges are inferred to occur during the equivalent of terrestrial Milankovitch cycles. Obliquity excursions are likely to have caused deposition of ice in the equatorial regions during its history. Migration of this ice from the equatorial regions to the poles occurs during the Martian summer at low obliquity periods, cycles of which are evidenced by NPLD and south polar layered deposits composition. Contemporary Mars is thought to possess minimal amounts of equatorial ice. In this context, sublimation of mid-latitude ice feeds accumulation of NPLD, with deposition becoming progressively thinner as ice sources move poleward (Vos et al. 2022).

9.1.2 Polar Layered Deposits

The polar layered deposits record ongoing climate change on Mars (Fig. 9.2). The stability of ice deposition and sublimation over the planet outside of the poles has been influenced by Mars obliquity excursions, similar to the Milankovitch cycles described above. Recent research has suggested that Mars may have had as many as 40 ice ages over the past 5 million years, depositing ice-rich material at the mid-latitudes through this process (Schorghofer 2007). The last major glacial period may have ended as recently as 400,000 years ago, with subsequent thickening of the poles by 300 m (Smith et al. 2016). NPLD measurements in particular indicate a reduction in Mars obliquity from 35 to 25° between 4 and 5 Mya, increasing sublimation rates in the mid-latitudes and making polar deposition more favourable.

Even with the ending of the last major glacial period, the abundance of ice-rich material on Mars has led to the discovery of an assemblage of ice-related landforms (Smith 2022).

9.1.3 Martian Ice Landforms

Remote sensing since Mariner 6 and 7 has identified numerous Martian ice-related features, particularly in the mid-latitudes. Terrain softening was identified from these and other spacecraft data, a process absent from the lava and impact-shaped surface of the Moon. Remotely sensed temperature measurements have shown that most

Fig. 9.2 HiRISE image of part of the polar layered deposits near the planet's north pole, showing layering of ice and mafic material (PIA18586, NASAS/JPL)

of Mars is permanently frozen to depths of several kilometres, making conditions favourable for ice preservation. Ground ice in these regions is likely in diffusive equilibrium with the Martian atmosphere, predominantly being cold-trapped within interstitial spaces in the regolith (Fisher 2005).

Recent Excavation Through Cratering

Direct observation of ice deposits from contemporary cratering were made by MRO, where impacts have excavated ice-rich material (Fig. 9.3; Byrne et al. 2009). One impact crater on young Amazonian flood lavas, 35° north of the Martian equator was imaged in 2022 and observations of bright material following the impact revealed the presence of excavated ice. Hyperspectral remote sensing from the CRISM instrument also identified spectra consistent with water ice, and this crater was the largest yet observed to excavate ice at this latitude. This impact was significant as it provided constraints on the presence of subsurface ice at lower latitudes. A past layer of buried ice of at least 30 m may have been deposited by past glacial activity and degraded over time, with remnants subsequently excavated by the crater impact. Climate models suggest this ice deposit is near the limits of stable sub-surface ice on Mars; warmer temperatures closer to the planet's equator would allow frozen water to escape (Dundas et al. 2023).

Mantling

Visible light observations of Martian regions of 30–60° show stippled, softened terrain that is rarely observed closer to the equator. This unit, overlays existing

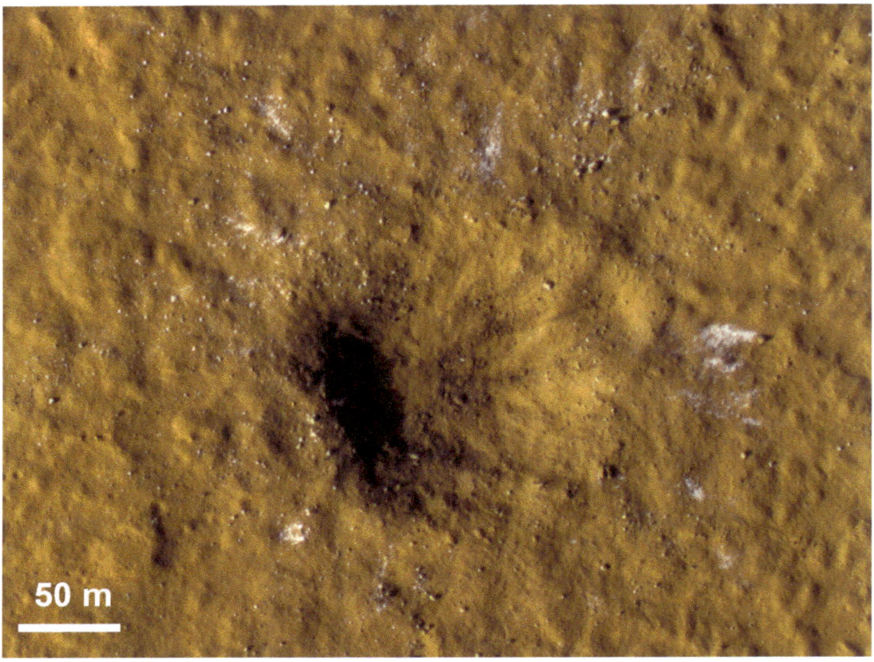

Fig. 9.3 Newly-discovered impact crater that has excavated ice-rich material just 35° north of the Martian equator. This discovery is near the limit of stability of subsurface ice under the current Martian climate (PIA25583, NASA/JPL)

topography and is interpreted to be ice-rich material deposited from the poles during recent obliquity excursions described above. Subsequent sublimation of this material leaves uncemented dust that is prone to deflation, producing the observed pitted texture. "Latitude dependent mantle"" (LDM), a < 10 Ma ice-rich layer estimated to average at least 10 m thick, covers approximately 23% of the northern and southern mid-latitudes of Mars, correlating with GRS detections of subsurface ice in these regions (Fig. 9.4a, b; Feldman et al. 2004; Conway and Balme 2014). LDM is affected by aspect, favouring deposition in pole facing slopes in the mid-latitudes and in many cases has experienced subsequent erosion by hillside gullies.

9.1.4 Ice Flows

The presence of water ice on Mars has been known since pre-space age times (Chap. 1), but the physical discovery and characterisation of ice flows had to wait until the space age. Mariner 9 and Viking orbital imagery from the early space age was of sufficient resolution to identify "fretted terrain" on the surface of Mars (Squyres

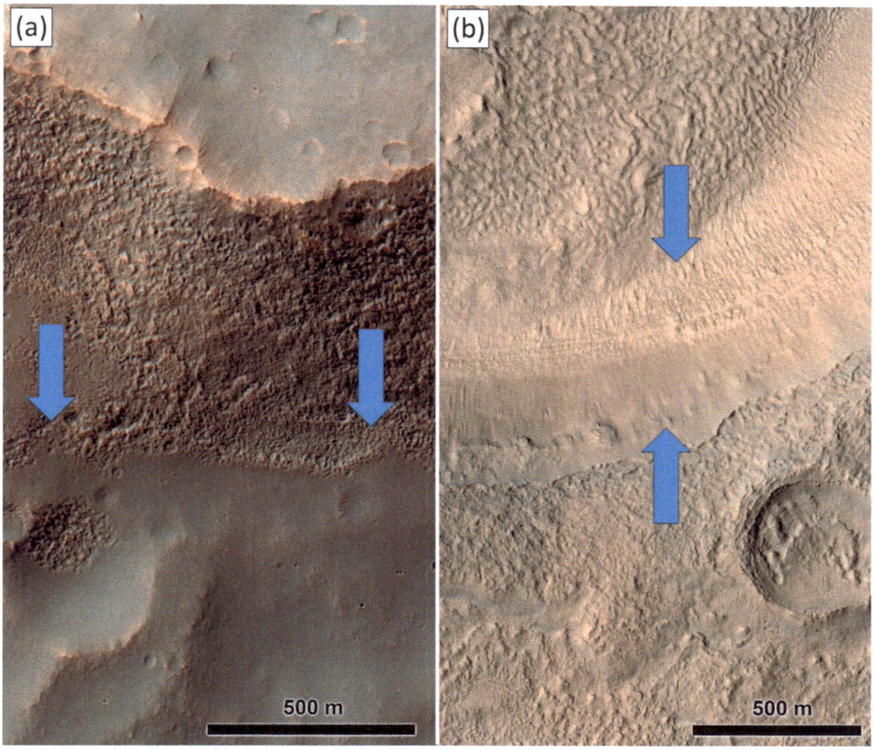

Fig. 9.4 **a** LDM deposits (arrows) covering terrain in Terra Sirenum (ESP037167-1445, NASA/JPL). **b** LDM covering a crater wall in Terra Cimmeria in the Hellas Basin region (ESP048063-1395, NASA/JPL)

1978). Fretted terrain was discovered by Mariner 9 in the northern lowland plains between 30–50° N and 0–60° E. It consists of numerous, angular mesas surrounded by flat-floored valleys.

The features were deposited on topographic lows and had lobate terminations and striations that paralleled valley walls and radiated outward from their source. Terrestrially, these geomorphic features are associated with indicative of viscous ice flows (Squyres 1978).

Viscous flow features (VFF) were imaged during the Viking orbiter mission, showing evidence of deformation of ice-rich material (Squyres 1978, 1979). VFF morphology includes lobate debris aprons (Fig. 9.5a), lineated valley fill (Fig. 9.5b) and concentric crater fill (Fig. 9.5c). Early interpretations postulated formation through small amounts of ground ice interactions with regolith (< 30%), though more recent multi-disciplinary work has shown substantial (90%) ice content was a better fit with observed geomorphology.

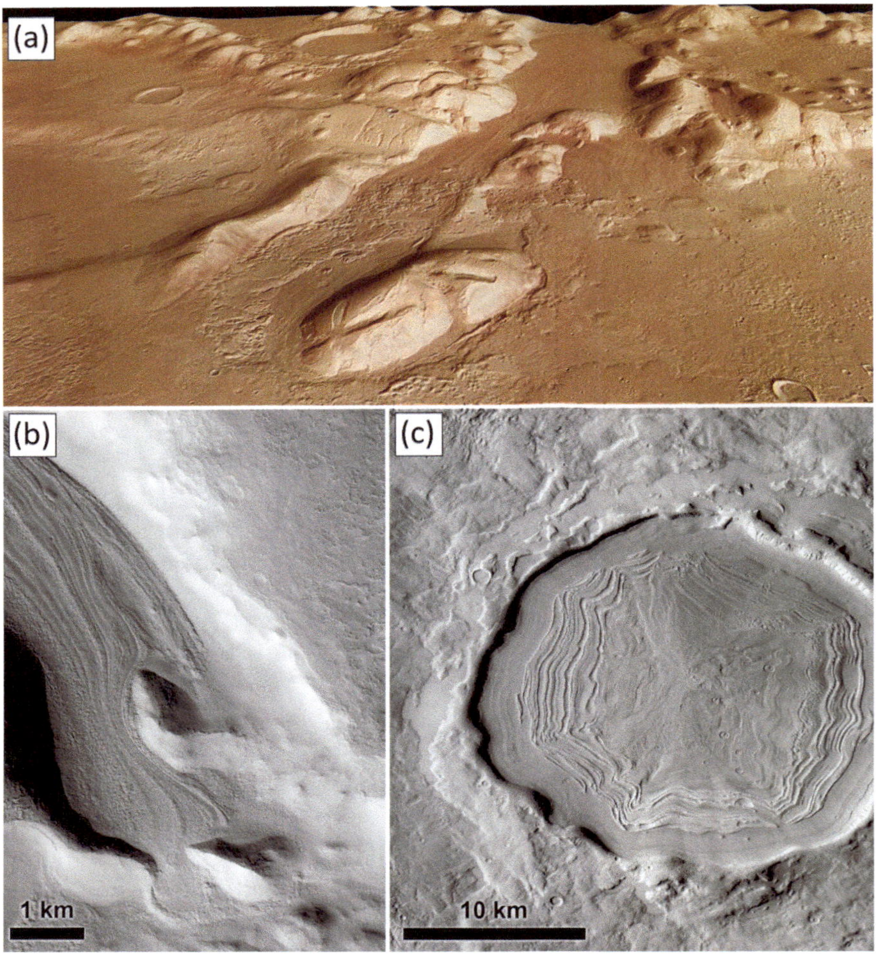

Fig. 9.5 **a** Mars Express perspective view of viscous flow features (lobate debris aprons) at Phlegra Montes in the northern plains of Mars (ESA/DLR/FU Berlin, CC BY-SA 3.0 IGO). **b** Lineated valley fill approximately 2.5 km across in the inside the Casius Quadrangle, an ice-rich region in the Martian northern plains (ESP_048658_2150, NASA/JPL). **c** Crater concentric fill in a 30 km wide impression inside the Casius Quadrangle (B22_018144_2182_XN_38N284W, NASA/JPL)

Lobate debris aprons extend up to tens of kilometres from the bottom of steep cliffs in the mid-latitudes. Evidence of viscous flow in these features is demonstrated by observations of flow around obstacles, and striations parallel to the direction of flow (e.g. Figure 9.5a). They also possess convex profiles that are characteristic of terrestrial ice flows and have low slopes of 1–3°. Orbital radar analysis indicates lobate debris aprons may be up to 1 km in thickness and be made of over 90% ice (Plaut et al. 2009). Recent analysis of boulder depositions has suggested that lobate debris apron formation, flow and sublimation are sensitive to Martian obliquity excursions. Pole-facing features favour more accumulation, being protected from direct sunlight for much of the Martian year than equator facing aprons. Increased accumulation would assist in higher flow rates at different periods, as evidenced by multiple boulder bands (Levy et al. 2021; Dundas et al. 2019).

Lineated valley fills occupy existing valleys and are common near the Martian dichotomy (e.g. Figure 9.5b). They are textured with linear patterns consisting of aligned pits and mounds along the direction of the fill. These striations are likely to be caused by convergence of flow fronts from depression walls, and downslope flow that can extend for tens of kilometres before terminating in convex lobes. Lineated valley fill is estimated to be 100–500 Ma in age, with deformation caused by an abundance of ice-rich material that is overlaid with debris deposits (Morgan et al. 2009).

Interior crater deposits, including concentric crater fill are not observed in equatorial regions and exhibit surface grooving oriented in a single direction or concentric to the host crater walls (e.g. Figure 9.5c). Initial hypotheses pointed to development through aeolian emplacement and removal, though subsequent ice-related discoveries, including radar detections, have identified these features as being a type of viscous flow, related to terrestrial glacial activity (Levy et al. 2010a). Dating as young as 700–10 Ma, interior crater deposits have estimated thicknesses of up to 1.7 km and can exist in craters as small as 2 km in diameter (Levy et al. 2014).

Glacial Flows

Glacial activity occurs on multiple bodies throughout the Solar System, such as the Earth, Pluto, and Mercury (Howard et al. 2017; Fastook et al. 2019). Higher-resolution imagery from subsequent orbiters have produced a wealth of evidence of the presence of small-scale ice flow features, corresponding to regions where the presence of water was identified by Mars Odyssey's gamma ray spectrometer (Bandfield 2007). These small-scale features included forms resembling terrestrial glaciers on Earth, with evidence of viscous deformation and downflow from cirque origins and lobate terminations (e.g. Fig. 9.6a). Termed "glacier-like forms" (GLF), 1309 of these features are known to exist, primarily between 30 and 60° bands north (56%) and south (44%) of the Martian equator (Hubbard et al. 2014).

GLFs exhibit a pole-facing preference for their location, where water ice deposition and preservation would be likely, often being shaded from direct sunlight. The precise composition of GLF is currently undetermined, as a covering of fine-grained regolith obscures direct spectral observations. The amount of water ice present in GLF is a popular topic of speculation by many researchers, as well as their age and present-day activity. It has been suggested that the arid Amazonian period was unfavourable for ice accumulation on Mars, although global obliquity changes of over 30° may have allowed deposition of ice-rich material closer to the equator. Local topographic environments, coupled with these excursions allow ice to accumulate faster than it can sublime (Brough et al. 2019).

The cold temperatures of Mars are also thought to "lock" many ice-rich features in place. On Earth, the pressure of overlying ice forms a liquid at a glacier's base,

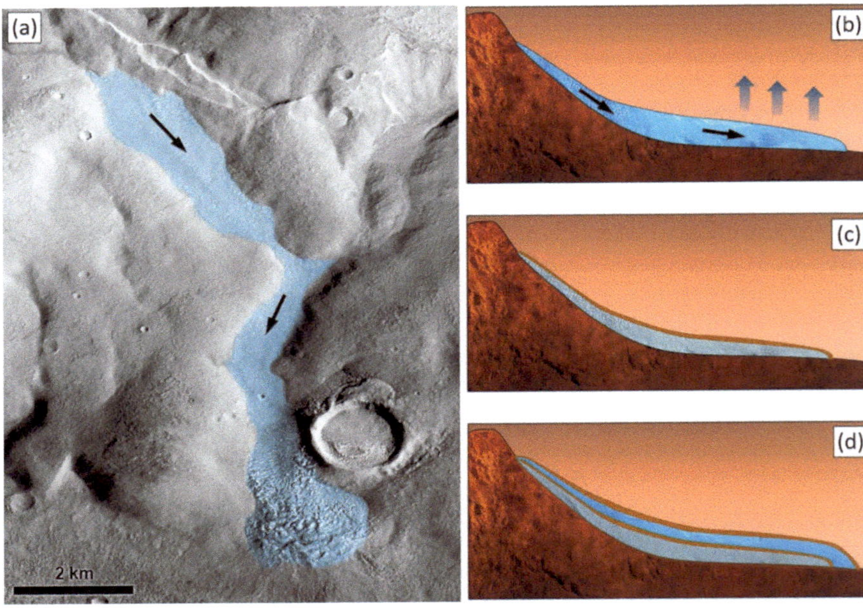

Fig. 9.6 **a** Example of an 8 km long glacier-like form (highlighted in blue) in Noachis Terra with downflow direction of ice-rich material highlighted by arrows (D06-029513_1401_XN_39S336W, NASA/JPL). **b** Accumulation and downslope flow of ice-rich material. This material sublimates into the Martian atmosphere over time (blue arrows). **c** Deposition of debris material may preserve the ice-rich layer. **d** Subsequent ice and debris depositions may occur over time, creating a layered feature

allowing it to creep downhill. So-called cold-based glaciers are unable to generate liquid water in this way and will move only through deformation of solid ice, a much slower mechanism that is thought to occur more readily on Mars (Hubbard et al. 2014).

SHARAD radar measurements have indicated GLFs composition are consistent with an abundance of water ice (Plaut et al. 2009). Recent research has also uncovered the presence of eskers associated with GLF. On Earth, eskers are ridges of sediments that are deposited by meltwater flowing in tunnels under glaciers. Their existence on Mars suggests viscous strain, possibly combined with geothermal heating provide enough energy for these features to flow (Butcher 2019). Chevron shaped crevasses have been observed where the centre of the GLF has flowed faster than the edges due to less friction with the surrounding walls, with downslope debris forming moraine-like ridges (Arfstrom and Hartmann 2005).

Are these features active today? Age estimates of GLF are difficult due to their small size, making it hard to estimate via crater counting methods. Nevertheless, age inferences have shed light into the past ice-age history of the late Amazonian period. Estimates indicate GLF started forming from 65 Mya, with major glacial wax and wane events occurring 45–65 Ma and again from as recent as 20–2 Ma ago

(Hepburn et al. 2020). During these cycles, GLFs would accumulate during glacial events (Fig. 9.6b), and leave ice-rich material protected by surface debris following sublimation and retreat (Fig. 9.6c). Subsequent iterations may have built multiple layers in the GLFs prior to the cessation of most major glacial activity from 2 Ma ago (Fig. 9.6d). Concurrent deposition occurring in the north polar cap from reduced obliquity excursions may have caused the overall shutdown of GLF activity (Smith et al. 2016; Hepburn et al. 2020).

Discoveries of remnants of glaciers near the Martian equator have caused a rethink of the extent of water ice and glacial activity on Mars (Fig. 9.7a). Equator-based regions were considered to be above the frost point (~ 200 K, Schorghofer 2020), the maximum temperature where water ice would remain stable on the Martian surface before sublimating (Mellon et al. 2004). Very recent analysis of HiRISE and CRISM remote sensing data revealed a 6 km long feature possessing crevasse fields and moraine bands associated with past glacial activity. Light toned deposits were identified as sulphate salts, suggesting a past interaction between ice and volcanic processes within Valles Marineris to the east of Noctis Labyrinthus (Lee et al. 2023). Along with the contemporary crater impact excavating ice described above, this finding indicates that ice has existed closer to the equator than previously thought.

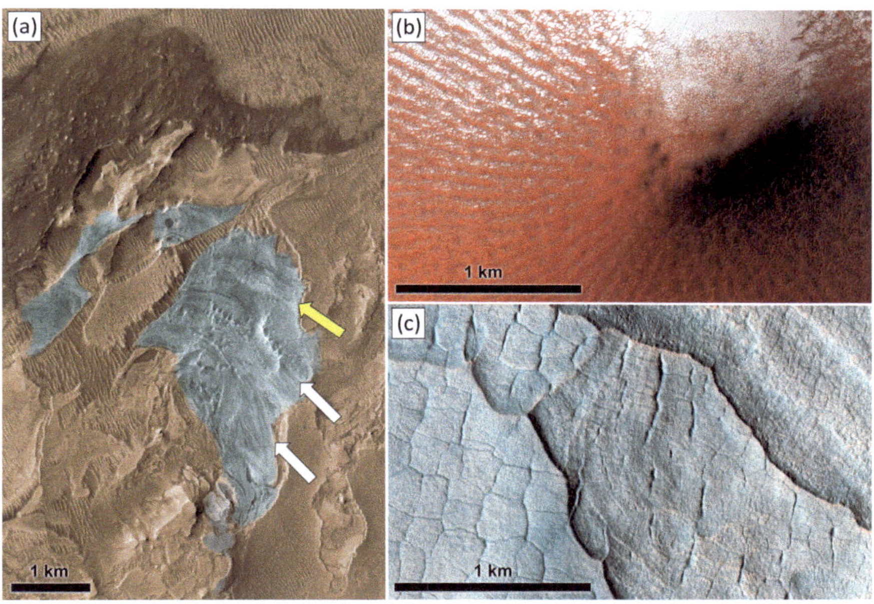

Fig. 9.7 **a** Likely remains of a ~ 1 km wide glacier near the Martian equator, colourised to highlight its location. White arrows denote the location of thrust bands and thrust moraines while the yellow arrow identifies crevassing of a type associated with glacial morphology (CTX and HiRISE composite, NASA/JPL). **b** HiRISE image of permafrost within the 36 km wide Louth Crater in the Martian north pole (ESP037551-2540, NASA/JPL). **c** Thermokarst features in the Martian northern hemisphere. The larger polygons are up to 100 m in size (PSP002070-2250, NASA/JPL)

Carbon Dioxide Glaciers

The presence of a CO_2 cycle on Mars, plus glaciation processes occurring on Pluto have triggered research into finding evidence for these types of glaciers. Recent modelling analysis identified discrepancies between expected and observed CO_2 deposition at the south polar cap, suggesting that glacial flow was required to distribute the material into its present position (Smith et al. 2022). Gravitational pressure in the polar regions is sufficient to liquify CO_2, enabling viscous flows to form. Visual observations of crevasses within the South Pole indicate that CO_2 ice has undergone glacier-like creep to pond in geographic lows (Cross et al. 2020).

Permafrost

In addition to near-surface ice that exchanges with the atmosphere, remote sensing has provided evidence for semi-permanent ice deposits below the surface in the Martian mid-latitudes (for example, terrain located within Louth Crater in the northern hemisphere, Fig. 9.7b). Following the Mars Odyssey mapping of subsurface ice from the mid-latitudes (Mellon et al. 2004), thermal inertia analysis indicated subsurface permafrost from 50 to 80° north and south of the equator. This layer may extend kilometres into the Martian crust, as suggested by petal-like impact craters whose ejecta appears to originate from fluidised material, possibly melting of subsurface ice. Subsurface radar analysis has also assisted in characterising the nature of subsurface ice on Mars (Orosei et al. 2015; Berman et al. 2021).

However, the highly reflective nature of near-surface ice makes deeper radar penetration and detection of deeper deposits difficult. Polygonal patterned ground, similar in appearance to that found in terrestrial periglacial settings of Earth, has been well documented on Mars and imaged by the Phoenix lander (Levy et al. 2010b). These occur at scales of metres to 20 km across, compared with 3–30 m range for Earth-based features (Lachenbruch 1962) but one interpretation is formation by ice-related processes in these areas (Mellon 1997). GIS analysis using techniques described in Chap. 6 of HiRISE imagery of Martian south pole polygons suggested ice-related plastic creep and deformation (Zhang et al. 2022).

Thermokarst

Terrestrial thermokarst are ice-related features that reveal the past presence of ice. When ice is removed, the host regolith is weakened and can locally subside and collapse, creating distinctive patterns that are visible from orbit. On Earth, ice content must be high enough to cause the ground to collapse, and this normally occurs through melting (French 2017). In this way, the presence of thermokarst serves as a marker for the past presence of ice-rich material and, where it is non-seasonal, can help provide information on past periglacial environments and climate change (Soare et al. 2011).

On Mars, features identified as thermokarst were large enough to be observed by the hundreds of metres resolution imagery of Mariner 9 and Viking (HiRISE example, Fig. 9.7c; Sharp 1973; Costard and Kargel 1995). Thermokarst features on Mars are distinguished from other features, such as impact craters, because they are rimless depressions with pole-facing slopes consistently steeper than those facing the equator, suggesting influence from insolation. Thermokarst are also present in southern Chryse Planitia, Hellas Basin and Utopia Planitia (Costard and Kargel 1995). The features in Utopia Planitia also exhibit polygonal overprinting and layering, providing further evidence for the presence of ice-rich material in the past. Recent landscape modelling of thermokarst formation showed a rarity in the availability of ice-melt under the past climate conditions leading to their formation. Instead, it is more likely that localised sublimation of ice up to tens of metres thick, caused collapse and development of the thermokarst. This differs from the liquid water-based process commonly occurring on Earth, showing that similar landforms can form through different processes (Dundas et al. 2015).

9.2 Water

The centuries old belief that Mars once contained oceans of water (Chap. 1) has heavily influenced remote sensing of the planet to the present day. Following the Viking missions, NASA has adopted a "follow the water" approach, with mission instruments and landing site selection tailored to identify evidence of past aqueous environments. While pre-space age notions of Mars canals feeding advanced civilisations have long since vanished, decades of exploration have demonstrated that liquid water has played a major role on Mars.

9.2.1 Valley Networks

Among the earliest indications for past water action on Mars were the discovery of extensive valley networks by Mariner 9. Up to 40,000 of these features have since been identified, some extending for thousands of kilometres and displaying evidence of deep valley erosion to bedrock. The branching valley networks exist on the oldest terrain on Mars, the Noachian, mostly between 60° south to 10° North (examples illustrated in Fig. 9.8a, b). They are characterised by branching (dendritic) channel networks that are similar to river networks on Earth. The total volume of liquid water required to erode the observed Noachian valley networks is estimated to be 6.86×10^{17} m^3, or a standing ocean of 5 km depth (Luo et al. 2017), with minimum formation timescales of up to 10 Ma (Hoke et al. 2011).

Proposed formation mechanisms for these Noachian-age valley networks range from rainfall to underground water. Observations of the headwaters of the valley networks is of high interest to researchers to help characterise the main processes for valley network formation. Some research has used the presence of amphitheatre heads of valleys to suggest groundwater flow, formed where material excavated by groundwater causes collapse of the overlying regolith and retreat by up-slope sapping (Gulick 2001). Some research suggests the highly dendritic nature of the

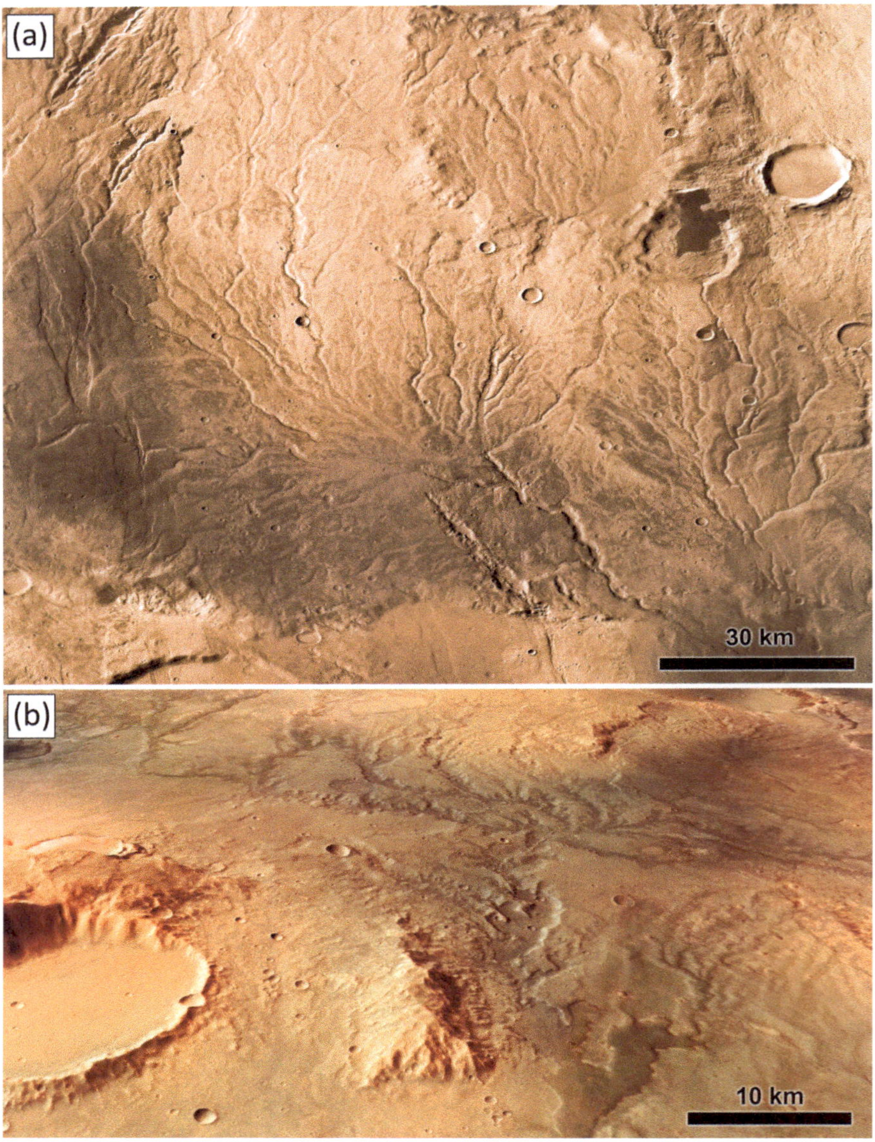

Fig. 9.8 **a** The heavily networked terrain of Warrego Valles, one of the large valley networks on Mars (colourised Viking mosaic, NASA/JPL). **b** Perspective view of another valley network to the east of Huygens Crater in the Martian southern highlands (ESA/DLR/FU Berlin, CC BY-SA 3.0 IGO)

valley networks and the presence of interior channels to indicate drainage by rainfall (Craddock and Howard 2002).

Past rainfall is further supported by the presence of degraded craters that lack central peaks, clearly-defined rims and have shallow floors (e.g. Figs. 1.12c and 8. 4a). As these craters are restricted to Noachian-aged terrain and do not appear in younger Hesperian or Amazonian landscapes, some researchers have argued that erosion caused by rain splash that declined early in Mars' history was the likely mechanism (Craddock and Lorenz 2017).

A local region exhibiting evidence of past rainfall is Mojave Crater. Heat generated by the impactor is thought to have produced localised, heavy rainfall, carving channels that have eroded to the crater rim (Fig. 9.9a; Goddard et al. 2014). Local environmental effects may assist with reconciling the "warm and wet" scenario with models of early solar formation. In these models a Noachian-aged Sun produced 70% of the current solar energy output, presenting difficulties in keeping much of Mars above freezing (Wordsworth et al. 2013).

Fluvial Estimations

Strahler's stream order is commonly used to identify the level of branching present in fluvial systems. First order features possess a single channel, whereas merging of two streams will create a resulting channel of a higher order (Shreve 1966). On Mars, as on Earth, the host slope of a fluvial system can influence the level of branching observed. Steeper slopes tend to produce lower order channels, as the higher velocities of water flows find the least resistive path to a drainage level. Smaller velocity flows of lower slopes increase the influence of resistive material to erosion, where the fluid may meander around obstacles, and changes in slope causes channels to merge into larger features. Strahler orders may also be used to infer the level of development of valley networks based on the hypothesis that, slopes being equal, larger and high-order systems would take longer to develop through more extended fluvial erosion than lower order, smaller networks (Mest et al. 2010; Shi et al. 2022).

Discharge calculations can provide an estimate of the amount of liquid water required to erode a feature, providing constraints on environmental and climate modelling of Mars' wetter past. Values calculated from valley networks and outflow channels provide upper and lower limits on the inventory of liquid water that needed to be available to erode these features during the Noachian and Hesperian eras. Stream discharge (Q), used for terrestrial hydrology, is also applicable for Mars geomorphic analysis and is a function of cross-sectional area and flow velocity:

$$Q = va \tag{9.1}$$

Manning's equation is the usual method for calculating flow velocity on Earth and has been used for Mars applications (Wilson et al. 2004). The equation provides a calculation of velocity v as:

$$v = (1/n)R^{2/3}S^{1/2} \tag{9.2}$$

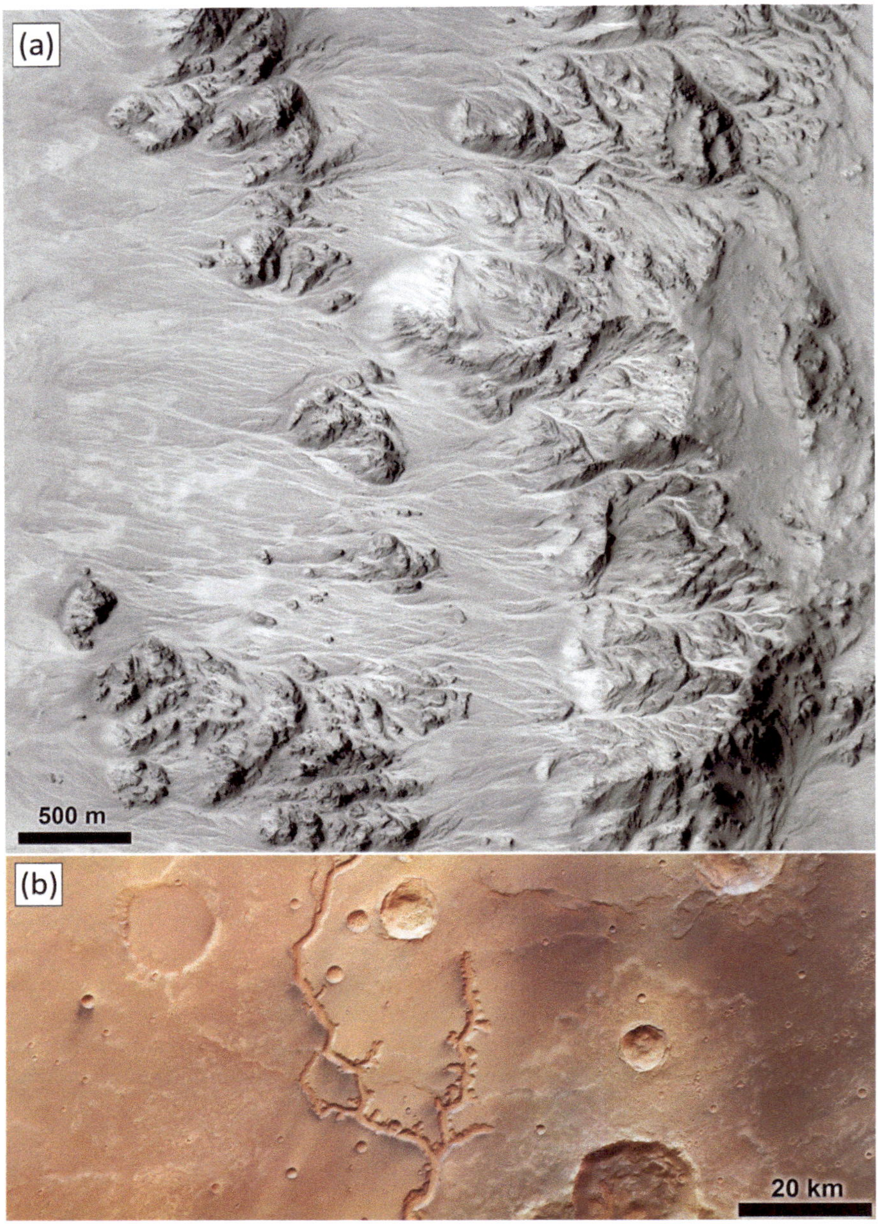

Fig. 9.9 **a** Fluvial features in Mojave Crater as imaged by HiRISE (PSP001415-1875, NASA/JPL). **b** The upper reaches of Nirgal Vallis, with the complete system stretching 700 km across the southern highlands (ESA/DLR/FU Berlin, CC BY-SA 3.0 IGO)

where

- v is the average velocity of the flow in m s^{-1},
- S is the tangent of the water surface slope,
- n is the Manning coefficient, and
- R is the hydraulic radius.

Hydraulic radius R represents the ratio between channel cross-sectional area of the wetted channel (A) and the portion of the channel in contact with water, being the 'wetted perimeter' (Pw) in the following relationship:

$$R = A/Pw \tag{9.3}$$

The properties of terrain are particularly important as roughness of a stream bed and side walls effects the friction acting on the flowing water and hence its velocity (Birot 1965), affecting the value of n. Terrestrial values range from 0.025 for smooth, low-friction streams to 0.75 for rocky, rough water courses.

A Manning's equation for Mars applications, adapted for Mars gravity, is as follows (Komar 1979):

$$v = (1/n)\left(g_m/g_e\right)^{1/2} R^{2/3} S^{1/2} \tag{9.4}$$

where the new variables are:

- g_m is the gravity of Mars (3.72 m/s), and
- g_e is the gravity of Earth (9.8 m/s).

Applying Manning's equation to an ancient water-carved Martian channel in Nirgal Vallis, south-east of Valles Marineris (Fig. 9.9b; Szilágyi-Sándor and Székely 2025), assuming the following values:

- Channel width w = 20 m,
- Flow depth d = 2 m,
- Slope S = 0.01,
- Manning's n = 0.03 (using a natural channel value).

To calculate R, first:

- Determine cross sectional area $A = w \times d = 20 \times 2 = 40\ m^2$,
- Wetted Perimeter $P = w + 2 \times d = 20 + 4 = 24$.

R is then A/P = 40/24 = 1.667 m.

Using Manning's equation provides:

- $V = (3.71/9.81)^{1/2} \times (1/0.03) \times (1.667)^{2/3} \times (0.01)^{1/2}$
- $V = 0.614 \times 33.33 \times 1.41 \times 0.1 = 2.88$ m/s.

9.2.2 Outflow Channels (Hesperian)

Large outflow channels are another large-scale feature eroded by liquid water. Dating from the Hesperian and Amazonian eras, they share some morphology with catastrophic floods on Earth, such as the Washington scablands. They have wider width-depth ratios when compared to valley networks, linear, longitudinal grooves and teardrop islands, and are often linked to chaotic terrain. These features were first imaged during the Mariner 6 and 7 flybys and were one of the first landform types that pointed to a more complex geology for Mars than Earth's Moon (Chap. 1; Sharp 1973).

Identified as the source of outflow channels, chaotic terrain features jumbled and hummocked geology, lacking an analogous process on Earth. Hydrologic analysis of Hesperian outflow channel morphology demonstrates a lack the dendritic catchment areas and the channels often originate from a single upland source. Hesperian-aged outflow channels and associated chaotic terrain sources are believed to have origins linked to localised volcanism events. Heating of ice-rich regolith would cause melting and releasing large amounts of subsurface ice, causing the remaining regolith to collapse into a series of 1 km or more deep hummocks and troughs.

An example is Aram Chaos, a 280 km wide basin located in eastern Chryse Planitia near the Martian dichotomy boundary. A 15 km wide and 2.5 km deep channel connects the depression to the larger Ares Vallis feature that drains to the northern lowlands (Lichtenberg et al. 2010). Following its formation at least 3.7 Ga, the impact crater was filled with sediments porous enough for accumulating subsurface water ice in voids between the deposited particles. This process was believed to have taken hundreds of millions of years, during which time modest geothermal heating was sufficient to destabilise the subsurface ice deposits. Localised ponding events from ice melt fed groundwater flow and small-scale erosion on the eastern rim to a point where it is breached. Collapsing overlying sediment and catastrophic water release through the breach in the eastern rim released an estimated 9.3×10^4 km^3 of liquid water, creating the connecting channel to Ares Vallis in tens of days (Fig. 9.10a, b; Roda et al. 2014).

The Aram Chaos fluvial event seems typical of the process by which Hesperian-related outflow channels were created. The low Martian temperature and atmospheric pressure estimations in the Hesperian period was less likely to support long term stability of liquid water on the surface; exposed water would freeze and sublime, meaning that outflow events carving channels of this period were likely short lived (Warner et al. 2010).

Inverted streams such as Antoniadi Crater in Syrtis Major and Medusae Fossae have also been discovered on Mars (likely example in Antoniadi Crater, Fig. 9.11) and are indicative of past fluvial activity. Processes for formation of these features on Mars are interpreted to be the same as those on Earth, where fluvial erosion leaves stream beds of more resistant material lying higher than the surrounding regolith. Over time, the overlying material is eroded, leaving the hardened stream bed exposed (e.g. Pain et al. 2007).

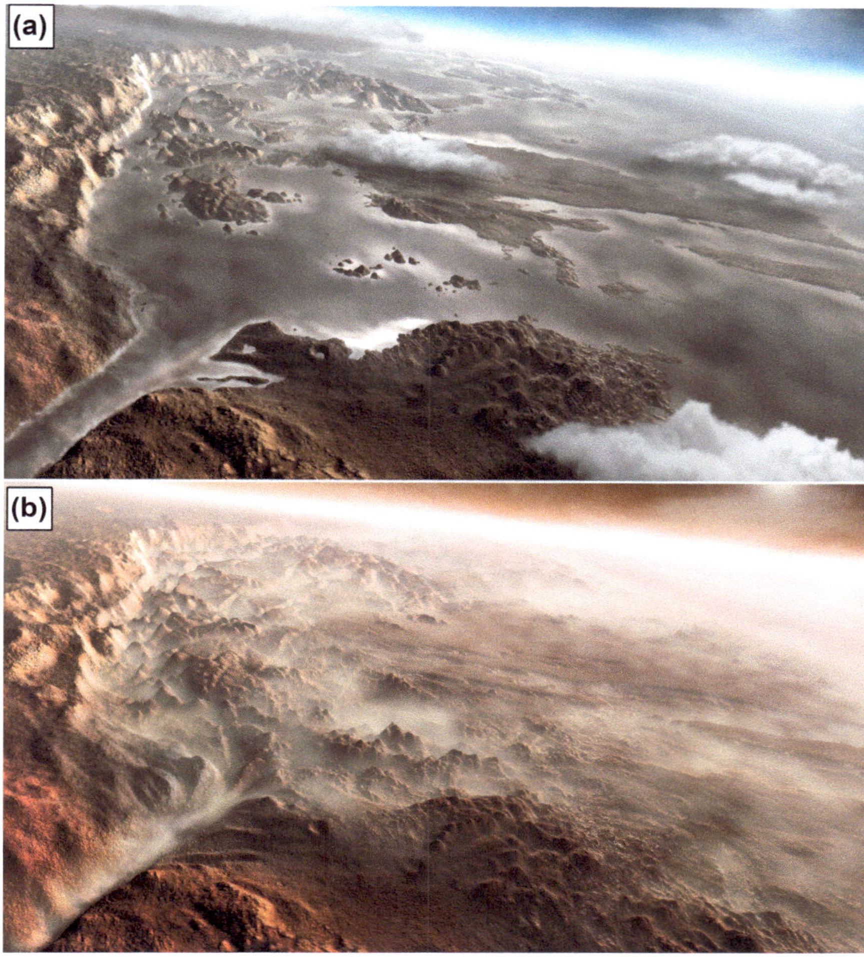

Fig. 9.10 **a** Illustration of Aram Chaos, following the point of collapse and releasing a short-lived flood of water to the surrounding landscape. **b** Illustration of Aram Chaos as it appears in the present day

9.2.3 A Case Study: The Kasai Valles System

Mars' largest outflow channel, at 2000 km long and 2.5 km deep, is in Kasei Valles in the northern hemisphere. It exhibits braided textures and teardrop formations characteristic of erosion by energetic fluid (Fig. 9.12a). This feature originates from the topographically high Tharsis Bulge region at the head of a Valles Marineris trough at 1° South, 80° West and north of Echus Chasma. The outflow channel flows north to Uranius Dorsum ridge before running east and debouching onto the Chryse Planitia plains north of the Martian Dichotomy. Estimates of discharge rates

Fig. 9.11 Branching features on the floor of Antoniadi Crater that are likely inverted channels (ESP012435-2015, NASA/JPL)

based on MOLA and hydrological analysis suggested rates between 8 × 10^4 and 2 × 10^7 m^3/s, comparable with large-scale, catastrophic flooding on Earth (Williams et al. 2000; Fig. 9.12b). Flooding was likely caused by a combination of volcanism, tectonics, collapse and subsidence in the Tharsis region that released massive amounts of groundwater.

The large size of Kasei Valles has caused it to cut across diverse terrains. The outflow system has cut through Hesperian-aged plains material and chaotic units, with the valley floor estimated to be Hesperian. Amazonian lava flooding has altered most of the norther half of Kasei Valles, and younger units covering most of the southern half to Echus Chasma (Tanaka et al. 2005). As a result, Kasei Valles is representative of interactions of different geologic processes that have occurred over the 3.7 Ga life of the feature. The oldest surfaces consisting of the Valles main path, north and south tributaries occurring from 3.7 to 3.5 Ga from two major flooding events. From 3.4 to 3.0 Ga, erosion was modified through topography uplift and subsequent subsidence, coinciding with late-stage valley network erosion elsewhere on Mars. Subsequent erosion to 2.0 Ga was less energetic, indicating a decline in the availability of liquid water and collapse of large-scale hydrology (Duran and Coulthard 2020).

As the hydrosphere shut down on Mars, subsequent alteration of Kasei Valles was driven by glacial ice-melt-related activity and volcanic processes. Distinct episodes

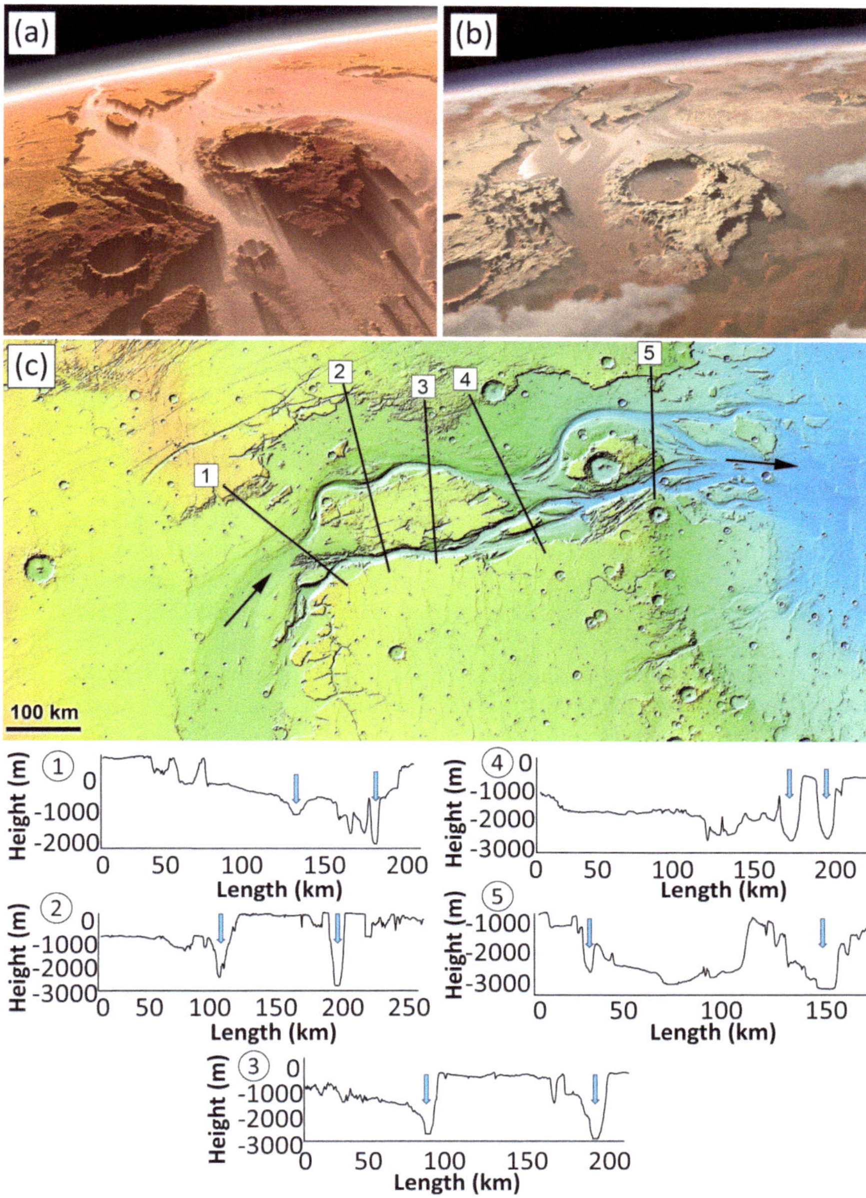

Fig. 9.12 **a** Perspective view illustration of Kasei Valles as it appears in the present day. **b** The outflow feature as it may have appeared during a catastrophic flooding event early in Mars' history. **c** Topographic overview of the Kasei Valles outflow channel, with flow originating north of Valles Marineris (bottom left arrow) and draining into Chryse Planitia (far right arrow). Five cross-profile samples for hydraulic analysis are numbered, with the cross profiles appearing below. Blue arrows denote the location of channels

occurred at least four times through the Amazonian, with the youngest lava flows dated to 100 Ma and near-surface ice and glacial melting persisting to recent periods as late as 70 Ma (Chapman et al. 2010). This has meant that present-day Kasei Valles is the product of alteration through multiple mechanisms of volcanism, flooding and ice-melt.

To help assist in interpreting past actions of liquid water on Kasei Valles, researchers have adapted terrestrial fluvial analysis methods to Mars applications. While it is impossible to directly measure flow rates in the absence of liquid water, values may be inferred with constraints using geomorphic analysis and adapted hydrological tools. Fluid velocity as it moves downhill is governed by a combination of viscosity, gravity and terrain roughness. Erosion by water normally contains enough energy to transport particles of regolith downstream, with debris aprons or alluvial fans being created from this debris. Evidence of fluvial deposition may be modified or erased by subsequent surface processes, such as lava flows or aeolian erosion or deposition; in the case of Kasei Valles, deposits are found to the north-east, where the feature debouches onto Chryse Planitia (Yousif 2018).

High-resolution topographical information from MGS MOLA, Mars Express HSRC and other missions has made calculating the cross-sectional area of Martian channels possible. Exploitation of this data entails using a customised GIS and deriving cross-profiles at intervals along the feature of interest (Fig. 9.12c). Estimations of water height are inferred from interpretation of changes in topography as measured from cross-profiles and imagery analysis. This is made difficult where the feature has undergone multiple episodes of erosion, as is the case of Kasei Valles. In Fig. 9.12c, five cross-profiles obtained from the MOLA data show the location of channels (blue arrows), and associated banks. Identification of bank slope changes may assist in inferring fluid height within the constraints of subsequent erosion activity described above.

Recent research uses high-resolution imagery to infer the chronological sequence of events through crater counting, mapping of superposition and geomorphic analysis. This research produced more accurate discharge estimates of Kasei Valles (2–4 orders of magnitude lower) when compared to analysis conducted with older Viking data (Williams et al. 2000). It is important to note that this and other research can only provide indicative estimates in inferring fluvial discharge through adaptation of terrestrial hydrological analysis. In these cases, results from the application of Manning's equation or other terrestrial analogue fluvial models should be interpreted as placing upper bounds on water flow.

While inherent uncertainties will persist in reconstructing ancient Martian fluvial activity, advances in high-resolution digital terrain models and multispectral remote-sensing datasets offer a path toward more reliable interpretations. These tools enhance our ability to quantify channel dimensions, assess the geological framework of valley systems, and identify minerals indicative of aqueous alteration.

9.2.4 *Crater Lakes*

On Mars, impact craters can form topographic lows, and breaching of their rims has led to crater lakes where valley networks have fed water to the crater floors. The Noachian highlands contain an abundance of crater lakes that often are associated with the valley networks described above (Fassett and Head 2008). Interest in these features is such that three dedicated surface missions have landed in Gusev (illustrated in Fig. 7.12a), Gale and Jezero Craters, all postulated to have had liquid water, to conduct surface geologic analysis. Gale Crater, investigated by orbital remote sensing and the Curiosity Rover (Fig. 9.13a), likely hosted at least two lakes, up to 1200 m deep. While the Gale Crater lakes likely formed during torrential rainfall periods (Heydari et al. 2023), other crater lakes may have been formed through a number of other processes, such as melting of surface or subsurface ice. As described above, landform analysis of the origin of the channels can be used to infer the water source causing the erosion.

Most of the crater lakes discovered on Mars are the closed basin type, where water has entered the crater through a rim breach, and ponded without draining away. Research has suggested that the crater lakes formed during Hesperian and Amazonian periods were relatively short lived. This is supported by orbital or and rover mineralogical analysis of crater floor material that suggested much of the chemical alteration by water, occurred during the Noachian periods (Ehlmann and Edwards 2014).

9.2.5 *Fluvial Deltas*

Fluvial deltas are present in at least 206 craters predominantly within 40° north and 54° south of the equator, likely where temperatures were more favourable for the existence of liquid water (Eberswelde Crater example, Fig. 9.13b). These features share the geomorphology of terrestrial fans, where sediment is deposited where fluvial action from an upland channel (Stock 2013). Estimations of formation rates for Martian alluvial fans range from tens to hundreds of thousands of years in conditions of increased water availability through snowmelt or runoff (Morgan et al. 2014). The hundreds of millions of years ranges in age of Martian alluvial fans suggest conditions favourable for their formation may have extended into the Amazonian era (Wilson et al. 2021).

Another prominent example of a fluvial delta is within Jezero Crater on the western edge of Isidis Planitia (Fig. 9.13c). The 49 km diameter crater is approximately 4 Ga old and is incised on its western rim with a channel that terminates in a fan-shaped delta deposit that formed over a period of up to 10 million years. Research into the evolution of the landscape suggests the crater experienced two distinct fluvial events, with water erosion ceasing by the Hesperian. Jezero was of particular interest to researchers as its location fit within the engineering constraints of a lander mission,

Fig. 9.13 **a** Perspective view of Gale Crater with Curiosity Rover landing ellipses shown (ESA/DLR/FU Berlin, CC BY-SA 3.0 IGO). **b** Eberswalde Crater delta-line fan, likely fluvial in origin (PIA25543, NASA/JPL). **c** Remains of a fluvial delta at Jezero Crater, Mars (ESA/DLR/FU Berlin, CC BY-SA 3.0 IGO)

plus the presence of clays detected in the vicinity of the delta made it a prime candidate for preserving evidence of past microbial life. Orbital analysis of Jezero using TES and CRISM instruments were consistent with olivine minerals formed by magma in the presence of water (Sun and Stack 2020). Syrtis Major, a shield volcano, produced lava flooding to the south of Jezero Crater, with glacial activity likely blocking this flow and facilitating water erosion through a series of channels draining into Jezero and other areas of Isidis Basin (Matherne 2019). Investigations by the Perseverance Rover suggest basalts from Syrtis Major were transported by water, to be deposited into the floor of the crater. Many deposits observed by Perseverance are rounded, indicating mechanical erosion by extended downstream water transport. The precise source of the water, such as glacial melt or catastrophic failure of an upstream dam, has not yet been determined (Mangold et al. 2023). The extreme age of the crater and deposits has led to subsequent crater floor erosion and alteration, likely including aeolian action (Paris and Morgan 2022). The sediments in Jezero Crater may provide the best possible chance of preserving microbial life, and Perseverance has cached core samples for later collection by a Mars Sample return mission (Bosak 2022).

9.2.6 Lakes

Remote sensing and surface observations have identified evidence for standing water within Martian topographic lows besides craters. Eridania Basin, at the boundary of Terra Cimmeria and Terra Sirenum, is the source of Ma'adim Vallis that fed Gusev Crater (Fig. 9.14a). It is an example of a topographically low area that may have held up to 562,000 km^3 of water. This water volume exceeded the total estimated to be in all other Martian craters and drainage areas. Elevation limits of Noachian valley networks, including Ma'adim Vallis, suggest the Eridania Basin may have hosted a lake of 1–1.5 km in depth (Pajola et al. 2016).

No surface mission has yet been to Eridania Basin. However, orbital remote sensing has identified deep basin deposits up to 400 m in elevation eroded into polygonal features and subsequently altered by Hesperian volcanic eruptions (Fig. 9.14b). Hyperspectral CRISM analysis supports the aqueous origin of the deep basin deposits by identifying Mg- and Fe-serpentine, Fe-rich mica, Fe-sulphide, saponites and Mg-Fe-Ca carbonates, minerals likely formed in a submarine environment in association with hydrothermal activity (Ojha et al. 2021). Research suggests that volcanic materials ejected into the basin floor superheated the water around it to dissolve and alter the mafic material and produce the above minerals in a similar process found near Earth's deep sea ocean vents. Climate change during the late Noachian period caused the lake to evaporate and divide to smaller water bodies before completely drying out. Additional chloride deposits discovered only at higher elevations of the basin supports the loss of Eridania Basin water through subsequent evaporation (Michalski et al. 2017).

9.2.7 A Martian Ocean

The northern plains of Mars have muted topography and extensive deposition and low elevations compared to the southern regions of the planet. Additionally, the large outflow channels of Kasei and Ares Vallis drain into this region and researchers suggest this area was originally the bed of an ancient ocean (Salvatore and Christensen 2014; Villanueva et al. 2015). Such an ocean likely covered up to 36% of the planet's surface at its peak and was present from the Noachian to Hesperian periods (4.1–3.0 Ga, Fig. 9.15; Nazari-Sharabian et al. 2020).

Indirect evidence for previously large amounts of water on Mars comes through studying deuterium/hydrogen (D/H) ratios of the planet's atmosphere. Deuterium, an

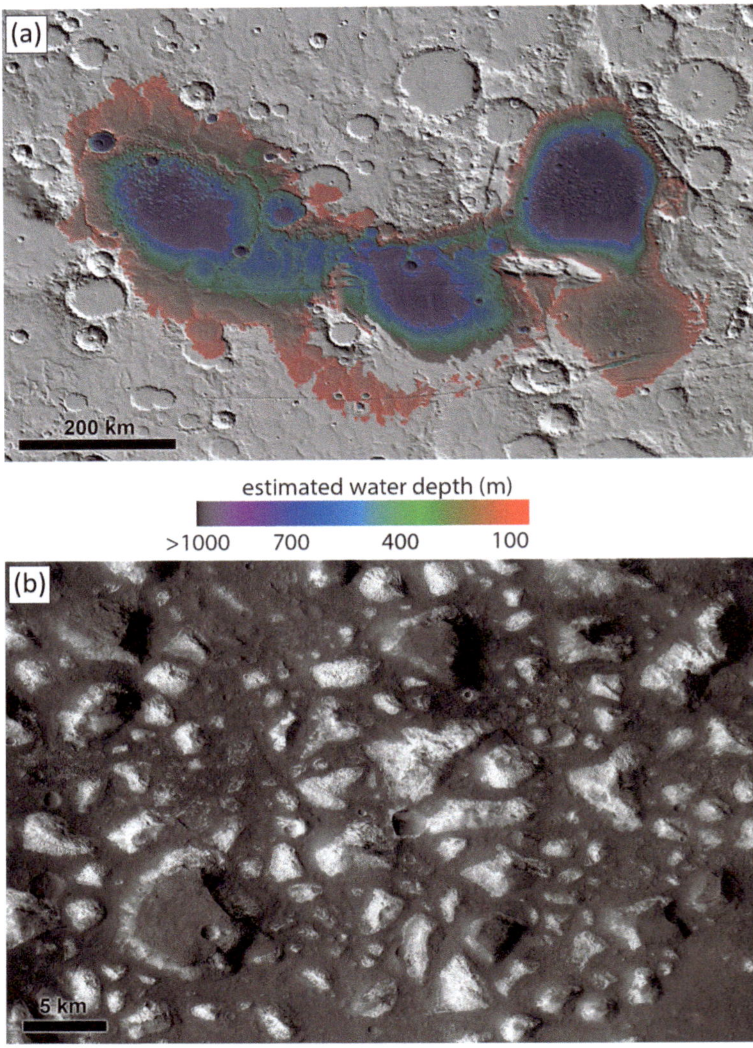

Fig. 9.14 **a** Proposed water depth analysis of Eridania Basin at the source of Ma'adim Vallis that drains into Gusev Crater (PIA22059, NASA/JPL). **b** CTX image of landforms within the basin and postulated to have formed through underwater hydrothermal activity (PIA22058, NASA/JPL)

isotope of hydrogen, is heavier and thus harder to escape into space when compared to the lighter element. Martian D/H rations were compared to those of Earth and found to be 5–6 times higher (Lunine et al. 2003). The high value suggests more Martian water was lost to space when compared to Earth, with solar radiation breaking the H_2O molecule, the lighter hydrogen escaping to space while oxygen bound to surface mineralogy in oxidising reactions (Vals et al. 2022).

Fig. 9.15 An illustration of early Mars abundant with liquid water as seen by a passing asteroid

Identifying the original and subsequent changes to the extent of an ancient ocean on Mars via remote sensing is made difficult by subsequent alteration and degradation of Noachian shorelines through to the present day (Cardenas and Lamb 2022). One such alteration process was likely tsunami action in the late Hesperian period. Geomorphic and thermal remote sensing analysis provided evidence of alteration by two tsunamis, likely caused by asteroid impacts on the northern ocean (Fig. 9.16). The first tsunami possessed sufficient energy to carry house-sized boulders and erode channels with wave backwash. The second, occurring several million years later, produced ice-rich lobes over runout distances of tens of kilometres (Rodriguez et al. 2016). Additional research attributed at least one late Hesperian (3 Ga) mega-tsunami to the meteorite that formed the 120 km Lomonosov crater. This impact may have caused at least one of the tsunamis that resurfaced coastal areas in northwestern Arabia Terra (Costard et al. 2019).

A Hesperian-aged boundary of constant elevation, as would be expected from a water boundary, is interpreted as a shoreline, formed at similar period to the major outflow channels. This feature was geologically mapped as the Vastitas Borealis Formation (VBF), however more direct analysis, was needed to reduce uncertainty as to whether it was not formed by other processes (Parker et al. 1989).

In 2021, China's Tianwen-1/Zhurong rover landed in the VBF and conducted surface-level investigations on its 12-month traverse towards the proposed shoreline. The vehicle was able to document 23 individual rocks using multispectral and microscopic imagery, identifying parallel-tabular bedding, sedimentary structures and cross-laminations, suggestive of opposing fluvial currents. The findings of the sedimentary structures and comparison with terrestrial analogues are consistent with the interpretation that the VBF was deposited in a low-energy marine environment and that an ancient ocean existed in this region (Xiao et al. 2023).

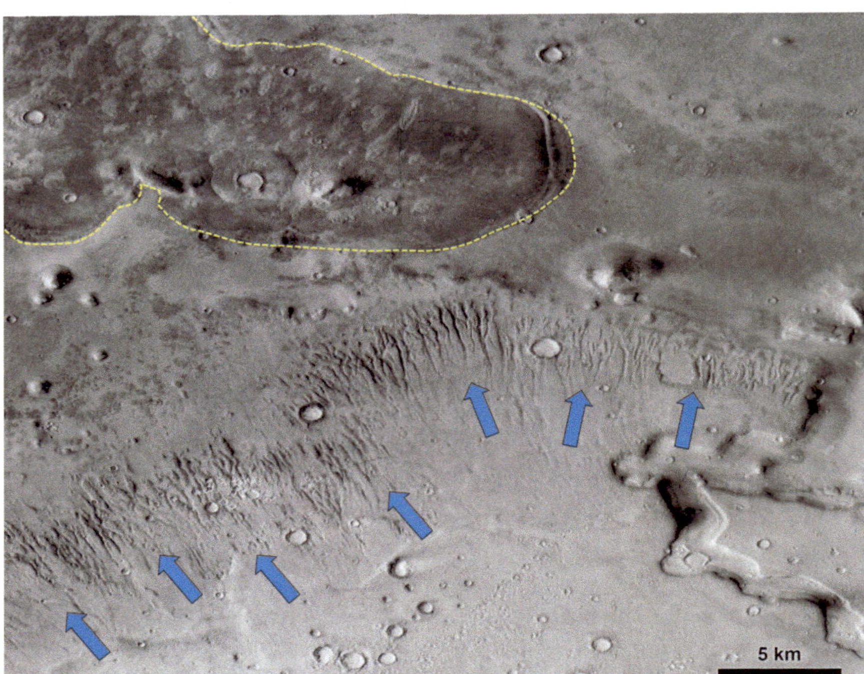

Fig. 9.16 Site of a likely Martian tsunami near the highland-lowland area of Arabia Terra, northeast of Kasei Valles. Dashed line denotes a lobate deposit while the blue arrows mark the direction of channel-carving backwash from the likely tsunami (CTX composite, NASA/JPL)

9.2.8 Case Study: Remote Sensing Reveals Complex Multi-process Interactions in Noachis Terra

The Noachis Terra highlands, west of Hellas Basin in the Mars' southern hemisphere are among the oldest geologic features on the planet. The rough terrain and higher elevations compared to the northern hemisphere have not made this region an attractive candidate for surface missions. All knowledge gathered from Noachis Terra originates from orbital remote sensing missions where increasingly sophisticated sensors have revealed a complex evolution of the region. This case study will focus on water and ice interactions within crater valleys of Choyr, Hunten, Le Verrier and an unnamed crater in a mid-latitude area on the highlands (Fig. 9.17a–c).

Noachis Terra has the most craters anywhere on Mars, formed during an early phase of Solar System evolution. These craters were among the first features to be identified during the early Mariner flyby missions (Chap. 1). Early analysis revealed them to be dissimilar to those of the Moon but were instead subject to different and more substantial erosive processes. Subsequent missions returning higher resolution data provided hints of water and ice-related processes acting on this region. The softening of craters, plus discovery of surficial materials with similar stippled characteristics of terrestrial ice-rich geology provided initial indications of ice-related processes.

Likely ice (Fig. 9.17b) and water-related activity (Fig. 9.17c) appeared to operate within many of the Noachis Terra craters themselves, carving valleys up to tens of kilometres long within the crater walls (defined as "crater valleys"). These are much smaller than the 10–100+ km long network valleys found on the plains (e.g. Figure 9.8a, b), or Hesperian outflow channels (e.g. Figure 9.10a, b). Research of fluvial and ice-carved features on Earth have provided an important baseline in this case study. Each have distinctive morphological features based on water or ice action that can be used to help identify their formation mechanism. For example, studies of terrestrial glacial valleys such as those Washington's Olympic Peninsula (Montgomery 2002), including cross-profiles, have provided useful information for characterising Martian features.

Glacial erosion leaves wide, U-shaped troughs with shallow gradients. Dividing the length by the width provides a useful metric for comparing these features with narrower, V-shaped water-carved valleys. Lower length/width ratios would result from ice-based erosion reflecting wide, U-shaped valleys as observed on imagery. In contrast, higher length/width ratios would be representative of narrower channels incised by liquid water (Bernhardt and Williams 2021). Terrestrial streams also tend to increase in width downstream, and have a high degree of sinuosity (a ratio between the valley's actual length and that of a straight line drawn down valley), depending on slope. They also typically have steeper walls, and a source consistent with fluvial origin such as branching valleys or an amphitheatre-like head (Leopold et al. 1964).

Lacking direct geological measurements, pure remote sensing and baseline information provided by terrestrial studies were used to separate and understand geomorphic processes operating in Mars' past. This included exploitation of CTX imagery, a

Fig. 9.17 **a** Noachis Terra study site west of Hellas Basin. The blue star denotes the location for the ice-related valley illustrated in part b, the green star denotes the location for a water-related valley shown in detail in part c. The white box denotes the crater shown in Fig. 9.18 (Viking mosaic, NASA/JPL). **b** Ice-flow valley highlighted in blue with arrows denoting downslope motion of ice-rich material (CTX mosaic, NASA/JPL). **c** Fluvial valley with arrows denoting main and tributary channels (CTX mosaic, NASA/JPL)

dataset that had coverage of the study area, as well as GSD needed to identify feature types, and elevation data provided by MOLA (Fig. 9.18a).

A custom GIS was developed by the author for successful exploitation of the disparate datasets used in this work (Hobbs et al. 2016). This entailed downloading and reprojecting the datasets to a standard format to enable distances and slopes to be measured (Chap. 6). Derivation of slopes was important for geomorphology interpretation (e.g. Figure 9.18b); gentler slopes tend to allow dendritic, branching valleys, where steeper slopes may only possess a single valley channel.

Measurements of crater valley length from origin to terminus were taken following development of the custom GIS. The valley's centreline was used as a reference (long red line in the representative crater valley shown in Fig. 9.18c), and this also enabled creation of a topographic profile. Sequential slope measurements from head to terminus were then sampled at equally-spaced intervals for each valley (intersection between long and short red lines, Fig. 9.18c). The course resolution (450 m/pixel) of the MOLA data required slope sampling of at least 600 m intervals to ensure different pixels and thus unique slope measurements being sampled. Valley width measurements were also taken (short red lines, Fig. 9.18c), to obtain the width/length ratio described above, used to infer glacial or fluvial activity.

Four Noachis Terra craters were studied in detail using the above remotely-sensed datasets, with two broad classifications of crater valleys ("ice" and "fluvial") identified (Fig. 9.19a–d). Nearly 40% of the valleys in the study site showed evidence of evolution through ice-flow. The 27 valleys possessed cirque-like origins, wide, U-shaped cross sections as inferred from imagery, and in many cases, evidence of ice-rich infill. Additional imagery analysis identified longitudinal ridges, inverted valleys and possible moraines at the foot of many of the "ice-flow" valleys, consistent with glacial activity.

The remaining crater valleys in the study site were also analysed, identifying distinctly different morphology to the ice valleys. These other valleys have much narrower widths and higher sinuosity values than the ice-flow features. They instead shared morphology with terrestrial hillside streams, where higher slopes forces water to take a more direct downhill route, reducing branching networks to lower slope areas.

Differences between the ice and fluvial valleys were identified, particularly widths and lengths (Fig. 9.19a, b). Fluvial valleys were longer and narrower than those carved by ice. While occurring on similar crater wall slopes to the studied ice valleys (Fig. 9.19c), they meandered across the landscape in a manner similar to terrestrial mountain streams, with higher sinuosity measurements than the ice valleys (Fig. 9.19d).

The origins of liquid water required to carve features interpreted to be fluvial in origin on Mars is of high interest to researchers. Rainfall, ice-melt and underground water flow have all been proposed as past mechanisms for fluvial erosion, with analysis of valley head and comparison with terrestrial features proving useful in inferring underlying process. Terrestrial groundwater sapping (subsurface flow) tends to form rectangular, flat-floored valley heads (Malin and Carr 1999), while surface runoff from rainfall may exhibit diffuse, dendritic drainage patterns that coalesce into

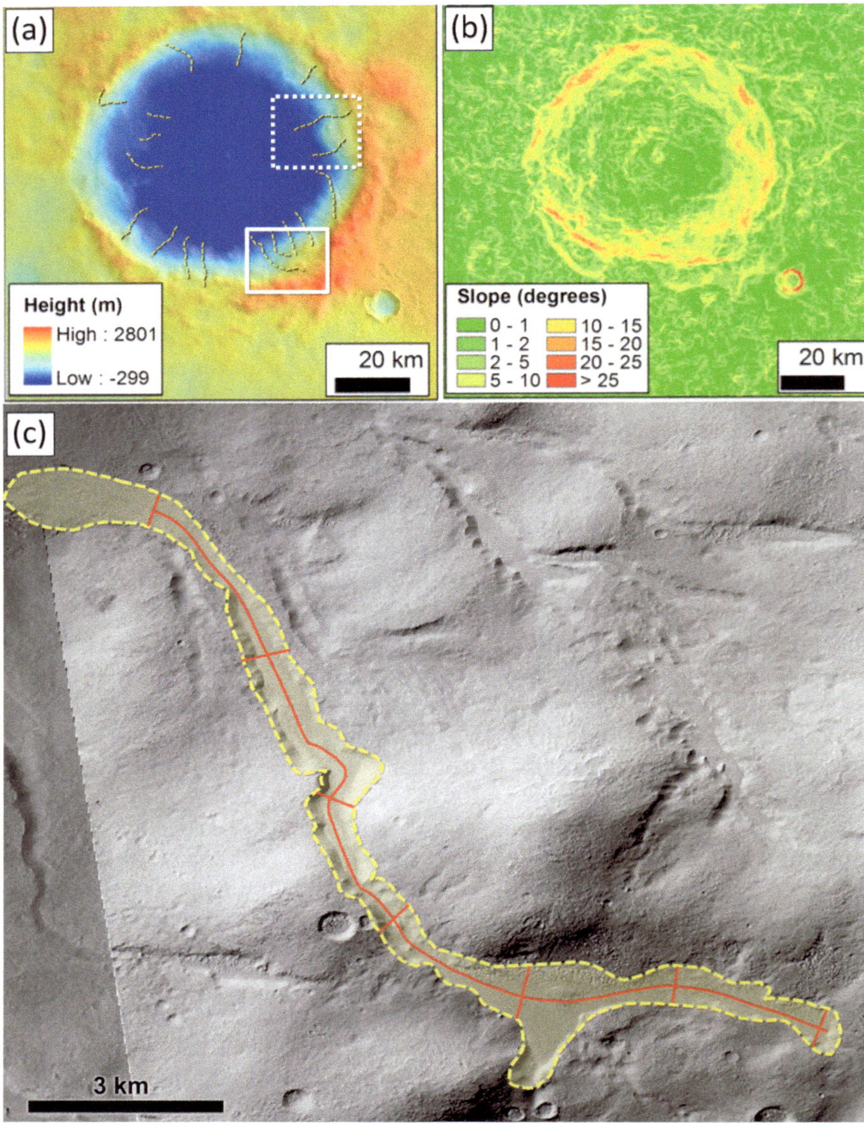

Fig. 9.18 **a** MOLA elevation image of an unnamed crater in Fig. 9.17a, with dashed lines showing the location of the sampled crater valleys. Dashed box shows the location of Fig. 9.20. White box shows the location of the valley in part c (NASA/JPL). **b** MOLA-derived slope map of the crater (NASA/JPL). **c** Detail of the measurements sampled from the study site crater valleys including length along the valley centre (long red line), and cross-profiles (short red lines). Slope measurements were taken at the intersection between the long and cross profiles (CTX mosaic, NASA/JPL)

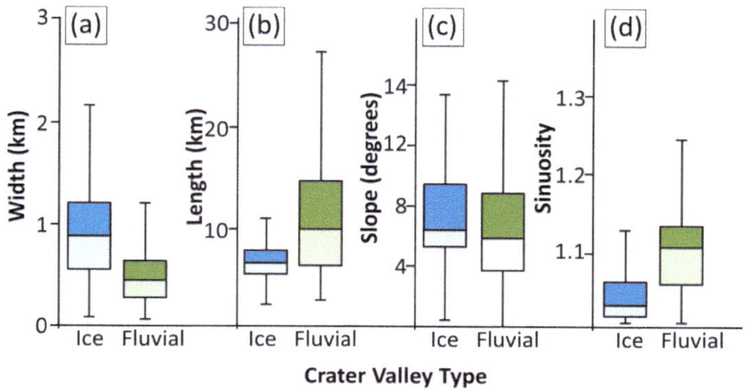

Fig. 9.19 Box whisker plot display of measurement results of "ice" and "fluvial" valleys from the GIS analysis of the study site. The bottom, middle and top box segments contain 25th percentile, median, and 75th percentile, covering 50% of the results. The "whiskers" cover the spread of the remaining data. **a** Valley width measurements. **b** Valley length. **c** Slope measurements. **d** Sinuosity measurements

more substantial channels downstream (Craddock and Howard 2002). The location of the valley heads near crater rims made underground, aquifer-fed erosion unlikely, with no clear mechanism for recharging subsurface reservoirs at these high elevations. These studied crater valleys consistently showed close association with textured material interpreted to be ice-rich in nature (examples shown in Fig. 9.20a). Many of the valleys were infilled with this material, possessing textures associated with ice sublimation (Fig. 9.20a inset). Examples of stubby and dendritic valley heads located with ice-rich material are shown in Fig. 9.20b, c. The dendritic valley head appears to have been fed by ice-melt, with a distinct change in surface texture occurring in the example valley head to the south (dashed line, Fig. 9.20c). The suggested ice-melt process is observed on Earth. Near the Earth's south pole, ADVs evolve through a process of freeze-thaw-melt processes that may have operated at the Martian study site in the past (Head and Marchant 2014). In addition, similar branching valley heads morphology are driven by snowmelt at Devon Island during the local spring (Clarke et al. 2019).

The findings identified in this case study, along with subsequent research, support the possibility of a cold, icy climate where liquid water erosion was dominated by ice-melt processes in this region of Mars (Bouquety et al. 2019). Martian global climate models (GCM) are also active areas of research using ever increasingly detailed data to create a "digital twin" of the planet's paleoenvironment. Using three-dimensional simulations, research attempted to correlate a "warm and wet" scenario, where average temperatures were above the freezing point of water with, rainfall and related runoff-related erosion on the planet's water-carved features (Palumbo and Head 2018). The GCM instead presented an early Mars similar to the case study described above, favouring snow accumulation, melting and runoff as the dominant water-based erosion.

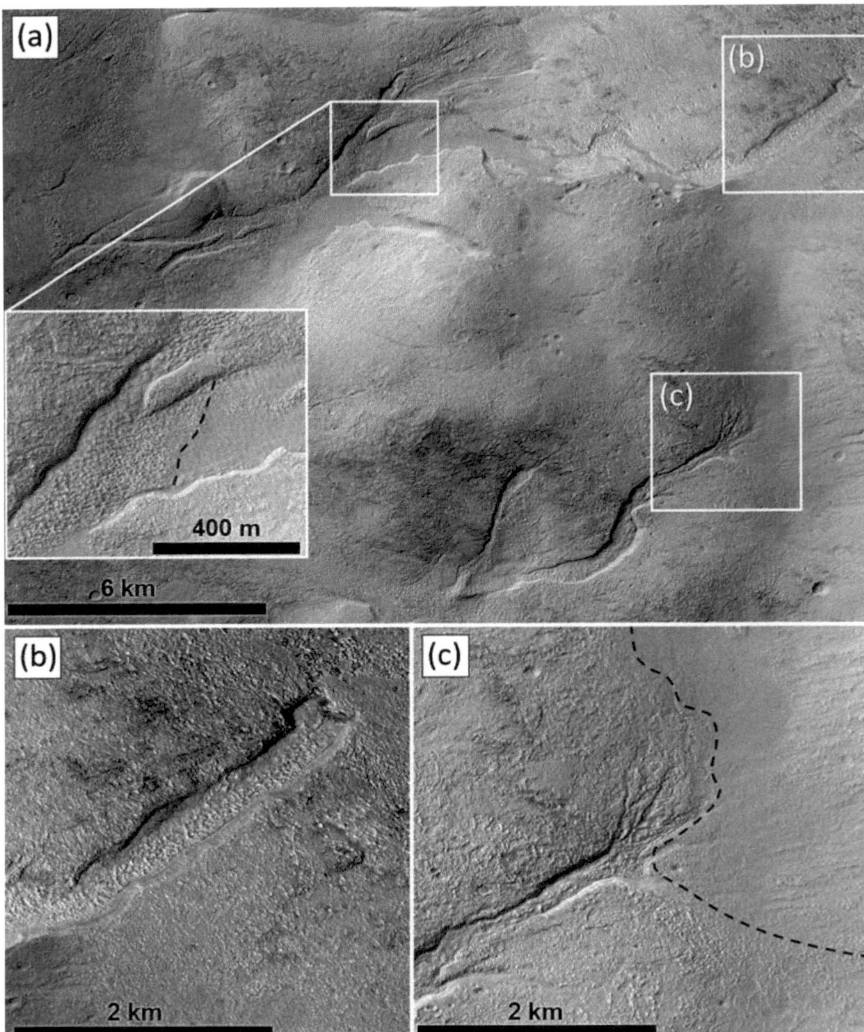

Fig. 9.20 **a** Overview image of two studied crater valleys. The inset highlights different textures in ice-rich infill material of the northern valley (P16_007255_1411_XN_38S340W, NASA/JPL). **b** Stubby head of the northern valley (P16_007255_1411_XN_38S340W, NASA/JPL). **c** Dendritic head of the southern valley with dashed line denoting change in ice-rich material texture (P16_007255_1411_XN_38S340W, NASA/JPL)

As an active area of research, other recent work has suggested that the Mars-wide, cold climatic periods described above were likely to have been interspersed with transient warmer eras. Episodic rises in temperature and humidity would have driven rain or snow precipitation to shape the fluvial landforms observed on Mars today. Asteroid impacts or volcanism-induced outgassing may have supported the production of these warmer, wetter periods that lasted long enough to cause fluvial erosion (Wordsworth et al. 2021; Steakley et al. 2023; Adams et al. 2025; Steckel et al. 2025).

This case study of remote sensing and baseline information from Earth analogue studies demonstrate how geomorphic history can be researched and inferred on Mars. The extreme age of this region of Mars has meant that it has been subject to volcanic activity, impact cratering, erosion by water and subsequent dry mass wasting events (Bernhardt and Williams 2021). While the ice and fluvial activity is of particular interest for astrobiological implications, it is important to note that Noachis Terra, along with much of Mars, has evolved through multiple-agent processes. Future surface missions would add a wealth of additional in-situ data, though engineering constraints make this unlikely in the immediate future.

9.3 Small-Scale Processes: Gullies and Recurring Slope Lineae

Identifying unambiguous evidence of smaller-scale water-based erosion on Mars has proven more difficult in recent years than previously thought. On Earth, seepage from water-rich regolith or snowmelt are quite common, along with water-lubricated debris flows or gullying. Research into equivalent features on Mars has tried to identify whether water played a dominant role in their morphology, or if it is a case of different processes behaving as a fluid.

9.3.1 Gullies

MGS provided surface imagery of sufficient resolution to enable analysis of small-scale geomorphologic features not seen in the decades-old Viking data. In 2000, sub-kilometre erosive features were discovered on hillslopes and crater walls consisting of small channels, triangular shaped depositional aprons and in many cases, a concave alcove (Malin and Edgett 2000). These features, found on the Martian mid-latitudes and polar regions, resembled terrestrial hillslope gullies cut by flowing liquid water or debris flows (Fig. 9.21). There are more gullies in the southern hemisphere than the northern hemisphere because of the higher incidence of cratering, and greater abundance of steeper (> 20°) slopes. Minimal impact cratering and overprinting,

Fig. 9.21 Colourised perspective view of Martian gullies embedded within the pole-facing Talu Crater wall in Noachis Terra, produced from HiRISE and Mars Express elevation data

plus direct observations of gully modification demonstrate the youthfulness (2 Ma to present day) of these features.

The gullies found on Mars are consistently geomorphologically different from dry mass-wasting flows found on Earth, Mars, and the Moon in that they possess U or V-shaped sinuous channels, fan-shaped depositional aprons, and are found on slopes below the angle of repose for dry material. The absence of gullies in the equatorial regions initially suggested water, possibly from melting of ice-rich material, was required for their formation. The changing nature of the erosive agents and temperature ranges required are also suggested by mid-latitude gullies (30–40°) being predominantly pole-facing while higher-latitude gullies (> 40°) in colder regions being mainly equator facing (Berman et al. 2005).

To account for the diversity of Martian gully location and pole versus equator-facing trends, researchers have proposed various mechanisms for their formation. Extensive research conducted in Meteor Crater, the ADV and other terrestrial areas were used as baseline references for inferring potential formation mechanisms for features observed on Mars. Fluvial-based processes, such as subsurface aquifer, melting of ice-rich material, brines and fluid-poor debris flows were proposed. However, additional remote sensors, such as subsurface radar soundings, have failed to correlate any subterranean water with gully activity, and many gullies occur on isolated topography, precluding the possibility of at least shallow subsurface aquifers in these areas (Conway et al. 2019).

Recent research has considered the radically different environment on Mars compared to Earth. The much colder temperatures on Mars allows CO_2 to act as an erosive agent, altering the planet's surface in processes absent on Earth. This is supported by direct observations of gully modification that have suggested more exotic processes for their formation. Channel incision and fan deposition was directly observed using HiRISE imagery, in conditions too cold for even water brines to remain liquid (Dundas et al. 2022). Repeated observations identified a correlation of active gully modification with seasonal CO_2 frost deposition and sublimation. This research suggests that sublimation of emplaced CO_2 above a water-ice permafrost layer is sufficient to lubricate regolith and behave in a fluid-like manner to form features typically associated with liquid water erosion, including lobate deposits and levee formation (de Haas et al. 2019). Erosion of more resistant material, such as bedrock, is likely facilitated by thermal cycle stresses, aeolian and freeze-thaw processes causing slope failure, explaining the formation of gully alcoves in this geology (Dundas et al. 2022).

Additional support for this mechanism came from attempts to identify minerals altered by water; analysis of 100 gullies failed to find evidence of hydrated minerals or brines that would have been indicated a fluvial origin (Núñez et al. 2016). Unlike larger water-related features, such as outflow channels or lakes described above, Martian gully erosion is likely transient in nature, involving small amounts of water as would be the case for debris flow mechanisms. Should this be the case, the water may not have interacted with the underlying regolith long enough to alter mineralogy (Sinha et al. 2023).

Could gullies instead evolve under more complex, multi-agent mechanisms? Analogous investigations of terrestrial gullies were helpful in characterising Martian gully morphology features. Some of these studies also revealed a complex, multi-agent process of gully erosion, where dry mass wasting, snowmelt and debris flows occurred simultaneously on some gullies (e.g. Hobbs et al. 2013). Recent Martian gully studies have highlighted the difficulty in identifying a singular explanation for gully evolution. It is likely that multiple dry, CO_2 and water-based processes of differing strength have acted on Martian gullies throughout their history (e.g. Dickson et al. 2023). Examples includes gullies in Talu Crater, where features usually associated with dry mass wasting are superimposed on larger gully channels, suggesting possible different erosive processes over time. Gullies and fine-scale channel features in this and other craters have varied with changes in underlying material, highlighting a dependence on underlying geology (Hobbs et al. 2015; Conway et al. 2019).

The lack of terrestrial analogues for CO_2-based landscape features, plus the difficulty in simulating such processes in laboratories has hampered more detailed analysis on fine-scale erosion characteristics from this mechanism. Future work in this area would provide better evidence for the ability of water and CO_2-based flows to reproduce features associated with water-based erosion on Earth.

9.3.2 Recurring Slope Lineae

Near-continuous observations of Mars by HiRISE has allowed identification of change on the planet's surface. Imagery of Newton Crater identified seasonal downslope flow of dark material that increased during the warmer seasons and slowed during the Martian autumn and winter. The material, 0.5–5 m wide traversed hundreds of metres downslope in temperatures possible for briny liquid water to exist. The recurring slope lineae (RSL), supported by spectral evidence of associated hydrated salts, are considered by some researchers to be the first direct evidence of contemporary water activity on Mars (Ojha et al. 2015). Other RSL were observed by HiRISE in Valles Marineris (Fig. 9.22), with this region of Mars possessing more of these features than the rest of the planet combined.

Similar to the Martian gullies, alternative processes have been suggested for RSL evolution. CO_2-related or dry granular flow activity were also considered due to some regions being too cold for pure liquid water, and perceived difficulty in identifying how aquifer-fed seepage would recharge under current Martian climate conditions. As part of identifying alternate processes, a recent study suggested that seasonal dust devil activity may destabilise deposits of fine sand that would flow downslope to form the RSL (McEwen et al. 2021). These processes fail to take into account detections of hydrated salts described above, and it is very possible that multiple processes may be involved in their formation. Along with Martian gullies, RSLs are an example of morphology that cannot be fully explained through remote sensing and are located in areas currently inaccessible to surface missions.

Fig. 9.22 Recurring slope lineae on a north facing Coprates Chasma slope in Valles Marineris (ESP_042228_1670, NASA/JPL)

Research into small scale activities described above places the importance of multi-agent surface processes, including those related to water and ice, into planetary context. Valley networks, gullies, layered ice deposits and other forms have recorded periods of ancient flooding, ice-melt and precipitation, long term groundwater activity and more recent climate-cycle-driven freeze-thaw and sublimation activity. Recent analysis has identified that ice and water have not operated in isolation but have operated and in some cases been driven by other processes such as volcanic eruptions or geothermal heating.

Remote sensing has enabled large scale discovery and analysis of ice and fluvial-based surface processes, providing planet-wide coverage to place specific surface mission analysis in greater context. The missions and sensors described in previous chapters have produced an abundance of datasets that make this analysis possible.

References

Adams D, Scheucher M, Hu R, Ehlmann BL, Thomas TB, Wordsworth R, Scheller E, Lillis R, Smith K, Rauer H, Yung YL (2025) Episodic warm climates on early Mars primed by crustal hydration. Nat Geosci 18:133–139. https://doi.org/10.1038/s41561-024-01626-8

Alwarda R, Smith IB (2021) Stratigraphy and volumes of the units within the massive carbon dioxide ice deposits of Mars. J Geophys Res Planets 126. https://doi.org/10.1029/2020JE006767

Arfstrom JD, Hartmann WK (2005) Martian flow features, moraine-like ridges, and gullies: terrestrial analogs and interrelationships. Icarus 174:321–335. https://doi.org/10.1016/j.icarus.2004.05.026

Bandfield JL (2007) High-resolution subsurface water-ice distributions on Mars. Nature 447:64–67. https://doi.org/10.1038/nature05781

Barlow N (2008) Mars: an introduction to its interior, surface and atmosphere. Cambridge University Press, New York, 264 pp

Becerra P, Sori MM, Thomas N, Pommerol A, Simioni E, Sutton SS, Tulyakov S, Cremonese G (2019) Timescales of the climate record in the south polar ice cap of Mars. Geophys Res Lett 46:7268–7277. https://doi.org/10.1029/2020JE006767

Berman DC, Hartmann WK, Crown DA, Baker VR (2005) Arcuate ridges and gullies in Martian craters: dependence on orientation and latitude. In: 36th annual lunar and planetary science conference, p 1213

Berman DC, Chuang FC, Smith IB, Crown DA (2021) Ice-rich landforms of the southern mid-latitudes of Mars: a case study in Nereidum Montes. Icarus 355. https://doi.org/10.1016/j.icarus.2020.114170

Bernhardt H, Williams DA (2021) Geology and history of the Malea Planum region: a new view of Mars' oldest large volcanic province. Icarus 366:114518. https://doi.org/10.1016/j.icarus.2021.114518

Birot P (1965) Fluvial processes in geomorphology. In: Leopold L, Gordon Wolman M, Miller J (eds) Annales de Géographie, vol 74. Armand Colin, pp 454–456

Bosak T (2022) Aqueously altered igneous rocks sampled on the floor of Jezero Crater, Mars. Science 377. https://doi.org/10.1126/science.abo2196hal-03812419

Bouquety A, Sejourné A, Costard F, Mercier D, Bouley S (2019) Morphometric evidence of 3.6 Ga glacial valleys and glacial cirques in Martian highlands: south of Terra Sabaea. Geomorphology 334:91–111. https://doi.org/10.1016/j.geomorph.2019.02.022

Bramson AM, Byrne S, Bapst J, Smith IB, McClintock T (2019) A migration model for the polar spiral troughs of Mars. J Geophys Res Planets 124:1020–1043. https://doi.org/10.1029/2018JE005806
Broquet A, Wieczorek MA, Fa W (2020) Flexure of the lithosphere beneath the north polar cap of Mars: implications for ice composition and heat flow. Geophys Res Lett 47. https://doi.org/10.1029/2019GL086746
Brough S, Hubbard B, Hubbard A (2019) Area and volume of mid-latitude glacier-like forms on Mars. Earth Planet Sci Lett 507:10–20. https://doi.org/10.1016/j.epsl.2018.11.031
Buhler PB, Piqueux S (2021) Obliquity-driven CO_2 exchange between Mars' atmosphere, regolith, and polar cap. J Geophys Res Planets 126. https://doi.org/10.1029/2020JE006759
Butcher FEG (2019) Wet-based glaciation on Mars. Ph.D. thesis, The Open University. https://doi.org/10.21954/ou.ro.0000ed1f
Byrne S, Dundas CM, Kennedy MR, Mellon MT, McEwen AS, Cull SC, Daubar IJ, Shean DE, Seelos KD, Murchie SL, Cantor BA (2009) Distribution of mid-latitude ground ice on Mars from new impact craters. Science 325:1674–1676. https://doi.org/10.1126/science.1175307
Cardenas BT, Lamb MP (2022) Paleogeographic reconstructions of an ocean margin on Mars based on deltaic sedimentology at Aeolis Dorsa. J Geophys Res Planets 127. https://doi.org/10.1029/2022JE007390
Carr MH, Head, James W (2003) Oceans on Mars: an assessment of the observational evidence and possible fate. J Geophys Res 108. https://doi.org/10.1029/2002JE001963
Chapman MG, Neukum G, Dumke A, Michael G, Van Gasselt S, Kneissl T, Zuschneid W, Hauber E, Mangold N (2010) Amazonian geologic history of the Echus Chasma and Kasei Valles system on Mars: new data and interpretations. Earth Planet Sci Lett 294:238–255. https://doi.org/10.1016/j.epsl.2009.11.034
Clarke J, Knightly P, Rupert S (2019) Melt-water formed dark streaks on slopes of Haughton crater as possible Mars analogues. Int J Astrobiol 18:518–526. https://doi.org/10.1017/S1473550418000526
Conway SJ, Balme MR (2014) Decameter thick remnant glacial ice deposits on Mars. Geophys Res Lett 41:5402–5409. https://doi.org/10.1002/2014GL060314
Conway SJ, de Haas T, Harrison TN (2019) Martian gullies: a comprehensive review of observations, mechanisms and insights from Earth analogues. Geol Soc Lond Spec Publ 467:7–66. https://doi.org/10.1144/SP467.14
Costard FM, Kargel JS (1995) Outwash plains and thermokarst on Mars. Icarus 114:93–112
Costard F, Séjourné A, Lagain A, Ormö J, Rodriguez JAP, Clifford S, Bouley S, Kelfoun K, Lavigne F (2019) The Lomonosov crater impact event: a possible mega-tsunami source on Mars. J Geophys Res Planets 124:1840–1851. https://doi.org/10.1029/2019JE006008
Craddock RA, Howard AD (2002) The case for rainfall on a warm, wet early Mars. J Geophys Res Planets 107:1–36. https://doi.org/10.1029/2001JE001505
Craddock RA, Lorenz RD (2017) The changing nature of rainfall during the early history of Mars. Icarus 293:172–179. https://doi.org/10.1016/j.icarus.2017.04.013
Cross AJ, Goldsby DL, Hager TF, Smith IB (2020) The rheological behaviour of CO_2 ice: application to glacial flow on Mars. Geophys Res Lett 47. https://doi.org/10.1029/2020GL090431
de Haas T, McArdell BW, Conway SJ, McElwaine JN, Kleinhans MG, Salese F, Grindrod PM (2019) Initiation and flow conditions of contemporary flows in Martian gullies. J Geophys Res Planets 124:2246–2271. https://doi.org/10.1029/2018JE005899
Dickson JL, Palumbo AM, Head JW, Kerber L, Fassett CI, Kreslavsky MA (2023) Gullies on Mars could have formed by melting of water ice during periods of high obliquity. Science 380:1363–1367. https://doi.org/10.1126/science.abk2464
Dundas CM, Byrne S, McEwen AS (2015) Modelling the development of Martian sublimation thermokarst landforms. Icarus 262:154–169. https://doi.org/10.1016/j.icarus.2015.07.033
Dundas CM, Mellon MT, Conway SJ, Gastineau R (2019) Active boulder movement at high Martian latitudes. Geophys Res Lett 46:5075–5082. https://doi.org/10.1029/2019GL082293

Dundas CM, Conway SJ, Cushing GE (2022) Martian gully activity and the gully sediment transport system. Icarus 386. https://doi.org/10.1016/j.icarus.2022.115133

Dundas CM, Mellon MT, Posiolova LV, Miljković K, Collins GS, Tornabene LL, Rangarajan VG, Golombek MP, Warner NH, Daubar IJ, Byrne S (2023) A large new crater exposes the limits of water ice on Mars. Geophys Res Lett 50. https://doi.org/10.1029/2022GL100747

Duran S, Coulthard TJ (2020) The Kasei Valles, Mars: a unified record of episodic channel flows and ancient ocean levels. Sci Rep 10. https://doi.org/10.1038/s41598-020-75080-y

Ehlmann BL, Edwards CS (2014) Mineralogy of the Martian surface. Annu Rev Earth Planet Sci 42:291–315. https://doi.org/10.1146/annurev-earth-060313-055024

Fassett CI, Head III JW (2008) The timing of Martian valley network activity: constraints from buffered crater counting. Icarus 195:61–89. https://doi.org/10.1016/j.icarus.2007.12.009

Fastook JL, Head JW, Deutsch AN (2019) Glaciation on mercury: accumulation and flow of ice in permanently shadowed circum-polar crater interiors. Icarus 317:81–93. https://doi.org/10.1016/j.icarus.2018.07.004

Feldman WC, Prettyman TH, Maurice S, Plaut JJ, Bish DL, Vaniman DT, Mellon MT, Metzger AE, Squyres SW, Karunatillake S, Boynton WV (2004) Global distribution of near-surface hydrogen on Mars. J Geophys Res 109:1–13. https://doi.org/10.1029/2003JE002160

Fisher DA (2005) A process to make massive ice in the Martian regolith using long-term diffusion and thermal cracking. Icarus 179:387–397. https://doi.org/10.1016/j.icarus.2005.07.024

French HM (2017) The periglacial environment. Wiley, Chippenham

Goddard K, Warner NH, Gupta S, Kim JR (2014) Mechanisms and timescales of fluvial activity at Mojave and other young Martian craters. J Geophys Res Planets 119:604–634. https://doi.org/10.1002/2013JE004564

Gulick VC (2001) Origin of the valley networks on Mars: a hydrological perspective. Geomorphology 37:241–268

Head JW, Marchant DR (2014) The climate history of early Mars: insights from the Antarctic McMurdo Dry valleys hydrologic system. Antarct Sci 26:774–800. https://doi.org/10.1017/S0954102014000686

Head JW, Pratt S (2001) Extensive Hesperian-aged south polar ice sheet on Mars: evidence for massive melting and retreat, and lateral flow and ponding of meltwater. J Geophys Res 106:12275–12299. https://doi.org/10.1029/2000JE001359

Hepburn AJ, Ng FSL, Livingstone SJ, Holt TO, Hubbard B (2020) Polyphase mid-latitude glaciation on Mars: chronology of the formation of superposed glacier-like forms from crater-count dating. J Geophys Res Planets 125. https://doi.org/10.1029/2019JE006102

Heydari E, Schroeder JF, Calef FJ, Parker TJ, Fairén AG (2023) Lacustrine sedimentation by powerful storm waves in Gale Crater and its implications for a warming episode on Mars. Sci Rep 13:18715. https://doi.org/10.1038/s41598-023-45068-5

Hobbs SW, Paull DJ, Clarke JDA (2013) The influence of slope morphology on gullies: terrestrial gullies in Lake George as analogues for Mars. Planet Space Sci 81:1–17. https://doi.org/10.1016/j.pss.2012.10.009

Hobbs SW, Paull DJ, Clarke JDA (2015) Analysis of regional gullies within Noachis Terra, Mars: a complex relationship between slope, surface material and aspect. Icarus 250:308–331. https://doi.org/10.1016/j.icarus.2014.12.011

Hobbs SW, Paull DJ, Clarke JDA (2016) Analysis of crater valleys, Noachis Terra, Mars: evidence of fluvial and glacial processes. Geomorphology 261:244–272. https://doi.org/10.1016/j.geomorph.2016.02.027

Hoke MR, Hynek BM, Tucker GE (2011) Formation timescales of large Martian valley networks. Earth Planet Sci Lett 312:1–12. https://doi.org/10.1016/j.epsl.2011.09.053

Howard AD (2000) The role of eolian processes in forming surface features of the Martian polar layered deposits. Icarus 144:267–288. https://doi.org/10.1006/icar.1999.6305

Howard AD, Moore JM, Umurhan OM, White OL, Anderson RS, McKinnon WB, Spencer JR, Schenk PM, Beyer RA, Stern SA, Ennico K (2017) Present and past glaciation on Pluto. Icarus 287:287–300. https://doi.org/10.1016/j.icarus.2016.07.006

Hubbard B, Souness C, Brough S (2014) Glacier-like forms on Mars. Cryosphere 8:2047–2061. https://doi.org/10.5194/tc-8-2047-2014

Komar PD (1979) Comparisons of the hydraulics of water flows in Martian outflow channels with flows of similar scale on Earth. Icarus 37:156–181

Lachenbruch AH (1962) Mechanics of thermal contraction cracks and ice-wedge polygons in permafrost. Geol Soc Am 70. https://doi.org/10.1130/SPE70-p1

Lange L, Forget F, Banfield D, Wolff M, Spiga A, Millour E, Viúdez-Moreiras D, Bierjon A, Piqueux S, Newman C, Pla-García J (2022) Insight pressure data recalibration, and its application to the study of long-term pressure changes on Mars. J Geophys Res Planets 127:1–26. https://doi.org/10.1029/2022JE007190

Lee P, Shubham S, Schutt JW (2023) A relict glacier near Mars' equator: evidence for recent glaciation and volcanism in eastern Noctis Labyrinthus. In: 54th lunar and planetary science conference, abstract 2998

Leopold LB, Wolman MG, Miller JP (1964) Fluvial processes in geomorphology. Dover Publications, New York, 544 pp

Levy J, Head JW, Marchant DR (2010a) Concentric crater fill in the northern mid-latitudes of Mars: formation processes and relationships to similar landforms of glacial origin. Icarus 390–404. https://doi.org/10.1016/j.icarus.2010.03.036

Levy JS, Marchant DR, Head JW (2010b) Thermal contraction crack polygons on Mars: a synthesis from HiRISE, Phoenix, and terrestrial analog studies. Icarus 206:229–252. https://doi.org/10.1016/j.icarus.2009.09.005

Levy JS, Fassett CI, Head JW, Schwartz C, Watters JL (2014) Sequestered glacial ice contribution to the global Martian water budget: geometric constraints on the volume of remnant, midlatitude debris-covered glaciers. J Geophys Res Planets 119:2188–2196. https://doi.org/10.1002/2014JE004685

Levy JS, Fassett CI, Holt JW, Parsons R, Cipolli W, Goudge TA, Tebolt M, Kuentz L, Johnson J, Ishraque F, Cvijanovich B (2021) Surface boulder banding indicates Martian debris-covered glaciers formed over multiple glaciations. Proc Natl Acad Sci 118. https://doi.org/10.1073/pnas.2015971118

Lichtenberg KA, Arvidson RE, Morris RV, Murchie SL, Bishop JL, Fernandez Remolar D, Glotch TD, Noe Dobrea E, Mustard JF, Andrews-Hanna J, Roach LH (2010) Stratigraphy of hydrated sulfates in the sedimentary deposits of Aram Chaos, Mars. J Geophys Res Planets 115. https://doi.org/10.1029/2009JE003353

Lunine JI, Chambers J, Morbidelli A, Leshin LA (2003) The origin of water on Mars. Icarus 165:1–8

Luo W, Cang X, Howard A (2017) New Martian valley network volume estimate consistent with ancient ocean and warm and wet climate. Nat Commun 8:1–7. https://doi.org/10.1038/ncomms15766

Malin MC, Carr MH (1999) Groundwater formation of Martian valleys. Nature 397:589–591. https://doi.org/10.1038/17551

Malin MC, Edgett KS (2000) Evidence for recent groundwater seepage and surface runoff on Mars. Science 288:2330–2335. https://doi.org/10.1126/science.288.5475.2330

Mangold N, Gupta S, Caravaca G, Dromart G, Gasnault O, Le Mouélic S, Quantin-Nataf C, Horgan B, Bell JF, Beyssac O, Maurice S (2023) From lake deposits to fluvial floods at the eastern delta front of Jezero Crater, Mars. In: 54th lunar and planetary science conference 2023, vol 54, p 2140

Matherne CM (2019) Role of glaciers in halting Syrtis Major lava flows to preserve and divert a fluvial system. Louisiana State University and Agricultural and Technical College, Baton Rouge, USA

McEwen AS, Schaefer EI, Dundas CM, Sutton SS, Tamppari LK, Chojnacki M (2021) Mars: abundant recurring slope lineae (RSL) following the planet-encircling dust event (PEDE) of 2018. J Geophys Res Planets 126:1–12. https://doi.org/10.1016/B978-0-12-813018-6.00002-9

Mellon MT (1997) Small-scale polygonal features on Mars: seasonal thermal contraction cracks in permafrost. J Geophys Res Planets 102:25617–25628

Mellon MT, Feldman WC, Prettyman TH (2004) The presence and stability of ground ice in the southern hemisphere of Mars. Icarus 169:324–340. https://doi.org/10.1016/j.icarus.2003.10.022

Mellon MT, Arvidson RE, Sizemore HG, Searls ML, Blaney DL, Cull S, Hecht MH, Heet TL, Keller HU, Lemmon MT, Markiewicz WJ, Ming DW, Morris RV, Pike WT, Zent AP (2009) Ground ice at the Phoenix landing site: stability state and origin. J Geophys Res Planets 114:1–15. https://doi.org/10.1029/2009JE003417

Mest SC, Crown DA, Harbert W (2010) Watershed modelling in the Tyrrhena Terra region of Mars. J Geophys Res Planets 115. https://doi.org/10.1029/2009JE003429

Michalski JR, Dobrea EZN, Niles PB, Cuadros J (2017) Ancient hydrothermal seafloor deposits in Eridania Basin on Mars. Nat Commun 8:15978. https://doi.org/10.1038/ncomms15978

Montgomery DR (2002) Valley formation by fluvial and glacial erosion. Geology 30:1047–1050. https://doi.org/10.1130/0091-7613(2002)030b1047:vfbfagN2.0.co

Morgan GA, Head JW III, Marchant DR (2009) Lineated valley fill (LVF) and lobate debris aprons (LDA) in the Deuteronilus Mensae northern dichotomy boundary region, Mars: constraints on the extent, age and episodicity of Amazonian glacial events. Icarus 202:22–38. https://doi.org/10.1016/j.icarus.2009.02.017

Morgan A, Howard AD, Hobley DEJ, Moore JM, Dietrich WE, Williams RME et al (2014) Sedimentology and climatic environment of alluvial fans in a Martian Saheki crater and a comparison with terrestrial fans in the Atacama Desert. Icarus 229:131–156. https://doi.org/10.1016/j.icarus.2013.11.007

Nazari-Sharabian M, Aghababaei M, Karakouzian M, Karami M (2020) Water on Mars—a literature review. Galaxies 8:40. https://doi.org/10.3390/galaxies8020040

Núñez JI, Barnouin OS, Murchie SL, Seelos FP, McGovern JA, Seelos KD, Buczkowski DL (2016) New insights into gully formation on Mars: constraints from composition as seen by MRO/CRISM. Geophys Res Lett 43:8893–8902. https://doi.org/10.1002/2016GL068956

Ojha L, Wilhelm MB, Murchie SL, McEwen AS, Wray JJ, Hanley J, Massé M, Chojnacki M (2015) Spectral evidence for hydrated salts in recurring slope lineae on Mars. Nat Geosci 8:829–832. https://doi.org/10.1038/ngeo2546

Ojha L, Karunatillake S, Karimi S, Buffo J (2021) Amagmatic hydrothermal systems on Mars from radiogenic heat. Nat Commun 12:1754. https://doi.org/10.1038/s41467-021-21762-8

Orosei R, Jordan RL, Morgan DD, Cartacci M, Cicchetti A, Duru F, Gurnett DA, Heggy E, Kirchner DL, Noschese R, Kofman W (2015) Mars advanced radar for subsurface and ionospheric sounding (MARSIS) after nine years of operation: a summary. Planet Space Sci 112:98–114. https://doi.org/10.1016/j.pss.2014.07.010

Pain CF, Clarke JDA, Thomas M (2007) Inversion of relief on Mars. Icarus 190:478–491. https://doi.org/10.1016/j.icarus.2007.03.017

Pajola M, Rossato S, Carter J, Baratti E, Pozzobon R, Erculiani MS, Coradini M, McBride K (2016) Eridania Basin: an ancient paleolake floor as the next landing site for the Mars 2020 rover. Icarus 275:163–182. https://doi.org/10.1016/j.icarus.2016.03.029

Palumbo AM, Head JW (2018) Early Mars climate history: characterizing a "warm and wet" Martian climate with a 3-D global climate model and testing geological predictions. Geophys Res Lett 45:10249–10258. https://doi.org/10.1029/2018GL079767

Paris AJ, Morgan K (2022) The transport of extrusive volcanic deposits on Jezero Crater through paleofluvial processes. J Wash Acad Sci 108:11–28

Parker TJ, Saunders RS, Schneeberger DM (1989) Transitional morphology in west Deuteronilus Mensae, Mars: implications for modification of the lowland/upland boundary. Icarus 82:111–145

Pelletier JD (2004) How do spiral troughs form on Mars? Geology 32:365–367. https://doi.org/10.1130/G20228.2

Plaut JJ, Safaeinili A, Holt JW, Phillips RJ, Head JW, Seu R, Putzig NE, Frigeri A (2009) Radar evidence for ice in lobate debris aprons in the mid-northern latitudes of Mars. Geophys Res Lett 36:1–4. https://doi.org/10.1029/2008GL036379

Putzig NE, Smith IB, Perry MR, Foss FJ 2nd, Campbell BA, Phillips RJ, Seu R (2018) Three-dimensional radar imaging of structures and craters in the Martian polar caps. Icarus 308:138–147. https://doi.org/10.1016/j.icarus.2017.09.023

Roda M, Kleinhans MG, Zegers TE, Oosthoek JH (2014) Catastrophic ice lake collapse in Aram Chaos, Mars. Icarus 236:104–121

Rodriguez JAP, Fairén AG, Tanaka KL, Zarroca M, Linares R, Platz T, Komatsu G, Miyamoto H, Kargel JS, Yan J, Gulick V (2016) Tsunami waves extensively resurfaced the shorelines of an early Martian ocean. Sci Rep 6:1–8. https://doi.org/10.1038/srep25106

Rodriguez JAP, Tanaka KL, Bramson AM, Baker VR, Zarroca M (2021) North polar trough formation due to in-situ erosion as a source of young ice in mid-latitudinal mantles on Mars. Sci Rep 11. https://doi.org/10.1038/s41598-021-83329-3

Salvatore MR, Christensen PR (2014) Evidence for widespread aqueous sedimentation in the northern plains of Mars. Geology 42:423–426

Scariah NV, Lala MGN, Krishna AP (2023) Seasonal variation of Martian surface temperature over gale crater and surroundings. Sol Syst Res 57:14–24. https://doi.org/10.1029/2019JE006175

Schorghofer N (2007) Dynamics of ice ages on Mars. Nature 449:192–194. https://doi.org/10.1038/nature06082

Schorghofer N (2020) Crocus melting on Mars. In: Europlanet science congress, vol 14. https://doi.org/10.5194/epsc2020-175

Schorghofer N, Aharonson O (2005) Stability and exchange of subsurface ice on Mars. J Geophys Res Planets 110. https://doi.org/10.1029/2004JE002350

Sharp RP (1973) Mars: fretted and chaotic terrains. J Geophys Res 78:4073–4083

Shi Y, Zhao J, Xiao L, Yang Y, Wang J (2022) An arid-semiarid climate during the Noachian-Hesperian transition in the Huygens region, Mars: evidence from morphological studies of valley networks. Icarus 373. https://doi.org/10.1016/j.icarus.2021.114789

Shreve RL (1966) Statistical law of stream numbers. J Geol 74:17–37

Sinha RK, Ray D, Haas TD, Conway SJ, Noblet A (2023) Morphologic and morphometric differences between gullies formed in different substrates on Mars: new insights into the gully formation processes, Earth surface. Dynamics 11:713–730. https://doi.org/10.5194/esurf-2022-64

Smith IB (2022) A retrospective on Mars polar ice and climate. Planet Sci. https://doi.org/10.1093/acrefore/9780190647926.013.242

Smith IB, Putzig NE, Holt JW, Phillips RJ (2016) An ice age recorded in the polar deposits of Mars. Science 352:1075–1078. https://doi.org/10.1126/science.aad6968

Smith IB, Schlegel NJ, Larour E, Isola I, Buhler PB, Putzig NE, Greve R (2022) Carbon dioxide ice glaciers at the south pole of Mars. J Geophys Res Planets 127. https://doi.org/10.1029/2022JE007193

Soare RJ, Séjourné A, Pearce G, Costard F, Osinski GR (2011) The Tuktoyaktuk Coastlands of northern Canada: a possible "wet" periglacial analog of Utopia Planitia, Mars. Geol Soc Am Spec Pap 483:203–218

Squyres SW (1978) Martian fretted terrain: flow of erosional debris. Icarus 34:600–613. https://doi.org/10.1016/0019-1035(78)90048-9

Squyres SW (1979) The distribution of lobate debris aprons and similar flows on Mars. J Geophys Res Solid Earth 84:8087–8096. https://doi.org/10.1029/JB084iB14p08087

Steakley KE, Kahre MA, Haberle RM, Zahnle KJ (2023) Impact induced H_2-rich climates on early Mars explored with a global climate model. Icarus 394:115401. https://doi.org/10.1016/j.icarus.2022.115401

Steckel AV, Tucker GE, Rossi M, Hynek B (2025) Landscape evolution models of incision on Mars: implications for the ancient climate. J Geophys Res Planets 130:e2024JE008637. https://doi.org/10.1029/2024JE008637

Stock JD (2013) Waters divided: a history of alluvial fan research and a view of its future. In: Shroder J, Wohl E (eds) Treatise on geomorphology, vol 9, pp 413–458

Sun VZ, Stack KM (2020) Geologic map of Jezero Crater and the Nili Planum region, Mars. [Map]. U.S. Geological Survey. https://doi.org/10.3133/sim3464

Szilágyi-Sándor A, Székely B (2025) The evolution of Nirgal Vallis (Mars) based on morphometric results: does tectonics play a role? Planet Space Sci 106128. https://doi.org/10.1016/j.pss.2025.106128

Tanaka KL, Skinner Jr JA, Hare TM (2005) Geologic map of the northern plains of Mars. U.S. Geological Survey. Miscellaneous Investigations Service. Map, I-2888, scale 1:15,000,000

Vals M, Rossi L, Montmessin F, Lefèvre F, Gonzalez-Galindo F, Fedorova A, Luginin M, Forget F, Millour E, Korablev O, Trokhimovskiy A (2022) Improved modelling of Mars' HDO cycle using a Mars' global climate model. J Geophys Res Planets 127. https://doi.org/10.1029/2022JE007201

Villanueva G, Mumma M, Novak R et al (2015) Strong water isotopic anomalies in the Martian atmosphere. Probing current and ancient reservoirs. Science 348:218–221

Vos E, Aharonson O, Schörghofer N, Forget F, Millour E, Rossi L, Vals M, Montmessin F (2022) Stratigraphic and isotopic evolution of the Martian polar caps from paleo-climate models. J Geophys Res Planets 127. https://doi.org/10.1029/2021JE007115

Warner N, Gupta S, Lin SY, Kim JR, Muller JP, Morley J (2010) Late Noachian to Hesperian climate change on Mars: evidence of episodic warming from transient crater lakes near Ares Vallis. J Geophys Res Planets 115. https://doi.org/10.1029/2009JE003522

Williams RM, Phillips RJ, Malin MC (2000) Flow rates and duration within Kasei Valles, Mars: implications for the formation of a Martian ocean. Geophys Res Lett 27:1073–1076

Wilson L, Ghatan GJ, Head Iii JW, Mitchell KL (2004) Mars outflow channels: a reappraisal of the estimation of water flow velocities from water depths, regional slopes, and channel floor properties. J Geophys Res Planets 109. https://doi.org/10.1029/2004JE002281

Wilson SA, Morgan AM, Howard AD, Grant JA (2021) The global distribution of craters with alluvial fans and deltas on Mars. Geophys Res Lett 48. https://doi.org/10.1029/2020GL091653

Wordsworth R, Forget F, Millour E, Head JW, Madeleine JB, Charnay B (2013) Global modelling of the early Martian climate under a denser CO_2 atmosphere: water cycle and ice evolution. Icarus 222:1–19. https://doi.org/10.48550/arXiv.2103.06736

Wordsworth R, Knoll AH, Hurowitz J, Baum M, Ehlmann BL, Head JW, Steakley K (2021) A coupled model of episodic warming, oxidation and geochemical transitions on early Mars. Nat Geosci 14:127–132

Xiao L, Huang J, Kusky T, Head JW, Zhao J, Wang J, Wang L, Yu W, Shi Y, Wu B, Qian Y (2023) Evidence for marine sedimentary rocks in Utopia Planitia: Zhurong rover observations. Natl Sci Rev 137. https://doi.org/10.1093/nsr/nwad137

Xiaoting X, Xu Y, Meng X (2022) SHARAD observations of temporal variations of CO_2 ice deposits at the south pole of Mars. Remote Sens 14(3):435. https://doi.org/10.3390/rs14030435

Yousif M (2018) Sensing the hydrogeomorphology and palaeohydrologic features of Mars planet with inference from the Egyptian desert: an insight from satellite images. Space Sci Int 4:16–36. https://doi.org/10.3844/ajssp.2018.16.36

Zhang L, Lu Y, Zhang J (2022) A polygonal terrain on southern Martian polar cap: implications for its formation mechanism. Remote Sens 14. https://doi.org/10.3390/rs14225789

Chapter 10
Mars Unveiled: Astrobiology

The belief that Mars once possessed or may even continue to support life is one of the main driving forces to explore of the planet. This is also fuelled by the pre-space-age discovery of polar water–ice caps which changed depending on Martian seasons, and also persistent persuasion of the existence of since-refuted canals to the start of the space age. As a result, many missions were sent to Mars to better investigate the planet beyond the turbulence of Earth's atmosphere (Chap. 1). Despite pre-space-age hopes, decades of remote sensing and robotic exploration have revealed a present Mars climate that is inhospitable to most forms of terrestrial life (e.g. Fig. 10.1).

The lack of appreciable magnetic field subjects the Martian surface to elevated levels of cosmic radiation and solar storms when compared to Earth. In addition, UV radiation, barely filtered by Mars' thin atmosphere, has sterilised the surface, forming perchlorates that react adversely with organic molecules. The CO_2 atmosphere is also unable to maintain thermal stability, and temperatures swing between 130 and 250 K. For context, a 235 K average temperature is the limit at which biological replication can occur on Earth. Atmospheric pressures close to water's triple-point (the point where water can coexist as a solid, liquid and gas) also prohibit the existence of standing water on most of the planet; ices exposed by the Phoenix lander near the south pole were observed to sublime over a period of days (see Fig. 7.8b).

10.1 Direct Search for Life

The first, and so far the only direct search for present life experiments were conducted by the Viking landers at Utopia and Chryse Planitias in 1976 (Fig. 10.2). Based on the best information available at the time, each lander's one cubic foot payload consisted of three independent experiments, gas exchange (Gex), pyrolytic release (Pr) and labelled release (Lr), supported by a fourth gas chromograph mass spectrometer

S. Hobbs, *Sensing Mars: Remote Sensing to Understand the Red Planet*,
https://doi.org/10.1007/978-3-032-24416-1_10

Fig. 10.1 The present-day Martian surface is hostile to almost all terrestrial life, and requires space-hardened robots, such as the Spirit Rover illustrated here, for exploration

(GCMS) instrument (Klein 1979). Each lander used a robotic arm to extract regolith from regions protected from direct sunlight and deposit into the experiments.

Gex searched for biologically released gasses through saturating a sample with water vapour at 0.2 atmospheres and heated to a range of 281–288 K. A second mode added nutrients along with water, and both experiments ran for up to 200 Martian sols. Gex detected oxygen release over this period, though the result was attributed to regolith oxidising reactions versus biological activity.

Pr incubated regolith in illuminated and dark conditions in a simulated Martian environment. Water without nutrients was added in this test, and CO_2 gasses, initially

Fig. 10.2 Trenches dug by the Viking 1 lander's scoop (in the foreground) in the first direct search for life experiment on Mars (Viking, NASA/JPL)

thought to possibly indicate biological metabolism, were detected. These CO_2 gasses were emitted following heating the sample to sterilisation temperatures, and experiment results were instead interpreted to originate from chemical processes. Lr incubated hydrated regolith at 283 K and 60 mbar for 13–90 sols. Gasses were detected in this experiment during incubation, with levels reducing following a sterilisation temperature of 433 K. Of the three experiments, Lr provided the only results that may have been consistent with biological processes, its Martian results consistent with terrestrial biology tests, though purely chemical origins of the observed results could not be excluded.

The results from the GCMS, which failed to find any evidence of organic compounds in the regolith at both lander sites, led researchers to conclude that the surface of Mars is sterile. The Viking experiments also highlighted the difficulty in obtaining compelling evidence for biological activity using a limited set of experiments at only two landing sites (Klein 1998). The ambiguous results from Viking influenced more "conservative" subsequent missions to Mars, commencing with a focus on the mineralogy, then habitability (water) and more recently organics, including pre-biotic markers and related compounds. Many NASA surface missions were also designed to "follow the water", searching for geological indicators of past liquid water playing a significant role in altering the environment (Hubbard et al. 2002).

10.2 Earth Analogues for Astrobiology

The significant advances in the Earth analogue studies have supplemented both the study of potential Martian fossils and that of astrobiology whose remote sensing and surface investigations remain very limited. Identifying regions where life can survive in extreme environments, "extremophiles"—began in earnest following the Viking missions. This type of research seeks to discover environments where water-based life can exist, with discoveries revealing that micro-organisms are capable of surviving in extreme environments.

An example is the ADVs, with average temperatures of -20 °C and precipitation of just 6 cm/year, making this region among the driest, coldest places on Earth (Fig. 10.3a). The ADVs are a popular Mars analogue site, hosting similar geomorphic features such as ice flows, gullying and frost activity, as found on Mars (Cowan et al. 2014). The extreme conditions severely limit the availability of liquid water, with transition from solid to liquid affected by insolation and local topography.

Viking pre-flight experiments in the early 1970s actually failed to unambiguously identify life in these valleys. Despite these limitations, biologists were surprised to discover numerous extremophiles ranging from desiccation and radiation-resistant bacteria within the organically poor soil, to photosynthetic forms residing inside light-toned rocks (Broady and Weinstein 1998, Fig. 10.3b). Experiments conducted where representative extremophiles were exposed to Mars-like conditions supported the

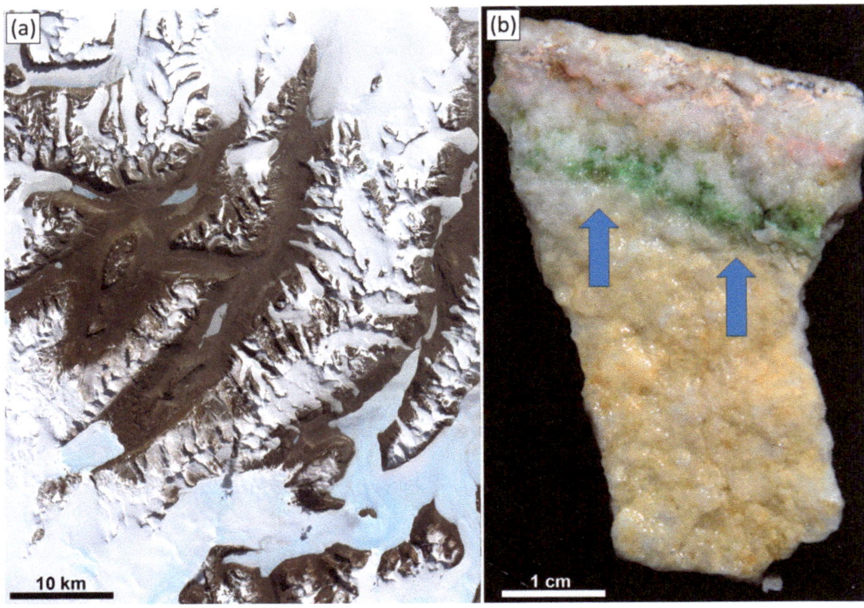

Fig. 10.3 **a** Landsat 7 image of the ADVs (NASA GSFC Oceans and Ice). **b** Fragment of an extremophile sample from a larger rock identified in the ADVs in 2014. Blue arrows highlight a layer of photosynthetic algae embedded within a light toned rock

possibility of survival on the Red Planet, though mostly in a dormant state (Musilova et al. 2015; Armstrong 2019).

Studying acidic terrestrial environments are also of interest to astrobiologists. The Opportunity Mars rover was targeted to land at the Meridiani Planum region as orbital remote sensing identified mineralogy consistent with liquid water (Chap. 7). During its 16-year operational life, the rover identified hematite-rich concretions, layering and ripples consistent with shallow water interactions at the site. Additional mineral investigations, such as the discovery of jarosite at El Capitan suggested the water was highly acidic and hostile to most forms of life (Fig. 10.4a, Elwood Madden et al. 2004).

The Rio Tinto in the Jaroso Ravine, Spain is considered a terrestrial analogue for the Opportunity site, and hosts microbial life forms existing in extremely acidic (< pH 2) conditions (Fig. 10.4b). The site consists of an iron and sulphur-rich hydrothermal system; and the utility of Raman spectroscopy for detecting indicators for biological activity was demonstrated in field research. A version of this spectroscopy awaits to be flown on the ESA ExoMars mission (Edwards et al. 2007).

Active hot spring and geyser discharge channels in El Tatio, Northern Chile are also of high interest as terrestrial analogues for environments hosting potential Martian astrobiology. Shallow outflow channels in this region have produced opaline silicate structures. These form through the leaching of metallic ions from

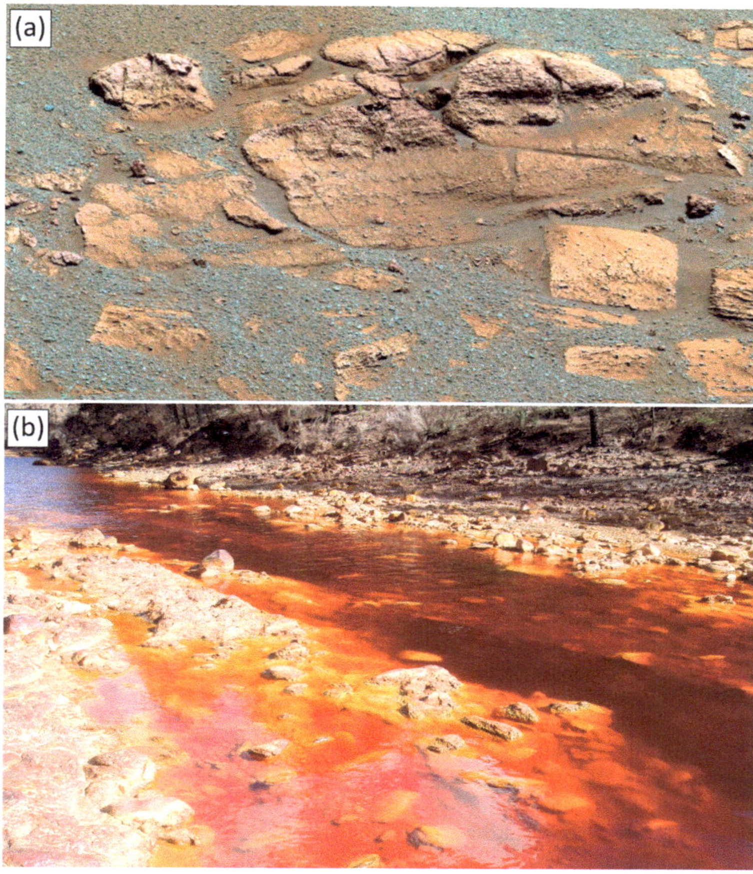

Fig. 10.4 **a** False colour image of the El Capitan from the Opportunity rover (PIA05457, NASA/JPL). **b** The acidic river in Rio Tinto, Spain (Public Domain)

basalt by acidic liquid, leaving behind silicate outcrops (Ruff and Farmer 2016). This is believed be similar to the past hot spring environment at Home Plate with associated mineralogy discovered by the Spirit Mars rover (Fig. 7.12a–c). Opaline silicate nodules were also discovered at the Home Plate site, appearing to be spectrally similar to those analysed at El Tato. Analysis of the rover data suggests the Home Plate silica rocks were formed through active geyser activity driven by hydrothermal activity. Rocks preserving evidence of siliceous microbial stromatolites or other biological activity are commonly found within terrestrial hot spring environments, including at the El Tato site. The similarity of the silicate structures found at Home Plate with terrestrial analogues that commonly preserve evidence for microbial life, has made the Gusev Crater site a potential candidate for a follow up surface mission (Ruff et al. 2020).

10.2.1 Case Study: Field Robotic and Laboratory Astrobiology Investigations

An example of terrestrial microfossil characterisation was the confirmation of fossil stromatolites in Australia as being biological in origin. These Precambrian cyanobacteria lived in shallow water. They accreted and cemented material in layers, with the living organisms on the top-most surface (Allwood et al. 2007). The physical appearance of fossil stromatolites bears remarkable similarity to natural, layered, sedimentary deposition. The ambiguity revealed the constraints of simple qualitative analysis of layered rock. Following the fossils discoveries, more complex analysis was required to identify the presence of organic globule clusters and aragonite nanocrystals that correlated with living stromatolites. This required extensive fieldwork, collection of substantial quantities of fossilised material, destructive and non-destructive testing by a battery of laboratory equipment. Multiple lines of research were then required to confidently identify the discoveries as biological in origin (Brasier et al. 2005; Allwood et al. 2009).

Recently, a field experiment to enhance understanding in reliable off-Earth fossil detection was conducted by the author using a Mars analogue rover at Arkaroola, South Australia. The test region in Arkaroola contains an abundance of fossil stromatolites. Discovery of similar fossils on Mars would be of high interest to astrobiologists, and the experiment was set up to quantify the utility and constraints of robotic astrobiology exploration with a limited set of instruments.

The Mars Society Australia's Little Blue Rover was built to the approximate dimensions of the Sojourner Rover flown on board Mars Pathfinder (Chap. 1) and represented a vehicle size that could be operated by space agencies on a limited budget. Little Blue's prime sensor was a 0.3 MPixel imaging camera able to capture three bands of visible and one NIR channel. Samples of stromatolite fossils and layered rocks of natural sedimentary origin were emplaced in an outdoor field setting brightly lit by natural sunlight. Little Blue was commanded to drive and photograph the target rocks at decreasing distances of 1 m, 50 cm and 25 cm respectively (output from the imager shown in Fig. 10.5a–c).

The NIR component failed to identify characteristics useful for assisting analysis that was dependant on shape recognition of fossil evidence and was subsequently discarded. Only the RGB bands were used to produce final images for analysis. Only broad features were identified at 1 m distance (Fig. 10.5a). At 50 cm distance, the target rocks began to show textured details and banding associated with stromatolite layering (Fig. 10.5b). Naturally occurring sedimentary processes were discernible at 25 cm distance (Fig. 10.5c), at which point experienced micropaleontologists were asked to assess the images and determine at which distance they were able to identify biological from natural activity.

Despite knowing that the imagery contained fossils, the micropaleontologists were unable to positively distinguish between the non-biologic sedimentary and fossil stromatolite samples. Extra sensor data, or actual samples for direct analysis were assessed by the micropaleontologists to be required for more a conclusive

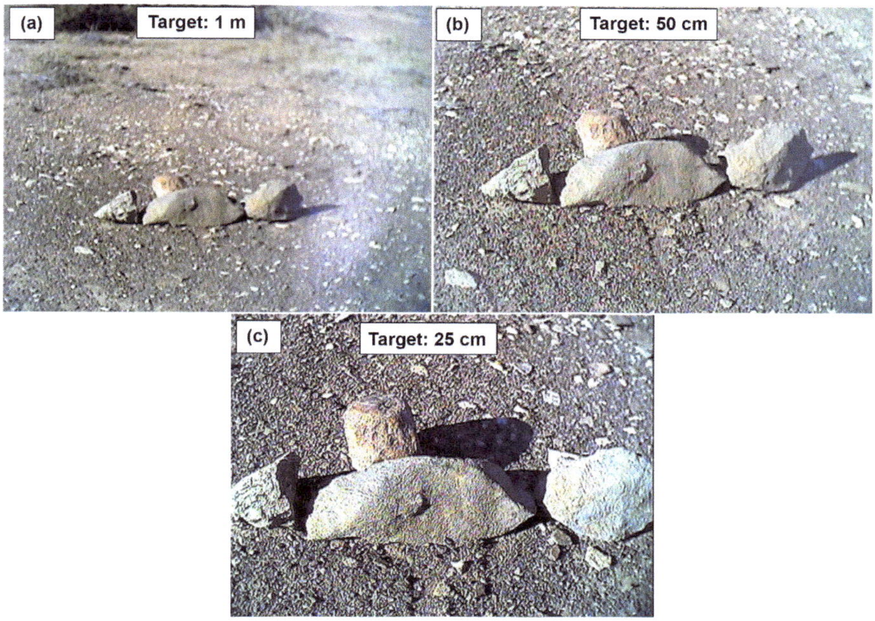

Fig. 10.5 **a** Little Blue Rover image of specimen samples from 1 m. **b** Image from 0.5 m distance. **c** 0.25 m image distance

determination of past biology. This simple experiment highlighted the constraints of off-Earth in astrobiology investigations.

Possible Martian Examples

Similar constraints demonstrated in the Little Blue Rover case study have appeared during Mars rover explorations, such as the Curiosity Rover microscopic imager investigations of Gale Crater sediments at Vera Rubin Ridge. Analysis of returned rover data identified evidence of lacustrine environments, potentially preserving organic compounds (Fraeman et al. 2020). In this context, discoveries of dark toned tubular features close to each other suggested a similarity to terrestrial fossil trace burrows (Fig. 10.6a; Elewa 2021). Similar to the Little Blue Arkaroola investigation, the analysis was based on imagery only, as the Curiosity Rover was unable to undertake chemical investigation with other instruments before traversing to the next location.

Natural explanations, such as crystallisation of the hydrated mineral gypsum were offered as credible alternatives. The restriction to comparative analysis at this site between terrestrial trace burrows and the Martian images precluded unambiguous assessment of biological or natural origin of these features.

Other discoveries of features resembling fossils within Gale Crater have led to suggestions to a mini-Cambrian explosion occurring at Curiosity's landing site

Fig. 10.6 **a** 3–5 mm wide, tube-like structures imaged by the Curiosity Rover that superficially resemble terrestrial fossil trace burrows. These features may be more likely formed through natural processes (1922MH0001520010703174C00-DXXX, NASA/JPL). **b** An outcrop imaged by Curiosity on Sol 553 that has been suggested by some researchers to contain microfossils 1–5 mm in size (0553MR0022370010303889E01-DXXX, NASA/JPL). **c** A portion of "Cheyaya Falls", a rock investigated by Perseverance in Jezero Crater (PIA26368, NASA/JPL). **d** Closeup of millimetre-sized "leopard spots" resembling terrestrial indications of past microbial life (PIA26368, NASA/JPL)

(Fig. 10.6b). This referred to the Terrestrial period where unprecedented species emerged between 541 and 530 Ma (Briggs 2015). The interpretations of parts of Gale Crater being a former lake bed were assisted by comparisons with analogues terrestrial environments, particularly Burgess Shale deposits formed during the Cambrian era (500 + Ma ago). Located in the Canadian Rockies and consisting of dark mudstones, the Burgess Shales host a wealth of well-preserved fossils of life forms existing in the period (Butterfield 2003). Specimens identified from Curiosity imagery resemble low-level microorganisms, including algae and stromatolites, or even marine organisms, such as tube worms, annelids and other larger fauna. Far more sophisticated analysis than was available for the Little Blue experiment (Fig. 10.5a–c) described above incorporated quantitative comparisons with similar terrestrial fossils. The authors suggested a possibility of the identified features being consistent with marine organisms becoming fossilised in mudstone as Mars suffered irreversible loss of water and the lake bed dried out (Joseph et al. 2023).

The above suggestions of microfossils remain controversial and are often contested amongst astrobiology peers. In the absence of additional data, more conclusive findings from the Gale Crater examples identified above are currently not possible. In order to obtain more robust findings, the most sophisticated robot ever sent to another world, Perseverance, is exploring a region in Jezero Crater believed capable of preserving evidence of past Martian life. This crater has a well-preserved fluvial delta that was active from the Noachian to Hesperian periods (Fig. 9.13c; Bell et al. 2022). The Fe/Mg smectite stratigraphy of the landing site has a high potential for preserving past biological activity. In July 2024 the Perseverance Rover completed a core sample of "Cheyava Falls", a probable mudstone rock in Jezero Crater (Fig. 10.6c, d). Unlike the two Gale Crater examples described above, imaged-based investigations were supplemented by spectroscopy and X-Ray elemental analysis, with results being consistent with terrestrial microbiology (Hurowitz et al. 2025a). Dark-toned rings enriched with organic molecules and calcium sulphate veins precipitated from past liquid water activity within the fractures are similar to terrestrial microbiological samples. In north-west England, microbes previously living within the veins of sedimentary rocks had precipitate chemicals that were later be detected through by Raman spectroscopy (McMahon et al. 2020). Cheyava Falls is located within Neretva Vallis, an ancient 400 m wide river valley formed through water draining into Jezero Crater. As of publication, the formations of the possible biomarkers in this rock remain unconfirmed, as natural processes are unable to be ruled out (NASA 2024; Hurowitz et al. 2025b).

Commencing from 2023, Perseverance collected and cached samples that included atmospheric, regolith, and rocky core for future return to Earth. A sample return mission, planned to fly no earlier than 2030, is expected to revisit the Jezero Crater site. The mission is likely to consist of flying drones, ground rovers or Perseverance itself, to collect the cached samples. Following launch and transport of these samples back to Earth, it is hoped the best laboratory equipment available may provide more conclusive analysis of the possible existence of Martian biological activity (Simon et al. 2023).

10.2.2 ALH84001—Mars Meteorite Analysed on Earth

As described above, remote sensing or in situ observations with a constrained suite of payload instruments limits their utility for the field of astrobiology. Terrestrial laboratory confirmation of astrobiology has also proven difficult, as demonstrated in the example of ALH84001 (Fig. 10.7). This Martian meteorite was discovered in Antarctica in 1984 and subjected to years of intensive analysis. In 1996 NASA announced that the 4.0 Ga meteorite contained chemical and microscopic evidence consistent with ancient bacteria. This included potential mineral biomarkers

Fig. 10.7 The ALH84001 Meteorite (S85-39565, NASA)

including large carbonate globules, polycyclic aromatic hydrocarbons, magnetite and iron sulphides. Electron microscopy also imaged tube-like features interpreted as microfossils (McKay et al. 1996).

The inability to identify where on Mars ALH84001 originated from placed constraints in understanding whether local geology may have driven inorganic processes. This made it difficult to understand the rock's history or identify process timings, such as aqueous alteration, that the meteorite had been subjected to. In addition, subsequent analysis identified purely chemical processes that could have produced the features observed in the meteorite. The proposed microfossils, while similar in shape to biological counterparts, were significantly smaller, and could be explained through pyroxene and carbonate crystal growth (Bradley et al. 1997). ALH84001 analysis thus demonstrated that, even with physical samples from Mars and sophisticated laboratory analysis, determination of whether Mars hosted biology remains difficult to answer.

10.3 Atmospheric Methane

Methane discovered in Mars' atmosphere through ground-based and orbital spectroscopy attracted interest as this gas is unstable, undergoing photochemical degradation to disappear in ~ 340 years or less without replenishment (Moores and Schuerger 2012). Terrestrial methane is produced through biological and volcanic processes, and its presence on Mars suggests transient processes capable of replenishment (Fig. 10.8a, b). The Tuneable Laser Spectrometer (TLS) on board Curiosity was able to detect small amounts of methane in Gale Crater that varied seasonally between 0.25 and 0.65 parts per billion (ppb). TLS operates by ingesting a sample of the Martian atmosphere and shining a laser beam through it to measure absorption at precise wavelengths. TLS provided measurements of methane, water and CO_2 isotopes respectively, the tuneable aspect of the spectrometer allowing it to operate in the Martian climate during the night. Following initial detection results, the TLS switched to daytime operations and results identified a diurnal cycle (Manne et al. 2017). Methane accumulated near the surface at nighttime is dispersed during solar heating during the day. The instrument may have also measured seasonal variations, suggestive of subsurface methane being released through changes in surface atmospheric pressure controlled by the Martian seasons (Ortiz et al. 2023).

Research conducted to identify the source of Martian methane is ongoing. Recent research identified a co-discovery of a methane spike by the Mars express orbiter one day after the Curiosity Rover recorded a methane spike on 16 June 2013. Modelling derived from understanding Earth-based processes suggested a possible explanation involving release of subsurface methane gas trapped in or below subliming ice. This explanation was supported by the identification of geological faults in Aeolis Mensae, east of Gale Crater, as the methane release source (Giuranna et al. 2019).

The above remote sensing and surface observations have assisted in inferring Martian methane properties. The gas is important for astrobiology as its creation through a combination of hydrogen and CO_2 gasses can be an energy source for chemosynthetic micro-organisms (Yamagishi et al. 2010). It also may be an indicator of active volcanic processes, supported by the recent volcanism activity at Cerberus Fossae (Chap. 8), or low temperature alteration of basalt by carbon-rich hydrothermal fluids (Lyons et al. 2005). Future detections and analysis of Martian methane are required to further characterise its origin and properties.

The above examples and investigations in this chapter suggest the possibility of past or even current biological activity. While the likelihood of Mars hosting past or present life has significantly driven exploration of the planet, hard evidence has yet to be found. Remote sensing and surface investigations have revealed an early environment compatible with life, with isolated regions likely capable of supporting biology to the present day. Terrestrial research has also identified biology capable of surviving in extreme environments, such as those found on Mars. Continued exploration of the planet with more sophisticated instruments, and missions to directly return Martian samples back to Earth are hoped to confirm whether Mars once hosted, or continues to support life.

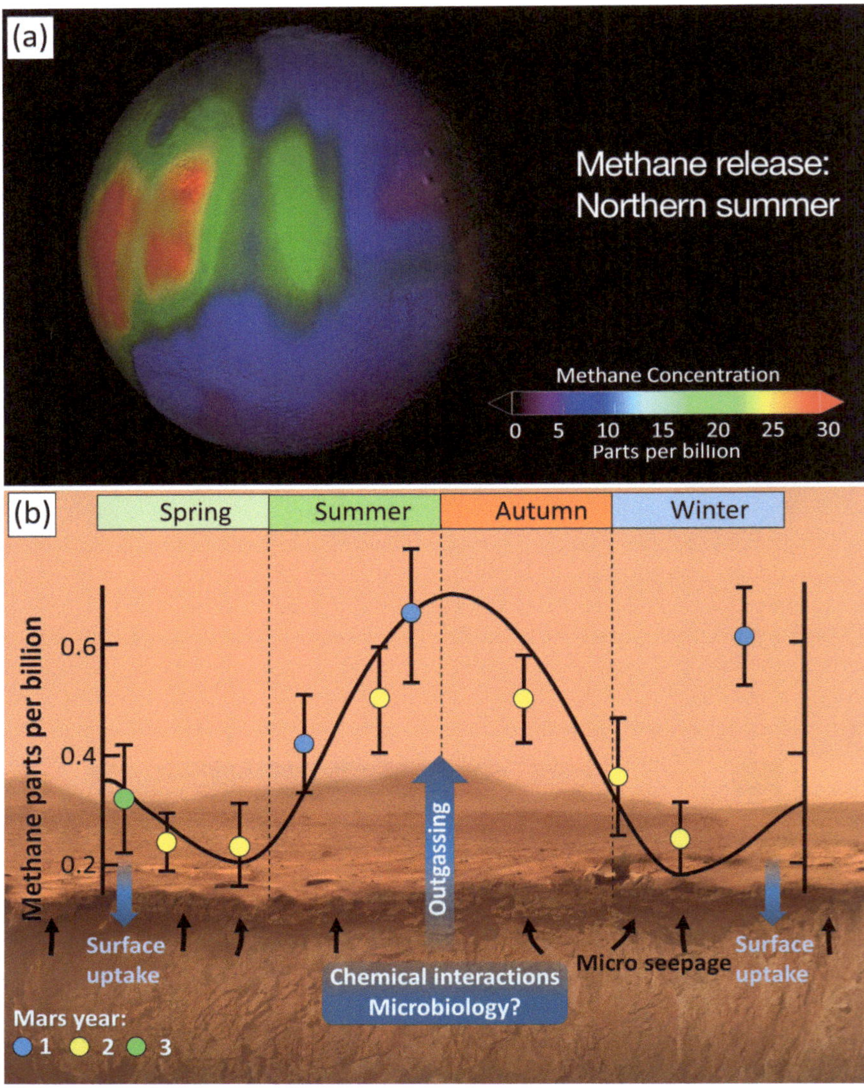

Fig. 10.8 **a** Global map of methane discovered on Mars (NASA). **b** Graphical representation of methane concentrations by season over a three Mars year period, showing the highest releases in late summer. Methane gas may be released through geological or potentially microbiological processes, with exchanges between the surface and atmosphere varying concentrations

References

Allwood AC, Walter MR, Burch IW, Kamber BS (2007) 3.43-billion-year-old stromatolite reef from the Pilbara Craton of Western Australia: ecosystem-scale insights to early life on Earth. Precambr Res 158:198–227. https://doi.org/10.1016/j.precamres.2007.04.013

Allwood AC, Grotzinger JP, Knoll AH, Burch IW, Anderson MS, Coleman ML, Kanik I (2009) Controls on development and diversity of Early Archean stromatolites. Proc Natl Acad Sci 106:9548–9555. https://doi.org/10.1073/pnas.0903323106

Armstrong RA (2019) The lichen symbiosis: lichen "extremophiles" and survival on Mars. J Astrobiol Space Sci Rev 1:378–397

Bell JF III, Maki JN, Alwmark S, Ehlmann BL, Fagents SA, Grotzinger JP, Gupta S, Hayes A, Herkenhoff KE, Horgan BH, Johnson JR (2022) Geological and meteorological imaging results from the Mars 2020 perseverance rover in Jezero Crater. Sci Adv. https://doi.org/10.1126/sciadv.abo4856

Bradley JP, Harvey RP, McSween HY Jr, Gibson E Jr, Thomas-Keprta K, Vali H (1997) No 'nanofossils' in Martian meteorite. Nature 390:454–455

Brasier MD, Green OR, Lindsay JF, McLoughlin N, Steele A, Stoakes C (2005) Critical testing of Earth's oldest putative fossil assemblage from the ~3.5 Ga Apex chert, Chinaman Creek, Western Australia. Precambr Res 140:55–102. https://doi.org/10.1016/j.precamres.2005.06.008

Briggs DE (2015) The Cambrian explosion. Curr Biol 25:R864–R868

Broady PA, Weinstein RN (1998) Algae, lichens and fungi in la Gorce Mountains, Antarctica. Antarct Sci 10:376–385

Butterfield NJ (2003) Exceptional fossil preservation and the Cambrian explosion. Integr Comp Biol 43:166–177

Cowan DA, Makhalanyane TP, Dennis PG, Hopkins DW (2014) Microbial ecology and biogeochemistry of continental Antarctic soils. Front Microbiol. https://doi.org/10.3389/fmicb.2014.0015

Edwards HG, Vandenabeele P, Jorge-Villar SE, Carter EA, Perez FR, Hargreaves MD (2007) The Rio Tinto Mars analogue site: an extremophilic Raman spectroscopic study. Spectrochim Acta Part A Mol Biomol Spectrosc 68:1133–1137

Elewa AM (2021) Fossils on Mars: a brief review of the evidence. J Astrobiol 7:29–37

Elwood Madden ME, Bodnar RJ, Rimstidt JD (2004) Jarosite as an indicator of water-limited chemical weathering on Mars. Nature 431:821–823. https://doi.org/10.1038/nature02934

Fraeman AA, Edgar LA, Rampe EB, Thompson LM, Frydenvang J, Fedo CM, Catalano JG, Dietrich WE, Gabriel TS, Vasavada AR, Grotzinger JP (2020) Evidence for a diagenetic origin of Vera Rubin ridge, Gale crater, Mars: summary and synthesis of Curiosity's exploration campaign. J Geophys Res Planets. https://doi.org/10.1029/2020JE006527

Giuranna M, Viscardy S, Daerden F, Neary L, Etiope G, Oehler D, Formisano V, Aronica A, Wolkenberg P, Aoki S, Cardesín-Moinelo A (2019) Independent confirmation of a methane spike on Mars and a source region east of Gale Crater. Nat Geosci 12:326–332. https://doi.org/10.5194/esurf-11-633-2023

Hubbard GS, Naderi FM, Garvin JB (2002) Following the water, the new program for Mars exploration. Acta Astronaut 51:337–350

Hurowitz JA et al (2025a) Redox-driven mineral and organic associations in Jezero Crater, Mars. Nature 645:332–340. https://doi.org/10.1038/s41586-025-09413-0

Hurowitz JA, Tice MM, Allwood AC, Cable ML, Hand KP, Murphy AE, Uckert K, Bell JF, Bosak T, Broz A, Clavé E (2025b) The detection of a potential biosignature by the perseverance Rover on Mars. In: 56th lunar and planetary science conference, p 2581, hal-04930754

Joseph RG, Rizzo V, Gibson CH, del Gaudio R, Sumanarathna AR, Armstrong RA, Ray JG, Elewa AMT, Bianciardi G, Duvall D, Wickramasinghe C (2023) Fossils on Mars? A "Cambrian Explosion" and "Burgess Shale" in Gale Crater? J Astrophys Aerosp Technol. https://doi.org/10.37421/2329-6542.2022.10.237

Klein HP (1979) The Viking mission and the search for life on Mars. Rev Geophys 17:1655–1662

Klein HP (1998) The search for life on Mars: what we learned from Viking. J Geophys Res Planets 103:28463–28466

Lyons JR, Manning C, Nimmo F (2005) Formation of methane on Mars by fluid-rock interaction in the crust. Geophys Res Lett. https://doi.org/10.1029/2004GL022161

Manne J, Bui TQ, Webster CR (2017) Determination of foreign broadening coefficients for methane lines targeted by the tuneable laser spectrometer (TLS) on the Mars Curiosity Rover. J Quant Spectrosc Radiat Transfer 191:59–66. https://doi.org/10.1016/j.jqsrt.2017.01.035

McKay DS, Gibson EK Jr, Thomas-Keprta KL, Vali H, Romanek CS, Clemett SJ, Chillier XD, Maechling CR, Zare RN (1996) Search for past life on Mars: possible relic biogenic activity in Martian meteorite ALH84001. Science 273:924–930

McMahon S, Parnell J, Reekie PBR (2020) Mars-analog calcium sulfate veins record evidence of ancient subsurface life. Astrobiology 20:1212–1223. https://doi.org/10.1089/ast.2019.2172

Moores JE, Schuerger AC (2012) UV degradation of accreted organics on Mars: IDP longevity, surface reservoir of organics, and relevance to the detection of methane in the atmosphere. J Geophys Res Planets. https://doi.org/10.1029/2012JE004060

Musilova M, Wright G, Ward JM, Dartnell LR (2015) Isolation of radiation-resistant bacteria from Mars Analog Antarctic dry valleys by preselection, and the correlation between radiation and desiccation resistance. Astrobiology 12:1076–1090. https://doi.org/10.1089/ast.2014.1278

NASA (2024) Perseverance finds a rock with 'Leopard Spots'. https://science.nasa.gov/resource/perseverance-finds-a-rock-with-leopard-spots/. Accessed 1 Aug 2024

Ortiz JP, Rajaram H, Stauffer PH, Lewis KW, Wiens RC, Harp DR (2023) Sub-diurnal methane variations on Mars driven by barometric pumping and planetary boundary layer evolution. J Geophys Res Planets. https://doi.org/10.22541/essoar.169160648.81638425/v1

Ruff SW, Farmer JD (2016) Silica deposits on Mars with features resembling hot spring biosignatures at El Tatio in Chile. Nat Commun 7:13554. https://doi.org/10.1038/ncomms13554

Ruff SW, Campbell KA, Van Kranendonk MJ, Rice MS, Farmer JD (2020) The case for ancient hot springs in Gusev crater, Mars. Astrobiology 20:475–499. https://doi.org/10.1089/ast.2019.2044

Simon JI, Hickman-Lewis K, Cohen BA, Mayhew LE, Shuster DL, Debaille V, Hausrath EM, Weiss BP, Bosak T, Zorzano MP, Amundsen HEF (2023) Samples collected from the floor of Jezero Crater with the Mars 2020 Perseverance rover. J Geophys Res Planets. https://doi.org/10.1029/2022JE007474

Yamagishi A, Yokobori SI, Yoshimura Y, Yamashita M, Hashimoto H, Kubota T, Yano H, Haruyama J, Tabata M, Kobayashi K, Honda H (2010) Japan Astrobiology Mars Project (JAMP): search for microbes on the Mars surface with special interest in methane-oxidizing bacteria. Biol Sci Space 24:67–82

Chapter 11
Mars of the Twenty-First Century: A Complex and Dynamic Planet

11.1 The Importance of Mars Missions

Humanity's understanding of Mars has changed dramatically within the space of one lifetime. Up to the first successful mission to Mars in 1964, the planet was thought to have hardened flora or even fauna living on its rust-red plains. Space age remote sensing has provided insight into the cold, hyper-arid surface conditions of present-day Mars with the development of increasingly sophisticated technology. As an example, the perception of the planet as a cratered, geologically dead world was as much due to engineering limitations as to analysis of returned data. These limitations constrained mission planners and operators to gather what data they could with primitive sensors using one-shot flybys able to target a fraction of the planet (Chap. 1).

The advent of more long-lived orbiters and landers vastly increased the amount of data received during their operation in the 1970s, revealing Mars as a once dynamic, active world, subject to volcanic, fluvial, ice and aeolian processes. Despite these missions employing a multitude of remote (and in the case of the Viking landers, direct) sensing techniques, research inferences were constrained by technology limitations. The best maps available of Mars to 1997 only showed objects greater than 150 m across. In addition, the highly successful Viking landers were able to directly sample only their immediate surroundings in Ares Vallis and Utopia Planitia respectively.

Numerous diverse and successful orbital and surface missions have returned large amounts of data from diverse locations on the planet, as well as from orbit. The abundance of multinational missions allows for collaborative sharing of results through exploitation of large portions of the electromagnetic spectrum. Active devices such as radar and laser range finding, and collectors of reflected or emitted energy such as thermal, optical and high-energy imaging and spectroscopy provide the most comprehensive data collection of any world besides Earth. Adaptations from terrestrial data management systems, and cartography, along with AI/ML processing are

S. Hobbs, *Sensing Mars: Remote Sensing to Understand the Red Planet*,
https://doi.org/10.1007/978-3-032-24416-1_11

being used to manage, process and exploit the hundreds of terabytes of data available to researchers worldwide.

The development and operation of Mars spacecraft begins with a clear understanding of mission goals that then drive robust space systems engineering practices. In this way, major development and testing requirements are considered early in the process, leading to a mission able to survive in the harsh Martian environment and return useful data. Recent advances in engineering simulations and "digital twins" have lowered the cost barrier to engineering and testing, while still maintaining the critical importance of assessing actual flight components.

The invention of the cubesat and "Space 2.0" intended to make access to space more affordable has had a positive impact on similar missions to Mars. In 2018, MarsCO demonstrated that a low-cost cubesat mission was able to survive and perform useful functions beyond Earth orbit. This success has led to the design and development of future missions possessing sensors once restricted to much larger and more expensive missions.

11.2 Remote Sensing as an Enabler for Understanding Mars

Mars has been the object of continuous orbital observations since 1997, allowing monitoring of seasonal changes and geomorphic processes over time. Impact cratering, wind, frost and ice are the main agents of present-day surface change. Slope failure, migration of sand dunes and other aeolian sediment transport are well documented, as is pit erosion from sublimating ice slabs. Much of this present activity is driven by seasonal heating and interaction of volatiles between the poles and the rest of Mars. Observations of fresh impact craters excavating near-surface ice deposits have also helped characterise the abundance of frozen water on the planet.

Present day Mars was shaped by multiple geologic processes that have occurred throughout its history, with abundant evidence of volcanic, ice, fluvial and aeolian-based erosion. As on Earth, Martian features created through one process were modified by later processes. Noachian aged fluvial channels that have hosted subsequent lava flows are a prominent example, and careful analysis of multiple remote sensing datasets or deployment of surface missions are essential for understanding the sequence of events. Identification and mapping of mineralogy using on-orbit and surface spectroscopy have identified distinct geology altering events aligning with the Noachian-Hesperian-Amazonian periods. Novel image-matching and computer vision techniques have revisited traditional crater counting techniques to further refine the history of Mars. This is achieved through automated processing of large data volumes and identifying primary vs secondary impact craters that may otherwise falsely inflate crater counts and hence surface age estimations.

Ultra-high-resolution imagery and elevation data has enabled detailed investigations of specific, small-scale regions of the planet. These include detailed analysis of Martian craters and slopes to characterise past ice, frost and water related activity such as in the case of glacier-like forms and hillside gullies. Concurrent orbiter and lander investigations also routinely occur with the latest generation of Mars rovers. Fine scale orbital imagery and spectral data is used to guide Mars rovers to scientifically interesting surface targets, enabling them to provide detailed measurements at ground level.

11.2.1 Multiple Processes Shaping Mars

An abundance of data from decades of spacecraft and lander data have revealed Mars to be a complex world that has evolved through interaction between multiple geologic processes. Formed and initially shaped by significant impacts along with the rest of the Solar System 4.5 Ga ago, the planet was at its most active in its first billion years. Powered by an active inner core and enhanced by a thick atmosphere, fluvial valley networks and volcanic provinces were formed. It was during the Noachian period, that the Martian environment was at its most hospitable to microbial life.

From 3.7 Ga ago impact cratering declined, and volcanic activity became the main geologic driver, forming lava plains and shield volcanoes in the Hesperian period. In many cases lava followed topographically low areas previously carved by fluvial activity, covering now inactive water-carved features, such as those discovered at Gusev Crater. Volcanic outgassing into the atmosphere interacted with liquid water to make it more acidic and precipitated out minerals of this class later discovered by the Spirit and Opportunity MER. As the planet began to cool, subsurface water began to freeze. This formed a cryosphere that was occasionally heated through volcanic activity to cause localised catastrophic flooding.

The Amazonian period, ~ 2 Ga to present was traditionally viewed as a major decline in Martian geologic activity, with the planet transitioning to a cold, hyperarid body predominantly modified by aeolian or seasonal polar cap freeze-thaw processes. Despite the reduction of Noachian and Hesparian-era surface processes, recent remote sensing and lander investigations have identified that Mars remains an active planet. Discoveries in Utopia Planitia have revealed salty water recently interacting with sand dunes, while remains of a modern glacier near the Martian equator suggest the preservation of ice in this region later than previously thought. In addition, volcanic activity in the Tharsis and Elysium regions continued throughout the Amazonian period, possibly to the present day.

11.3 Benefit to Earth

Remote sensing and exploration of Mars has produced benefits for Earth. Besides inspiring generations of science, technical engineering and mathematics professionals; studying the planet is providing a baseline for understanding surface and atmospheric processes and climate change that are directly relevant to terrestrial applications. Engineering advancements developed for Mars missions have supported and grown mission team expertise for Earth observation satellites and airborne sensors. In addition, international collaboration, fostered through multi-instrumented missions, ensures engineering and scientific data are shared to a wide stakeholder set. This has resulted in accelerated advances in remote sensing methods and shared responsibilities for environmental management and custodianship.

11.4 Future Exploration

A Mars sample return mission is an immediate future goal for Mars exploration for NASA and ESA. Despite decades of payload development and mission sophistication, resolution of whether Mars hosted microbial life is likely to be solved only with analysis of returned samples. Cached material deposited by the Perseverance Rover may be collected by Ingenuity-like Mars helicopters and launched from the planet's surface. On reaching orbit an ESA-designed transfer vehicle will fly the samples to Earth for analysis by the most advanced techniques available (Meyer et al. 2022). Mars sample return represents an extremely costly endeavour for Mars exploration; at the time of publication, funding has been significantly cut and launch dates for the venture remains in question.

The ExoMars Rosalind Franklin Rover is intended to directly search for Martian biosignatures. Equipped with a sophisticated drill, Rosalind Franklin is designed to extract samples from depths up to 2 m for analysis using an on-board laboratory. The six-wheeled, solar powered rover is planned to become the first European vehicle to safely land on the surface of the planet, targeted for eastern Chryse Planitia. Rosalind Franklin has suffered a series of design and test challenges during its development and a confirmed launch date remains unclear (Quantin-Nataf et al. 2021).

The Japanese Martian Moons eXploration (MMX) mission is planned to undertake detailed investigations of both Martian moons, Phobos and Deimos. MMX will conduct detailed flybys of the outer moon Deimos while dedicated landings and sample extraction will be made from Phobos. A return module is planned to transport collected Phobos samples to Earth for investigation (Usui et al. 2020).

In November 2025 NASA launched the Escape, Plasma and Acceleration Dynamics Explorers (ESCAPADE) mission. Two identical spacecraft were launched in November 2025 on a two-year journey to Mars. Unlike other missions, ESCAPADE was not launched during a favourable opposition window (Chap. 2) but instead will study space weather near Earth. On reaching Mars, ESCAPE will study the interactions between the planet's atmosphere and the solar wind to help understand gas loss over time (Lillis et al. 2024; Parker et al. 2024).

Along with mainstream ventures, the emergence of Space 2.0 will continue to foster the design and launch of low-cost cubesat missions to Mars. These will exploit ongoing availability of low-cost sensors and militarised supporting hardware such as power and propulsion systems capable of surviving the harsh environment of deep space. Missions of this type will lower the entry-point cost for Mars exploration, allowing countries with smaller space budgets to conduct their own exploration. This will allow more of the planet's surface to be explored, including landing sites that have not been considered a priority for visitation to date.

So far, the Martian moons have received limited attention and are recently being considered as an attractive target for cubesat-sized missions. Perseus is an example of a cubesat mission proposal that is planned to image Phobos through a series of close flybys (Nallapu et al. 2020). Once flown, Perseus would supplement data obtained from the MMX mission.

11.4.1 Future Plans for Human Habitation

Of all planets in our Solar System, Mars remains the most likely destination for future crewed missions (Fig. 11.1). As the second closest planet to Earth, Mars allows for transfer between the two planets in as little as six months using conventional propulsion systems, or less time if more exotic methods are employed. The Martian environment, although harsh compared to terrestrial standards, is also one of the least toxic to humans in the Solar System. The planet possesses abundant ice and other volatile reserves that can be exploited by future missions for generating rocket fuel, breathable oxygen and water needed for sustainment.

Development of multiple crewed proposals have occurred since the 1940s, though none has yet led to the construction of physical flight hardware. A popular mission concept exploits a nine-month Hohmann transfer orbit (Chap. 2) to Mars, stay time of

Fig. 11.1 A future return to Gale Crater may one day involve a crewed mission. Human adaptability and ingenuity would vastly increase the scientific data available for this area of Mars

16 months to await a transfer window to Earth, and an additional nine-month return. Shorter missions that would be launched ahead of Mars, then enter orbit once the planet catches up to the spacecraft, are also possible, at the expense of higher initial launch energy and reduced surface stay times.

Adaptations of a "Mars Direct" method are also being considered. Versions of this method involve sending space hardware and habitats ahead of a crewed landing, where automated or teleoperated systems would produce propellant for ascent to Earth, and construct living quarters for a subsequent crewed mission. Such crews would only be sent to Mars once confirmation of fully operational habitats and ascent vehicles is received. The Mars Direct method would save transporting enormous quantities of fuel and other supplies from Earth, instead using in situ resources on Mars (Impey 2019).

The private sector is also interested in going to Mars. Of the contenders, SpaceX is perhaps the most advanced in terms of planning and development of hardware. Founded in 2002, SpaceX has developed reusable launchers as well as commercial cargo and crew transport to the International Space Station. The company's current plans entail using a super heavy rocket booster to lift reusable "Starships" into Earth orbit. Following refuelling, the Starship would travel to and land on Mars, where in situ resources would be used to refill the rocket in a manner similar to Mars Direct (Heldmann et al. 2022; Young and Docherty 2025).

As of publication, neither government or commercial funded crewed Mars proposals have committed to firm launch dates, with the ventures heavily dependent on significant funding, technology development and political motivation. As planning and designs progress, future crewed missions will be shaped by the ongoing analysis of remotely sensed data of Mars.

References

Heldmann JL, Marinova MM, Lim DS, Wilson D, Carrato P, Kennedy K, Esbeck A, Colaprete TA, Elphic RC, Captain J, Zacny K (2022) Mission architecture using the SpaceX Starship vehicle to enable a sustained human presence on Mars. New Space 10:259–273. https://doi.org/10.1089/space.2020.0058

Impey C (2019) Mars and beyond: the feasibility of living in the solar system. In: The human factor in a mission to mars: an interdisciplinary approach, pp 93–111. https://doi.org/10.1007/978-3-030-02059-0_5

Lillis RJ, Curry SM, Curtis D, Taylor E, Courtade S, Ma Y, Parker J, Hara T, Luhmann JG, Barjatya A, Espley J (2024) ESCAPADE: a twin-spacecraft mars mission launching in 2024. In: 55th lunar and planetary science conference, pp LPI-contribution

Meyer MA, Kminek G, Beaty DW, Carrier BL, Haltigin T, Hays LE, Agree CB, Busemann H, Cavalazzi B, Cockell CS, Debaille V (2022) Final report of the Mars sample return science planning group 2 (MSPG2). https://doi.org/10.1089/ast.2021.0121

Nallapu RT, Dektor G, Kenia N, Uglietta J, Ichikawa S, Herreras-Martinez M, Choudhari A, Chandra A, Schwartz S, Asphaug E, Thangavelautham J (2020) Trajectory design of Perseus: a cubesat mission concept to Phobos. Aerospace 7. https://doi.org/10.3390/aerospace7120179

Parker JS, Ott C, Koehler A, Baskar S, Rosen M, Sullivan T (2024) Escape, plasma and acceleration dynamics explorers (ESCAPADE) mission design. In: Proceedings of the 44th annual American astronautical society guidance, navigation, and control conference, Mar 2022. Springer International Publishing, Cham, pp 187–206

Quantin-Nataf C, Carter J, Mandon L, Thollot P, Balme M, Volat M, Pan L, Loizeau D, Millot C, Breton S, Dehouck E (2021) Oxia Planum: the landing site for the ExoMars "Rosalind Franklin" rover mission: geological context and prelanding interpretation. Astrobiology 21:345–366. https://doi.org/10.1089/ast.2019

Usui T, Bajo KI, Fujiya W, Furukawa Y, Koike M, Miura YN, Sugahara H, Tachibana S, Takano Y, Kuramoto K (2020) The importance of Phobos sample return for understanding the Mars-moon system. Space Sci Rev 216:1–18. https://doi.org/10.1007/s11214-020-00668-9

Young D, Docherty N (2025) An anticipatory regime of multiplanetary life: on SpaceX, Martian colonisation and terrestrial ruin. Sci Cult 34:168–193. https://doi.org/10.1080/09505431.2024.2393096

Appendix A
List of Major Successful Mars Instruments

Instrument	Host mission	Host agency	Operation start/end
TV Camera System	Mariner 4	NASA	Jul-65/Jul-65
TV Camera System	Mariner 6	NASA	Jul-69/Aug-69
TV Camera System	Mariner 7	NASA	Aug-69/Aug-69
Imaging Science Subsystem (ISS)	Mariner 9	NASA	Nov-71/Oct-72
Imaging System (TV Camera)	Mars 3	Soviet Union	Dec-71/Mar-72
Infrared Radiometer	Mars 3	Soviet Union	Dec-71/Mar-72
Plasma Detector	Mars 3	Soviet Union	Dec-71/Mar-72
Imaging System (TV Camera)	Viking Orbiter	NASA	Aug-76/Jul-80
Infrared Thermal Mapper (IRTM)	Viking Orbiter	NASA	Aug-76/Jul-80
Ultraviolet Spectrometer (UVS)	Viking Orbiter	NASA	Aug-76/Jul-80
Cosmic Ray Telescope (CRT)	Viking Orbiter	NASA	Aug-76/Jul-80
Pressure Modulator Infrared Radiometer (PMIR)	Viking Orbiter	NASA	Aug-76/Jul-80
Surface Composition Experiment (SCE)	Viking Orbiter	NASA	Aug-76/Jul-80
Mars Atmospheric Water Detector (MAWD)	Viking Orbiter	NASA	Aug-76/Jul-80
Magnetometer	Viking Orbiter	NASA	Aug-76/Jul-80

(continued)

S. Hobbs, *Sensing Mars: Remote Sensing to Understand the Red Planet*,
https://doi.org/10.1007/978-3-032-24416-1

(continued)

Instrument	Host mission	Host agency	Operation start/end
Television Imaging System	Mars 5	Soviet Union	Dec-73/Feb-74
Gamma Spectrometer	Mars 5	Soviet Union	Dec-73/Feb-74
Plasma Detector	Mars 5	Soviet Union	Dec-73/Feb-74
X-Ray Spectrometer	Mars 5	Soviet Union	Dec-73/Feb-74
UV Spectrometer	Mars 5	Soviet Union	Dec-73/Feb-74
Radiometer	Mars 5	Soviet Union	Dec-73/Feb-74
Television camera system	Phobos 2	Soviet Union	Jul-88/Mar-89
NIR Imaging Spectrometer	Phobos 2	Soviet Union	Jul-88/Mar-89
ISM thermal infrared spectrometer/ radiometer	Phobos 2	Soviet Union	Jul-88/Mar-89
Thermal Imaging Camera	Phobos 2	Soviet Union	Jul-88/Mar-89
Gamma Ray Spectrometer	Phobos 2	Soviet Union	Jul-88/Mar-89
X-Ray telescope	Phobos 2	Soviet Union	Jul-88/Mar-89
Automatic Space Plasma Experiment with Rotating Analyzer (ASPERA)	Phobos 2	Soviet Union	Jul-88/Mar-89
Mars Orbiter Camera (MOC)	Mars Global Surveyor	NASA	Sep-97/Nov-06
Thermal Emission Spectrometer (TES)	Mars Global Surveyor	NASA	Sep-97/Nov-06
Mars Orbiter Laser Altimeter (MOLA)	Mars Global Surveyor	NASA	Sep-97/Jun-01
Gamma Ray Spectrometer (GRS)	2001 Mars Odyssey	NASA	Mar-02/09
Thermal Emission Imaging System (THEMIS)	2001 Mars Odyssey	NASA	Feb-02+
High Resolution Stereo Camera (HRSC)	Mars Express	ESA	Jan-04+
Visible and Infrared Mineralogical Mapping Spectrometer (OMEGA)	Mars Express	ESA	Jan-04+

(continued)

(continued)

Instrument	Host mission	Host agency	Operation start/end
Mars Advanced Radar for Subsurface and Ionosphere Sounding (MARSIS)	Mars Express	ESA	Jul-05+
High Resolution Imaging Science Experiment (HiRISE)	Mars Reconnaissance Orbiter	NASA	Nov-06+
Compact Reconnaissance Imaging Spectrometer for Mars (CRISM)	Mars Reconnaissance Orbiter	NASA	Nov-06/Dec-22
Mars Climate Sounder (MCS)	Mars Reconnaissance Orbiter	NASA	Nov-06+
Mars Color Imager (MARCI)	Mars Reconnaissance Orbiter	NASA	Nov-06+
Context Camera (CTX)	Mars Reconnaissance Orbiter	NASA	Nov-06+
Shallow Radar (SHARAD)	Mars Reconnaissance Orbiter	NASA	Nov-06+
Mars Atmospheric and Volatile EvolutioN (MAVEN) Instruments	MAVEN	NASA	Nov-14/Dec-25
Colour and Stereo Surface Imaging System (CaSSIS)	ExoMars Trace Gas Orbiter	ESA	Apr-18+
Methane Sensor for Mars	Mars Orbiter Mission (Mangalyaan)	ISRO	Sep-14/Oct-22
Mars Exospheric Neutral Composition Analyzer (MENCA)	Mars Orbiter Mission (Mangalyaan)	ISRO	Sep-14/Oct-22
Thermal Infrared Imaging Spectrometer (TIS)	Mars Orbiter Mission (Mangalyaan)	ISRO	Sep-14/Oct-22
Lyman-Alpha Photometer (LAP)	Mars Orbiter Mission (Mangalyaan)	ISRO	Sep-14/Oct-22
Mars Colour Imager (MCI)	Mars Orbiter Mission (Mangalyaan)	ISRO	Sep-14/Oct-22
Emirates eXploration Imager (EXI)	Emirates Mars Mission (Hope)	UAE Space Agency	Feb-21+
Emirates Mars InfraRed Spectrometer (EMIRS)	Emirates Mars Mission (Hope)	UAE Space Agency	Feb-21+
Emirates Mars Ultraviolet Spectrometer (EMUS)	Emirates Mars Mission (Hope)	UAE Space Agency	Feb-21+

(continued)

(continued)

Instrument	Host mission	Host agency	Operation start/end
Mars Orbiter Subsurface Investigation Radar (MOSIR)	Tianwen-1	CNSA	Feb-21+
Mars Magnetometer (MOMAG)	Tianwen-1	CNSA	Nov-21+
Mars Mineralogy Spectrometer (MMS)	Tianwen-1	CNSA	Feb-21+
Medium-Resolution Camera (MRC)	Tianwen-1	CNSA	Feb-21+
High-Resolution Camera (HRC)	Tianwen-1	CNSA	Feb-21+

Surface instrument	Surface mission	Host agency	Operation start/end
TV Camera	Mars 3 Lander	Soviet Union	2-Dec-71
Radiometer	Mars 3 Lander	Soviet Union	2-Dec-71
Barometer	Mars 3 Lander	Soviet Union	2-Dec-71
Thermometer	Mars 3 Lander	Soviet Union	2-Dec-71
Imaging System	Viking Landers	NASA	Jul-76/ Nov-82
Meteorology Instrument System (MET)	Viking Landers	NASA	Jul-76/ Nov-82
Seismometer	Viking Landers	NASA	Jul-76/ Nov-82
Gas Chromatograph-Mass Spectrometer (GCMS)	Viking Landers	NASA	Jul-76/ Nov-82
Biology Experiments	Viking Landers	NASA	Jul-76/ Nov-82
Imager for Mars Pathfinder (IMP)	Mars Pathfinder	NASA	Jul/Sep-97
Atmospheric Structure Instrument/Meteorology Package (ASI/MET)	Mars Pathfinder Lander	NASA	Jul/Sep-97
Rover Instrumentation (MO)—Soil Mechanics	Sojourner Rover	NASA	Jul/Sep-97
Imaging Camera	Sojourner Rover	NASA	Jul/Sep-97
Alpha Proton X-Ray Spectrometer (APXS)	Sojourner Rover	NASA	Jul/Sep-97
Panchromatic and Color Cameras (Pancam)	Mars Exploration Rovers	NASA	Jan-04/ Mar-10/ Jun-18

(continued)

(continued)

Surface instrument	Surface mission	Host agency	Operation start/end
Alpha Particle X-Ray Spectrometer (APXS)	Mars Exploration Rovers	NASA	Jan-04/ Mar-10/ Jun-18
Mini-Thermal Emission Spectrometer (Mini-TES)	Mars Exploration Rovers	NASA	Jan-04/ Mar-10/ Jun-18
Microscopic Imager	Mars Exploration Rovers	NASA	Jan-04/ Mar-10/ Jun-18
Surface Stereo Imager (SSI)	Phoenix Lander	NASA	Jul/Nov-08
Robotic Arm (RA)	Phoenix Lander	NASA	Jul/Nov-08
Robotic Arm Camera (RAC)	Phoenix Lander	NASA	Jul/Nov-08
Thermal and Evolved Gas Analyzer (TEGA)	Phoenix Lander	NASA	Jul/Nov-08
Microscopy, Electrochemistry, and Conductivity Analyzer (MECA)	Phoenix Lander	NASA	Jul/Nov-08
Mars Descent Imager (MARDI)	Phoenix Lander	NASA	Jul/Nov-08
Mastcam	Curiosity Rover	NASA	Aug-12+
Alpha Particle X-Ray Spectrometer (APXS)	Curiosity Rover	NASA	Aug-12+
Chemistry and Mineralogy (CheMin)	Curiosity Rover	NASA	Aug-12+
Sample Analysis at Mars (SAM)	Curiosity Rover	NASA	Aug-12+
Rover Environmental Monitoring Station (REMS)	Curiosity Rover	NASA	Aug-12+
Seismic Experiment for Interior Structure (SEIS)	InSight Lander	NASA	Nov-18/ Dec-22
Heat Flow and Physical Properties Package (HP3)	InSight Lander	NASA	Nov-18/ Dec-22
Rotation and Interior Structure Experiment (RISE)	InSight Lander	NASA	Nov-18/ Dec-22
Instrument Deployment Arm Camera (IDC)	InSight Lander	NASA	Nov-18/ Dec-22
Instrument Context Camera (ICC)	InSight Lander	NASA	Nov-18/ Dec-22
Rotation and Interior Structure Experiment (RISE)	InSight Lander	NASA	Nov-18/ Dec-22
Mastcam-Z	Mars 2020 (Perseverance)	NASA	Feb-21+
SuperCam	Mars 2020 (Perseverance)	NASA	Feb-21+
SHERLOC	Mars 2020 (Perseverance)	NASA	Feb-21+
RIMFAX	Mars 2020 (Perseverance)	NASA	Feb-21+

(continued)

(continued)

Surface instrument	Surface mission	Host agency	Operation start/end
Rover Environmental Monitoring Station (REMS)	Mars 2020 (Perseverance)	NASA	Feb-21+
PIXL (Planetary Instrument for X-Ray Lithochemistry)	Mars 2020 (Perseverance)	NASA	Feb-21+
MOXIE (Mars Oxygen ISRU Experiment)	Mars 2020 (Perseverance)	NASA	Feb-21+
MEDA (Mars Environmental Dynamics Analyzer)	Mars 2020 (Perseverance)	NASA	Feb-21+
Ground Penetrating Radar (RoPeR)	Tianwen-1 (Zhurong Rover)	CNSA	May-21/ May-22
Mars Rover Magnetometer (RoMAG)	Tianwen-1 (Zhurong Rover)	CNSA	May-21/ May-22
Mars Climate Station (MCS)	Tianwen-1 (Zhurong Rover)	CNSA	May-21/ May-22
Mars Surface Compound Detector (MarSCoDe)	Tianwen-1 (Zhurong Rover)	CNSA	May-21/ May-22
Multispectral Camera (MSCam)	Tianwen-1 (Zhurong Rover)	CNSA	May-21/ May-22
Navigation and Topography Cameras (NaTeCam)	Tianwen-1 (Zhurong Rover)	CNSA	May-21/ May-22
Surface Composition Detector (Spectrometer)	Tianwen-1 (Zhurong Rover)	CNSA	May-21/ May-22

Appendix B
Acronym List

Acronyms	Full name
ADV	Antarctica Dry Valleys
AEGIS	Autonomous Exploration for Gathering Increased Science
AI	Artificial Intelligence
APXS	Alpha Proton Spectrometer
ATM	Atmospheric Chemistry Suite
D/H	Deuterium Hydrogen Ratio
CCD	Charge-Coupled Device
CFL	Compact Fluorescent Lamp
CHeMin	Chemistry and Mineralogy X-Ray Diffraction
CHEMCAM	Chemistry and Camera
CMOS	Complementary Metal Oxide Semiconductor
CNN	Convolutional Neural Network
CO_2	Carbon Dioxide
COTS	Commercial Off-The-Shelf
CRISM	Compact Reconnaissance Imaging Spectrometer for Mars
CTX	Context Camera
DN	Digital Number
DTM	Digital Terrain Model
EM	Ejecta Mobility
EMM	Emirates Mars Mission
EMS	Electromagnetic Spectrum
EMUS	Hope's Emirates Mars Ultraviolet Spectrometer
EPS	Electrical Power System
ESA	European Space Agency

(continued)

S. Hobbs, *Sensing Mars: Remote Sensing to Understand the Red Planet*,
https://doi.org/10.1007/978-3-032-24416-1

(continued)

Acronyms	Full name
FiDO	Field Integrated Design and Operations
FPGA	Field-Programmable Gate Array
FPI	Fabry–Pérot Interferometer
FWHM	Full Width Half Maximum
Ga	Billion Years
GAN	Generative Adversarial Networks
GCM	Global Climate Model
GCMS	Gas Chromatograph Mass Spectrometer
GEO	Geosynchronous Earth Orbit
Gex	Gas Exchange
GIS	Geographic Information Systems
GLF	Glacier-like Form
GPS	Global Positioning System
GRS	Gamma Ray Spectrometer
GSD	Ground Sample Distance
HAB	High-Altitude Balloon
HiRISE	High-Resolution Imaging Science Experiment
HRSC	High Resolution Stereo Camera
IAU	International Astronomical Union
JMARS	Java Mission-Planning and Analysis For Remote Sensing
Ka	Thousand Years
LDM	Latitude Dependent Mantle
LEO	Low Earth Orbit
LIDAR	Light Detection and Ranging
Lr	Labelled Release
Ma	Million Years
MarsCO	MarscubeOne
MEO	Medium Earth Orbit
MAHLI	Mastcam/Z, Hand Lens Imager
MARSIS	Mars Advanced Radar for Subsurface and Ionosphere Sounding
MASI	Mastcam-Z Analog Spectral Imager
MAVEN	Mars Atmosphere and Volatile Evolution
MCDM	Multi-Criteria Decision Making
MCT	Mercury Cadmium Telluride
MER	Mars Exploration Rovers
MFF	Medusae Fossae Formation
MGS	Mars Global Surveyor

(continued)

(continued)

Acronyms	Full name
ML	Machine Learning
MMS	Mars Mineralogical Spectrometer
MOC	Mars Orbiter Camera
MOLA	Mars Orbiter Laser Altimeter
MOM	Mars Orbiter Mission
MRO	Mars Reconnaissance Orbiter
NDVI	Normalised Difference Vegetation Index
NIR	Near Infra-Red
NOMAD	Nadir and Occultation for Mars Discovery
NPLD	North Polar Layered Deposits
ODE	Orbital Data Explorer
OGC	Open Geospatial Consortium
OMEGA	Observatoire pour la Minéralogie, l'Eau, les Glaces et l'Activité
PDS	Planetary Data System
PNT	Position, Navigation and Timing
Pr	Pyrolytic Release
RGB	Red, Green, Blue
RIMFAX	Radar Imager for Mars' Subsurface Experiment
RMI	Remote Micro Imager
RSL	Recurring Slope Lineae
SAR	Synthetic Aperture Radar
SHARAD	Shallow Radar
SHERLOC	Scanning Habitable Environments with Raman and Luminescence for Organics and Chemicals
SNR	Signal to Noise Ratio
SPICAM	Spectroscopy for the Investigation of the Characteristics of the Atmosphere of Mars
SWIR	Shortwave Infra-Red
TES	Thermal Emission Spectrometer
TIR	Thermal Infra-Red
TLS	Tuneable Laser Spectrometer
TRL	Technology Readiness Level
THEMIS	Thermal Emission Imaging System
VIS/NIR	Visible/Near Infra-Red
UV	Ultraviolet
VBF	Vastitas Borealis Formation
VFF	Viscous Flow Features
WFS	Web Feature Service

(continued)

(continued)

Acronyms	Full name
WMS	Web Map Service
XRD	X-Ray Diffraction

Index

A
Alba Mons, 201, 209
Albedo, 1, 13, 22, 166, 212
ALH84001, 283, 284
Alpha Proton X-ray Spectrometer (APXS), 121
Amazonian era, 25, 175, 193, 199, 207, 209, 211, 221, 233, 246, 251, 291
  minerals, 184
Antarctic Dry Valleys, 52, 261, 264, 277
Antoniadi Crater, 246
Arabia Terra, 199, 207, 256
Aram Chaos, 246
Artificial intelligence, 152
  Autonomous Exploration for Gathering Increased Science (AEGIS), 158
  edge-based, 158
  image segmentation, 152
Atmosphere, 165
  column, 166
  dew point, 172
  Hadley cells, 168
  ionosphere, 111, 112, 114
  Mie scattering, 66
  pressure, 165
  rayleigh scattering, 66
  snowstorm, 174
  sputtering, 169
  thermosphere, 168
  troposphere, 168

B
Band ratios, 90
Binning
  pixel, 73
Bloemfontein Observatory, 8
Brines, 111, 175–177, 264, 265

C
Canals, 6–8, 241, 275
Cassini, J.D., 4
Cerberus Fossae, 3, 203, 285
Chaotic terrain, 3, 208
Charge Coupled Device, 39, 81, 83, 84, 87, 95
Chromatic aberration, 69
Chryse Planitia, 194, 241, 246, 247, 250, 275, 292
Cloud
  orographic, 174
  particle size, 173
Colbert, Elias, 6
Communication
  distance, 33
  X-band, 34
Compact Reconnaissance Imaging Spectrometer for Mars (CRISM), 68, 73, 93, 95, 96, 175, 207, 233, 253
Complementary Metal-oxide-Semiconductor, 83, 84, 87, 95
Convolutional neural network, 155, 157, 199
  pooling layers, 155
Crater
  breccia, 189
  central peak, 189, 195, 219
  degradation, 194
  excavation, 189
  formation process, 190

S. Hobbs, *Sensing Mars: Remote Sensing to Understand the Red Planet*, https://doi.org/10.1007/978-3-032-24416-1

ghost, 195
present impact rate, 198
shockwave, 189
Cubesat, 3, 40, 47, 48, 50, 97, 290, 293

D
Data processing and filters, 134
Dataset processing levels, 133
Dating
crater age, 196, 211
crater size frequency, 196
Deep Space Network, 33–35
Deimos, 6, 20, 21, 292
De Vaucouleurs, Gerard, 7
Digital number, 66, 74, 77, 133, 134
Digital Terrain Model, 140, 143
HiRISE, 141
volume calculation, 143
Dune
barchan, 143, 215
barchanoid, 215
composition, 213, 219
fossilised, 221
inter-crater, 221
intra-crater, 217
linear, 215
star, 215
Dust storms, 171

E
Electromagnetic spectrum, 61
Elevation data, 138
Ellipsoid
Mars, 132
terrestrial, 132
Engineering
alternative options workflow, 49
phase A, 31
phase B, 31
phase C, 32
phase D, 32
phase E, 32
phase F, 32
pre-phase A, 30
trade study analysis, 49
European Space Agency, 15, 30, 34, 278, 292
ExoMars, 292

F
Flat fielding, 71
Fretted terrain, 235

G
Gale Crater, 91, 152, 175, 219, 230, 251, 281, 282, 285
Galileo Galilei, 4
Gamma Ray Spectrometer, 122
hydrogen abundance, 122
radiation detection, 122
Generative adversarial networks, 157
Geographic information system, 143, 150
Georeferencing, 135
GIS - management, 143
Glacial flow, 237
Glacier flow
carbon dioxide, 240
near equator, 239
Glacier-like form, 237
age estimate, 238
Global mineralogy, 180
Goodness of fit, 154
Graben, 200, 209
Gusev Crater, 279

H
Hall, Asaph, 6
Hellas Basin, 171, 209, 241
Hellas Planitia, 175, 218
Herschel, Sir William, 6
Hesperian era, 23, 175, 208, 212
erosion, 195
minerals, 183
ocean, 253
outflow channel, 246
volcanism, 202, 209, 291
water activity, 246, 251
wind activity, 221
High Resolution Imaging Science Experiment (HiRISE), 14, 31, 39, 266
calibration, 77
elevation model, 150
optical design, 69
pushbroom operation, 83
Hohmann transfer, 37
Home Plate, 18, 182, 183, 279
Hubble Space Telescope, 8
Huygens, Christen, 4
Hydrated minerals, 182
Hyperspectral
calibration methods, 96, 101
designs, 92

detector considerations, 95
fabry-perot interferometer, 95
full width half maximum, 96
photon flux, 99
sensors, 92
surface mission spectrometers, 97

I
Ice
impact excavation, 233
movement, 234
Imaging
Bayer filter, 53, 84, 85, 89, 90, 95
multispectral, 53, 68, 84, 86, 87, 89, 90
Insight, 18, 40, 84, 89, 147, 149, 192, 203
Interior crater deposits, 237
Inverted stream, 246
Iron oxide, 87, 91, 101, 177

J
James Webb Telescope, 8
Jezero Crater, 18, 90, 114, 117, 121, 152, 182, 251, 252, 283

K
Kasei Valles, 248, 250
Kuiper, Gerald, 6

L
Landing site selection, 147
Landing site selection – example methods, 147
Latitude dependent mantle, 234
Lava, 3, 12, 23, 179, 181, 194, 199–203, 207, 209, 211, 233, 248, 252, 290, 291
tube, 201, 202, 211
Lineated valley fill, 237
Lobate debris aprons, 236
Lowell, Percival, 6, 7

M
Manning's equation, 243
Mariner 4, 8, 9, 81, 166, 189
Mariner 6 and 7, 10, 117, 135, 232, 246
Mariner 9, 11–13, 21, 41, 68, 117, 129, 138, 139, 145, 148, 199, 207, 212, 234, 235, 240, 241
Mars
diameter, 19
mass, 19
sample return, 292
seasons, 19
solar orbit, 19
Mars 1 mission, 9
Mars 3, 46
Mars 4, 5, 6 and 7, 13
Mars Advanced Radar for Subsurface and Ionosphere Sounding (MARSIS), 13, 112
Mars Atmosphere and Volatile Evolution (MAVEN), 13, 120, 132, 168
MarsCubeOne (MarsCO), 40, 48, 89
Mars Direct, 294
Mars Exploration Rovers, 2, 18, 50, 84, 149, 171, 181–183, 191, 278
Mars Express, 13, 93, 96
Mars Global Surveyor, 13, 83, 115, 132, 219, 250, 263
Mars Odyssey, 13, 115, 117, 122, 125, 175, 177, 211, 237, 240
Mars' orbit, 37
Mars Orbiter Laser Altimeter (MOLA), 13, 64, 132, 139, 141, 146, 150, 152, 229, 250, 259
Mars Orbiter Mission, 15, 39
Mars quakes, 203
Mars Reconnaissance Orbiter (MRO), 13, 31, 33, 34, 41, 177, 233
attitude control, 39
orbit, 45, 95
Martian Moons eXploration (MMX), 292
Mastcam-Z, 51, 85, 87, 88
Medusae Fossae Formation, 170, 194, 213, 246
Meridiani, 18, 65, 91, 117, 150, 179, 183, 278
Methane, 284
Michelson interferometry, 117
Microfossil, 280, 283, 284
Mission design
background radiation, 33
electrical power system, 35
nuclear heating, 37
power availability, 35
thermal considerations, 36
thermal cycling, 36
Mojave Crater, 243
Mosaicking
GIS data, 135
Mössbauer spectrometer, 121

N
NASA, 13, 34, 283
  engineering phases, 30
  landing site selection, 148
  Stereo pipeline, 137
  technology readiness level, 29
Navigation
  aerobraking, 41
  magnetorquer, 39
  orbit type, 41
  perturbation forces, 41
  position, navigation and timing, 37
  propulsion, 39
  solar radiation pressure, 41
  star tracker, 39
Noachian era, 23, 175, 212, 291
  atmosphere, 169
  cratering, 195, 196
  minerals, 182
  ocean, 253
  valley network, 241, 253
  volcanism, 199, 200, 219
Noise
  shot, 72
  thermal, 72
Normalized Difference Vegetation Index (NDVI), 52
Northern polar cap, 7, 230
North Polar Layered Deposits, 230

O
Observatoire pour la Minéralogie, l'Eau, les Glaces et l'Activité (OMEGA), 68, 92, 93, 96, 175
Olivine, 87, 97, 101, 179, 181, 201, 252
Olympia Undae, 221
Olympus Mons, 96, 133, 199, 200, 203
Open Geospatial Consortium, 146
Optical
  components, 69
  design, 68
  diffraction limit, 69
  orbiter imager types, 81
  popular sensor types, 83
Orbit
  aerostationary, 45
  elliptical, 44
  frozen, 45
Orthorectification
  GIS data, 137
Outflow channel, 246
  subsurface heating, 246
Ozone layers, 169

P
Pathfinder, 18
Pavonis Mons, 200, 201
Pedestal crater, 193
Perchlorate, 16, 18, 175, 275
Phobos, 6, 20, 292
Phobos 2 mission, 13, 22
Phoenix, 2, 18, 61, 149, 173, 175–177, 229, 240, 275
Photogrammetric processing (HRSC example), 141
Planetographic and planetocentric coordinates, 132
Polar Layered Deposits, 232
Prime Meridian, 129
Processing
  histogram stretch, 134
  image enhancement, 134
Proctor Crater, 219
Projection
  lambert conformal conic, 132
  mercator, 130
  polar stereographic, 130
  robinson, 130
Pushbroom, 94
Pyroxine, 181

R
Radar
  factors affecting, 110
  ground penetrating, 110
  remote sensing, 109
  synthetic aperture, 109
Radiative transfer models, 76
Radiometric calibration, 74
Rampart crater, 192
Recurring slope lineae, 266
Regression techniques, 153
Resolution
  ground sample distance, 66
  spatial, 66
  spectral, 68
  temporal, 68
Rio Tinto, 278, 279

S
Saltation, 215
Schiaparelli, 129
Schiaparelli, Giovanni, 6

Secondary craters, 189, 191, 196
Secondary payload, 40
Sensing
high-energy, 120
noise, 72
noise mitigation, 73
passive and active, 64
pre-processing, 71
signal to noise, 72, 99
Sensor
Mars Advanced Radar for Subsurface and Ionosphere Sounding (MARSIS), 44, 112, 177
RIMFAX, 113
Rover Penetrating Radar (RoPeR), 113
SHallow RADar (SHARAD), 111–113, 238
SHERLOC, 120
Slipher, E.C., 8
Solar flux, 63
Solar spectrum, 62, 74
Solar wind, 41
Southern polar cap, 230
composition, 230
Space 2.0, 47
commercial-grade components, 50
high altitude balloon, 51
Science, Technology, Mathematics and Education (STEM), 50
Spectral discrimination, 87
Spectral resolution, 96
Spectroscopy for the Investigation of the Characteristics of the Atmosphere of Mars (SPICAM), 121
Stereoscopic imagery, 140
Stromatolite, 279, 280, 282
Sun-synchronous, 45
Surface mission localisation, 149
Syrtis Major, 4, 213, 246, 252

T
Tharsis volcanoes, 200
Thermal inertia, 11, 115–119, 146, 148, 150, 151, 211, 240
Thermal infra-red sensing, 114
Thermal microbolometer, 118
Thermal sensor applications, 115
Thermal sensors, 64
Thermal sensors flown to Mars, 117
Tianwen-1, 45
Tie points
GIS data, 135

U
Ultraviolet
atmospheric interactions, 120
sensors flown to Mars, 120
Utopia Planitia, 15, 18, 50, 152, 175, 176, 207, 221, 241, 289, 291

V
Valles Marineris, 2, 12, 170, 175, 182, 207, 221, 239, 247, 266
Valley network, 241
Ventifacts - aeolian, 213
Vignetting, 71
Viking, 8, 9, 13–16, 18, 21, 33, 41, 45, 50, 81, 82, 84, 86, 117, 121, 129, 132, 143, 145, 147–149, 219, 234, 235, 240, 241, 250, 263, 275–277, 289
search for life, 16, 275, 277
Viscous flow features, 235
Visible-light imagers, 81
Volcanism, 199
basalt, 199
mud, 207
pyroclastic flows, 199
shield, 201, 252, 291
terrestrial studies, 199

W
Water
present stability, 175, 176
rainfall, 241
subsurface flow, 23, 241, 246, 248, 259
Web Feature Service, 146
Web Map Service, 146
Wind streak, 212

Y
Yardang - aeolian, 212, 213
Yellow clouds, 170

Z
Zero longitude
Airy Crater, 129
Zhurong, 18, 152, 221, 256

MIX
Papier aus verantwortungsvollen Quellen
Paper from responsible sources
FSC® C105338

If you have any concerns about our products,
you can contact us on
ProductSafety@springernature.com

In case Publisher is established outside the EU,
the EU authorized representative is:
Springer Nature Customer Service Center GmbH
Europaplatz 3, 69115 Heidelberg, Germany

Printed by Libri Plureos GmbH
in Hamburg, Germany